Space Weapons

Space Weapons

Deterrence or Delusion?

Rip Bulkeley and Graham Spinardi

Foreword by Brian Aldiss
Edited by Christopher Meredith

BARNES & NOBLE BOOKS
TOTOWA, NEW JERSEY

First published in the USA 1986 by
Barnes & Noble Books
81 Adams Drive
Totowa, New Jersey, 07512

Library of Congress Cataloguing in Publication Data

Bulkeley, Rip
  Space weapons
  1. Space weapons   I. Spinardi, Graham.   II. Title.
  UG1530.B84   1986   358'.8   86-10880

ISBN 0–389–20640–7
ISBN 0 389–20641–5 (Pbk.)

Typeset by Oxford Publishing Services, Oxford
Printed in Great Britain by TJ Press, Padstow, Cornwall

Auschwitz continues to haunt, not the memory but the accomplishments of man – the space flights; the rockets and missiles . . . Herbert Marcuse

This book is for our families, without whom . . .

List of Contents

List of Figures

List of Tables

List of Boxes

Note to the Reader

This book has been written as an account and an evaluation of space weapons and the issues they involve, for readers who may know little about them. We have tried to explain some fairly technical matters in plain language. Sometimes, however, plain language has to be technical to do its job properly. Terms peculiar to the field of space weapons, such as 'endo-atmospheric', have been introduced with some degree of explanation, if only the context itself. Sometimes 'scare-quotes' are used to do this, just as this sentence demonstrates the notion of scare-quotes to anyone not familiar with them. Where such an explanation is not given, either because it belongs, properly, to a later section of the book, or because the expression is expected to be familiar to many readers already, the term will be found in the glossary (p. 315), together with most acronyms (DEW, NATO etc), however familiar; the exceptions are explained on the spot. Readers feeling lost at any point should consult the glossary, if they think the term is being used for the first time, or the index, if they think it has already been explained and they wish to refer back to that passage. Page references to such explanations are given in **bold type** in the index, and most expressions in the glossary are also in the index. Some material which is either slightly more technical, or else basically explanatory, or both, has been placed in 'boxes' within the text, for people to read or skip as they choose.

Nevertheless, there is no avoiding the fact that anyone who wants to get on more or less equal terms with military and political decision-makers in the space weapons field, but who sets out from little knowledge of the subject, may have to work very hard with initially unfamiliar concepts and arguments that bring together aspects of weapons technology, strategy, law, and international relations. We have done our very best to help, but we have not left crucial issues undiscussed because they were 'too difficult for the general reader'. Three years ago we were general readers ourselves.

Foreword

Over the last ten years, a significant revolution in thought has taken place. Whereas the populations of the West – during the period of the Cold War and after – showed an almost united fear of nuclear attack from the Soviet Union, those populations now show a greater fear of the nuclear weapons themselves. They fear those weapons whether held in Soviet or Western (American) arsenals.

Fear is a wretched but not unfamiliar state of mind. It has probably prevailed over many populations during most periods of history. Napoleon, the Turks, the Mogul invasion, the Black Death, famine, etc. – each have, in their time, been generators of fear; and sometimes, possibly, of correction, as the Black Death eventually brought about better conditions for peasant labourers through their resulting scarcity.

If fear has changed minds in our day, so be it. Where the survival, not merely of mankind, but of the entire biosphere, the living planet itself, is at stake, it is wiser to fear the proliferation of nuclear weapons than another nation state.

Nowhere has the nuclear arms race been less controlled than in its intrusions, actual or potential, into outer space. There it has shown rapid evolutionary development. As the authors of this book point out, the world's first 'flying bomb', the V-1, had a maximum speed of about 400 mph and a constant altitude of between 2000 and 3000 feet. Its immediate successor, the V-2, was quite another animal. It climbed on an elliptical curve to altitudes of 50 to 60 miles, to the fringes of space, before falling back at a speed which reached over 2300 mph before impact.

V-2s thus reached target five minutes after launch, where the V-1s had taken 30 minutes or more. Since then, the space weapons, as Eisenhower used to call them, have grown faster, bigger, hungrier. Complexity has greatly increased with the entry of computerization into this field: *weapon systems* are now the order of the day. They hem in our planet as barbed wire once protected the trenches of World War 1.

These machines are so far only deployed. When they are employed, they will be launched against us, against human beings wherever we may live. This book effectively comes to terms with this situation, working its way steadily through the welter of ugly acronyms, the ABMs, STEWs, ICBMs, ASMSs, SALTs and SDIs, towards our greater understanding.

Is this a situation in which understanding can help us? I believe it is, as it is in most situations. It is useless to run from the current facts; there is nowhere to run to. Nor are many of us, as I see it, exactly guiltless. For the instincts of war – the reflexes which make us hate our neighbour instead of loving him or her – are not the monopoly of politicians, the Politburo or the Pentagon.

One of the excellences of this book lies in its understanding that space weapons are products of our lifestyle. Soviet and Western perceptions, with all their cultural and psychological biases, are analysed. We are shown clearly the complexities of the whole technological ethos, which goes well beyond any sloganeering.

Until a few years ago, I was a proponent of the 'deterrence' or Balance of Terror theory. It had, after all, worked for thirty years, or so it could be argued. But nothing remains the same. The original uneasy balance is in any case being destabilized by the availability of nuclear weapons to countries all round the world. Two other factors served to change my mind, scientific discovery and personal experience.

Under the first heading has come a growing understanding of what a nuclear exchange of even limited power might entail. Recent research has uncovered the strong possibility that most life in the northern hemisphere might be extinguished and, if in the northern hemisphere, then the southern too, since the hemispheres are not autonomous. Only such creeping things as the ants and cockroaches would survive. This is the nuclear winter.

As for personal experience, I visited Greenham Common and saw the women camping there in primitive conditions outside the base, besieging it as if it contained a hostile power. As possibly it does. In my boyhood, Greenham was an RAF base. The RAF during the Battle of Britain were the nation's heroes. Here was as dramatic a reversal as could be imagined. The women shook the wire and shouted and sang carols, for the year was dying towards Christmas. '. . . Peace on Earth and mercy mild . . .' It must have been borne in on everyone there that time had come for a reversal of loyalties. Our understandings have to evolve as rapidly as our missile systems.

Otherwise, there will be only the ants and cockroaches, who on the whole lead boring lives and cannot name Shakespeare or Sagittarius as the unfeeling 'eyes and ears' of weapons systems still unused glitter high in the darkening sky overhead.

This book, I believe, offers an alternative for which we have to fight, a Balance of Understanding. Those brilliant technological devices already or soon to be poised above us, the higher hardware, come there of our generally undirected but communal will. By communal will, we can get them down again. Or, better still, direct them towards more positive ends, as swords were once beaten into ploughshares: towards the exploration of our beckoning galaxy.

Brian Aldiss
Oxford, November 1985

Part 1
Origins

1
Just Another Place

Space is just another place where wars will be fought.
Senator Goldwater, 1984

This is a book about an old idea which has acquired a new and disturbing importance in recent years, and seems sure to retain it: the idea of what are popularly but not very precisely termed 'space weapons'. Most types of actual or possible space weapon will be straightforwardly described and explained. But the book is not meant as a comprehensive guide to every space weapon there has ever been, or that may be devised in the next, say, twenty years. On balance, more attention is given to assessing the controversial implications of such armaments for world peace and the future of human society. Where evaluative conclusions have seemed justified, we have not hesitated to state them.

Human beings are capable of amazing, if depressing, ingenuity and persistence when it comes to fighting. Almost anything can and has been used as a weapon, so that the concept cannot really be applied to objects in themselves, but only to how they are used. Earth satellites are fragile objects, travelling at high speeds, and very vulnerable to destruction by simple collision. Any manœuvrable space object might therefore become a 'weapon' in relation to any other it is able to approach. Nevertheless, some space objects are, or would be, weapons in the fullest sense of the word, being purposely designed to cause the types of destruction listed in the definition which follows.

The expression 'space weapon' is used here to refer to any specialized destructive device built to operate or take effect in space, which means for present purposes the region of 'near' or 'contiguous' space surrounding Earth. At about 80 kilometres above sea level the remaining traces of the atmospheric gases

3

become too thin to support aerodynamic flight at any feasible speed. A little further 'out', and their individual atoms and molecules go into free-falling orbits of their own around the planet. From that point everything that can happen in the zone of Earth's neighbouring space is dominated by one major factor, the gravitational field or 'pull' of the planet itself. Our topic therefore comprises all destructive devices moving under the influence of gravity in near or 'planetary' space that can be aimed either against other space objects or against targets on Earth below; also, any machines built to attack space objects from down here on the planet.

This covers a great deal. For one thing, the operation of space weapons is so demanding – at the present level of human technology – that it requires the simultaneous use of many different and separate machines, such as enormous radars, computers, rockets, and satellites. Often such interconnected parts of a 'weapon system', as it is termed, can be hundreds or thousands of kilometres apart, which makes their effective combined operation completely dependent on their communication links. (This point is worth remembering when trying to judge how well they might work, how much they might cost and how they might themselves be accidentally or deliberately disrupted.) Also, weapon systems intended for one purpose often have features and even actual components in common with those intended for another.

In particular, it will be necessary to divide our attention between two main types of space weapon. First, there are those designed to destroy artificial satellites in orbit around Earth, many of which are not themselves weapons, though they are increasingly employed to give direct assistance to various uses of armed force down on the planet. These anti-satellite weapons are commonly referred to as 'asats' (pronounced 'aysats').

The second main type of space weapon comprises all those designed to destroy long-range missiles, which climb above the atmosphere into the nearer fringes of 'outer space' during much of their journey from one point on Earth to another. We have to stretch our definition slightly to cover all such weapons, because long-range missiles can be attacked not only in space but also during the two short atmospheric phases of their flight, at the beginning and at the end. Although they pass through space, however, the missiles themselves are not covered by the definition, because both their launch-points and their targets would be on Earth.

Box 1.1   Ballistic and Cruise Missiles

Ballistic missiles are so called because for most of their flight they are unpowered. Like a thrown or kicked ball, or a bullet, they follow a path determined by their initial impulse, gravity and any air friction they encounter. Ballistic missiles can be capable of travelling distances as great as 10,000 kilometres (ICBMs). Such intercontinental ranges are achieved by initial rocket thrusts that hurl the missile up out of the atmosphere, so that it only meets with air resistance in the few minutes at the beginning and end of its flight. The launcher and the warheads, or 're-entry vehicles', are protected from overheating and from aerodynamic drag by streamlining and by the use of special materials.

Cruise missiles, on the other hand, are essentially unmanned aircraft, remaining within the atmosphere throughout their powered flight. They depend on air resistance for aerodynamic lift, and on atmospheric oxygen for their turbo-jets. Their jet engines resemble those of other aircraft, with which they are sometimes bracketed in nuclear strategic discussions as 'air-breathing' weapon-delivery systems. To date, all cruise missiles have had speeds below the speed of sound, though supersonic versions are now being developed.

The first anti-missile weapons were missiles themselves, armed with nuclear warheads much like the ones they were intended to intercept. They were known at first as 'anti-ballistic-missile missiles', which was shortened to 'Anti-Ballistic Missiles' or 'ABMs'. Later, as different ways of attacking long-range missiles came to be more feasible, from swarms of steel rods to very intense beams of light, such weapons began to be referred to by a more general phrase, 'Ballistic Missile Defence' or 'BMD'. However, the 1972 ABM Treaty between the Soviet Union and the United States (appendix 1) defines 'ABM' to mean *anything* able 'to counter strategic ballistic missiles . . . in flight. . .'. So the difference between the two expressions is very slight, with 'ABM' pointing a little more towards the weapons and 'BMD' towards their military role.

A third category of space weapon has long been imagined and even planned for, but has not been much developed beyond some prototypes in the early 1960s. This is the 'Space-To-Earth' weapon, sometimes abbreviated to STEW. Thirty years ago these were usually conceived as orbital atomic bomb carriers. Today the US Air Force has technical studies well in hand for a shuttle-like 'space fighter' known as the Trans-Atmospheric Vehicle (TAV) (p. 193).

Such programmes help us to understand what the phrase 'space weapons' does *not* mean. It does not mean that a kind of weapon is coming into fashion, the effects of which would be remote from our planet, and so to speak exiled far off into space. Today, and for as far ahead as it is reasonable to forecast, space weapons are also planetary weapons, using or passing through the space close to Earth to play their part in terrestrial conflict. They are no more able to lift the threat of war from our land-dwelling species than were the naval or the aerial weapons which preceded them (p. 199).

Much of the current world-wide interest in space weapons has resulted from President Reagan's initiation, in March 1983, of an intensive US development programme aimed at producing weapons capable of destroying nuclear missiles, if these were ever launched against the United States or any of her allies. That programme, the Strategic Defense Initiative (SDI), will be described in chapters 6, 7 and 8 below, and some of its main organizational and budgetary aspects are summarized in appendix 2. The technical and political history of space weapons, from which the SDI emerged, is surveyed in chapters 2 to 5. Chapter 4, which describes current military satellites and the first really effective asats of the United States and the Soviet Union, shows that by no means all US space weapon programmes are included in the SDI.

The United States has not been alone in developing space weapons. Its historical lead in this field has been followed for over twenty years by the Soviet Union. That rival military space effort is described and discussed, in such detail as is possible, in chapters 3, 4 and 9, and in appendix 3. Regrettably, Soviet sources are usually less forthcoming than American. Whereas in the American case we are sometimes left in the dark over crucial details, the Soviet authorities occasionally give the impression that they have no space weapon programmes whatsoever. In a recent letter to the Boston-based Union of Concerned Scientists (UCS), Soviet General Secretary Mikhail Gorbachev declared that, at present: 'The Soviet Union is not *developing* attack space weapons or a *large* ABM system' (1985 – emphases added). But other Soviet representatives go further, declaring that their country: 'does not have any space weapons, [and] is not engaged in research on space attack weapons' (*International Herald Tribune*, 1 October 1985).[1] Much can, however, be said about Soviet programmes, with reasonable probability, and we have done the best we can in the circumstances.

We have not always been able to separate the description of space weapons, and of possible devices to counteract them, from the discussion of their military value, or lack of it, and of the prospects for international agreement to restrict their development or deployment. But the major issues in these important areas, which became subjects of world-wide debate during 1985, are treated separately in chapters 10 and 11. In chapter 12 some other political effects of current military developments in space are considered.

This book seeks to provide the basis for an overall understanding of space weapons, covering not just their technical problems and possibilities but most aspects of their present significance for humanity. We are confident that the important technical, military and political issues concerning space weapons are not too heavy or complex for most sensible people to understand.

The world's first ever 'space weapon' experiment was a series of three high-altitude nuclear test explosions carried out above the South Atlantic by United States forces in August and September 1958, known as Project Argus.

The idea behind Project Argus was conceived before Dr James Van Allen's theory about the existence of natural radiation 'belts' around our planet had been confirmed by the United States' first satellite (the world's third), Explorer I, launched on 31 January 1958. The Argus experiment was intended to find out whether nuclear explosions on the edge of space could use properties of the Earth's magnetic field to produce an artificial shell of rapidly moving electrons around the planet, that would remain in place for some time. As an experiment, Project Argus succeeded brilliantly. The explosions managed to create just such an effect, and to do so at an altitude between the major areas of natural radiation, where it could be carefully mapped by another satellite, Explorer IV. A report was published a few months later by the scientist at the University of California's Livermore Radiation Laboratory whose idea it had all been, Mr N. Christofilos (1959). It gives the impression that there had never been anything more to Project Argus than a very expensive and environmentally irresponsible probing of one of the less fascinating secrets of nature.

In reality, however, Project Argus was conceived from the start as an experiment to study a possible kind of weapon. To the extreme vexation of the American government officials concerned, details of the experiment became known to journalists at the *New York Times*, which published a detailed account of it all (19 March 1959) before an official version of events could be issued. There is little doubt that but for the *New York Times* and their unknown informants there might never have been an official version at all. As things fell out, it was obliged to be a fairly detailed one.

The Livermore Laboratory's Director, Dr Herbert York, was the first person to whom Mr Christofilos confided his idea, and was also largely responsible for getting it tested by means of the costly and complex experiment already referred to. At a Congressional Hearing on 10 April 1959, Dr York explained the Project's origins quite frankly:

> In the fall of 1957, as we all very acutely know, the Russians launched their first sputnik. Mr Christofilos was really personally affected by this as a great many people were, and he was considerably stirred up by the fact that the Russians had pulled off quite an accomplishment and set himself to thinking, trying to think up what he could do about this.
>
> As a result, he invented what has become known as the Argus effect or the Argus phenomena or what have you . . .
>
> What Christofilos had in mind at first – when he first got the idea for the use of this – was that this would produce on any object which might be orbiting the earth and pass through this region, it would produce high-intensity radiation inside. And therefore this could be used as a means of stopping or preventing manned satellites. That was an original idea. That is what he invented it for, a means for prohibiting manned satellites.
>
> As I said, he was sort of personally upset by sputnik and he decided he was going to do something about it and came up with this invention. (H. Science and Astronautics, 1959: 1–2)[2]

The Argus effect did not in fact lead to a workable anti-satellite weapon, nor to an anti-missile weapon either, for which it and related phenomena were also considered. But that does not affect the point of the example, which is to illustrate the tendency, throughout the first decades of human dabbling in the shallow margins of space, for many an 'accomplishment' to be seen by someone else as a threat which it was vital 'to do something about'. Those 'somethings' are space weapons.

2
The Beginning of Space Weapons

Even if it were never actually to be used, the strategic value of such a space mirror is so great that one of the advanced states will surely undertake the enterprise in the foreseeable future, despite having to finance the necessary investment during peacetime.

Hermann Oberth, 1928

What is usually considered the first comprehensive technical study of the way rockets could be used to launch space vehicles, including artificial earth satellites with people on board, was published in Munich in 1923, when Ronald Wilson Reagan was just twelve years old. Better known to scholars in its later, revised version as *The Way to Space Travel*, it had been written by a Romanian trainee mathematics teacher called Hermann Oberth. Its interest for present purposes derives from the third section, in which Oberth suggested certain applications for the principles he had earlier explained with great technical thoroughness. Amongst them was a proposal for placing an enormous sectional solar mirror, 1000 kilometres across, in orbit around Earth. Oberth suggested such a device could either serve benign economic purposes, or else function as a massively destructive and invincible weapon of war (1929: 355).[1] Despite the space industry's public relations emphasis on the peacefulness of their products, the possible use of space machines of all kinds for military purposes has seldom been out of consideration by their designers ever since. Thirty years later, just as the first US satellite project was announced, Wernher von Braun brought Oberth to America as an adviser to the US rocket programme.

In the two decades before 1945, all those working on long-range rockets, including the German scientists who developed the 'V' weapons of World War 2, were well aware of their eventual potential as launchers for space vehicles. But war confined the work of scientists on both sides to weapons that might be rapidly available. Articles about Oberth's orbital mirror-weapon could be published in wartime German newspapers only because it was in no way a feasible project at the time. Potential military uses of space were shelved for the duration.

After 1945 the technology of rocket propulsion made rapid strides. Progress was especially fast in the United States, which was able to combine the results of its own extensive wartime studies with the experience of leading German rocket scientists, brought to the country in 'Operation Paperclip' (McGovern, 1965). Sounding rockets, carrying instruments straight upwards to collect data on the upper atmosphere and the fringes of 'outer space', went higher and higher. Then in February 1949 the US Army fired a two-stage rocket to a height of 415 kilometres, a record which stood unbroken for the next eight years. It began to seem it would soon be possible to place machines into orbit around Earth as artificial satellites, or 'moons' as they were sometimes called by analogy with Earth's only natural satellite.

Early post-war studies and proposals for artificial satellites focused largely on their potential scientific role, and on such non-destructive and auxiliary military uses as reconnaissance, communication, and navigation – all of which predictions have been amply fulfilled in the event. However, one former senior official is on record to the effect that consideration was given from the outset to possible anti-satellite devices (Stares, 1985: 106). By contrast, the influential 1946 study of possible applications for military satellites, carried out by the think-tank within Douglas Aircraft that would shortly become the RAND Corporation, claimed that a military satellite would be 'an observation aircraft which cannot be brought down' (Douglas Aircraft, 1946: 17).

In the early 1950s US rocket programmes began to languish under the restraints of tight budgeting, both because of the lack of interest of senior USAF officers, headed by General Curtis LeMay, and because of the requirements of the Korean War. Enthusiasts for the 'new frontier' of space turned for support to the American public, which had acquired and alarmingly combined a typically post-war liking for science fiction, and a fascination for revelations about 'flying saucers', with an intense Cold War anxiety about the 'Communist menace' of the Soviet Union. Thus many Americans were ready to be thrilled and enthused by preparations for the real thing, provided they would take place beneath the emblem of the Stars and Stripes.

One very influential publication was the special 'Space-Flight' issue of *Collier's* magazine (22 March 1952). Wernher von Braun contributed a long and detailed account of the large manned satellite he had been offering his new homeland since the day of his arrival in 1945 (1952a). It would be a 76 metre wide ring-shaped structure (later derided by one critic as a 'space-dunking doughnut'), placed in a two-hour polar orbit (figures 2.1; 4.2) at an altitude of 1730 kilometres. Not only would such a space station be 'a watchdog of the peace', by virtue of the constant surveillance of the planet by its 'technicians . . . using . . . powerful telescopes'. It would also, asserted von Braun, be a very effective atom bomb carrier. He explained in Oberthian tones that:

> Small winged rocket missiles with atomic warheads could be launched from the station . . . [and] could be accurately guided to any spot on earth.
> [This] would offer . . . the most important tactical and strategic advance in military history. (1952a: 74)

Von Braun gave more details of his space weapons proposal in a book based on the magazine articles (Ryan (ed.), 1952). There, he alternated the carrot and stick approaches to security in a manner that has since become all too familiar:

> [W]hether in the hands of a single peace-loving nation, or in the hands of the United Nations, the space station would be a deterrent which might cause a successful outlawing of war . . .
> The space station, with all its potentialities for exploration of the universe, for all kinds of scientific progress, for the preservation of peace or for the destruction of civilization, can be built . . . Perhaps the military reasons for establishing such a station are in the long run among the least significant, but in the existing state of the world they are the most urgent. Unless a space station is established with the aim of preserving peace, it may be created as an unparalleled agent of destruction – or there may not be time to build it at all.
> Under the impetus of these considerations, therefore, perhaps the space station will become a reality, not a generation hence, but in – say – 1963. (1952b: 56,68–9)[2]

The *Collier's* 'Space-Flight' issue appeared three years before the world's first experimental satellite project was announced, and more than five years before the first such object was actually

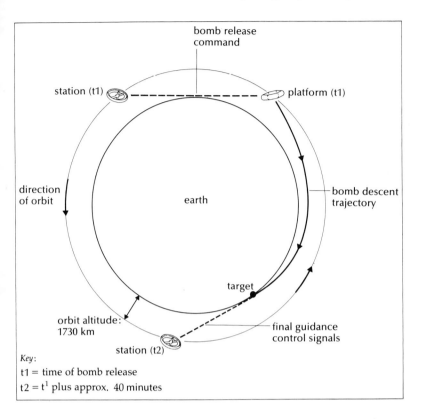

bomb release
command

station (t1)

platform (t1)

direction
of orbit

earth

bomb descent
trajectory

target

orbit altitude:
1730 km

final guidance
control signals

station (t2)

Key:

t1 = time of bomb release

t2 = t^1 plus approx. 40 minutes

Figure 2.1   The von Braun Space Station and Bomb Platform

placed in orbit. Nor had decision-makers yet realized the full
implications, for missile technology, of the 1950 US decision to
develop the thermonuclear (hydrogen) bomb. Previously, the
inaccuracy of early missiles had prevented their adoption by the
United States, though not by the Soviet Union, as means for the
delivery of atomic bombs. But by the mid-1950s it became
accepted that the vast and unlimited increase in explosive power,
represented by thermonuclear warheads, would make even pretty
inaccurate missiles a militarily suitable method of delivery. Once
development of long-range missiles was made a priority in the
United States, military satellites naturally followed suit. The
USAF was authorized in 1954 to begin the studies that led to its
first reconnaissance satellites, SAMOS and Discoverer, in the
early 1960s.

Thus von Braun's ideas about nuclear bombardment from space

were aimed, in part, at providing a technical solution for a particular problem – the military unwieldiness of inaccurate long-range rockets in combination with merely atomic-sized warheads – that was on the eve of being tackled far more directly and effectively from both ends, through improved missile guidance and far larger warhead yields. Still, his article firmly established the idea of possible deployments of nuclear weapons in space orbit as a long-term policy option, both with public opinion and amongst official decision-makers.

In the immediate debate von Braun's critics argued, in a pattern that would later become familiar, that any such space station would be a highly vulnerable and valuable sitting target, more threatened by than threatening to an enemy. And just as with the idea of setting up a nuclear missile base on the Moon, which began to be canvassed by military space enthusiasts around the same time, it was objected that any technology capable of hoisting missiles accurately up to and down from such a station would surely also be capable of sending them with equal precision, and with far greater promptness and less expense, from one point on the planet's surface to any other. Von Braun responded that:

> the station could defend itself in the case of attack and . . . could prevent rival stations from being established. (1952b: 56)

He does not seem, in retrospect, to have won the military argument. But he did leave people with an abiding impression that one or other side might before long feel it had a use for nuclear weapons in space. During both the Eisenhower and the Kennedy administrations it would remain quite usual for hedging references to be made to such a possibility, right alongside the reasoned decisions in official policy papers to the effect that it would make no real military sense (US NSC, 1958; Parson, 1962: 204–7).

When the space age actually dawned in October 1957 with the successful orbiting of Sputnik 1 by Russian scientists, there were two distinct strands to the negative response felt by so many Americans. The first was jealousy. It had been widely expected that this achievement would be an American one. Its importance had been emphasized to the point of exaggeration for at least three years, precisely because it was assumed by all concerned to be a prize almost certainly within the American, and only the American, grasp. For example, a writer in *Scientific American* at the end of 1955 might begin his account of the coming great historic event of the first satellite launch in general terms:

> Some time in the next few years a new object will appear in the heavens . . . It will be mankind's first feeler on the space frontier.

But soon enough the obligatory patriotic note was sounded:

> The man-made satellites that the U.S. is planning to launch for the International Geophysical Year . . . (Newell, 1955)

Even more revealing, for by then a careful study of the actual state of the 'race' should have prompted rather more caution, was an account of the US satellite programme which appeared only a few months before Sputnik 1. The book opened with a quasi-fictional narrative of the launching of 'the first man-made moon', set firmly at Cape Canaveral. Even if several attempts should fail, the author claimed, 'Satellite 1' would be launched from Florida. Not from Tyuratam, a town in Kazakhstan the world had not yet heard of. Little can he have relished the irony with which his only caveat would later read:

> This is reality. Perhaps not one hundred per cent reality, for some of the things may not happen exactly as described. (Buedeler, 1957: 12)

Such examples of American over-confidence on the threshold of space could be multiplied many times over from these years.

By contrast, Soviet announcements about a similar project came later, and usually in a lower key. As (rarely) laid before the international press, they had little convincing detail. The advanced state of Soviet preparations, revealed by several open references in domestic publications, had been picked up in the West, both in military intelligence estimates and in open but relatively specialist journals (*Aviation Week*, 29 October 1956; Parry, 1960: 183–7). But such warnings were ignored at higher levels of the US government and were largely unknown outside it.

The second element in the dismayed American reaction was fear. Before the United States had even put so much as a 'tin grapefruit' (as the 1.5 kg experimental Vanguard 1 satellite came to be derided) into space, the first Sputnik was followed by a second larger one (at least 3150 kg), this time with a live passenger, the bitch Laika.[3] As a direct result of the public fascination a few years before with the possibility of satellites armed with nuclear weapons, Americans now viewed the Russian achievement with growing apprehension. For one thing, the Sputniks meant that an earlier Soviet claim that year, to have successfully tested the first intercontinental-range ballistic missile, would now have to be accepted as the unwelcome truth. This brought the nuclear threat directly and inescapably to every Main Street for the very first time.

The technical and political post-mortems were intense and thorough. They resulted in a two-fold American determination to 'do something' about the Russian lead in space. First, of course, was a simple resolve to catch up and surpass the rival superpower in space technology and exploration. But second, the fear of military applications of satellites, either for nuclear bombardment or for less far-fetched purposes such as intelligence gathering, led to renewed interest in possible techniques for their destruction. There was also the thought that the Soviet Union might make progress with such anti-satellite weapons, in view of the resentment against Western spy-planes which they had from time to time expressed by shooting them down.[4]

Before considering actual space weapon programmes, however, it is important to notice that a vital part of any such system is always the ground-based detection and tracking capability on which both its development and its possible use are completely dependent. Part of the angry humiliation of the American response to the Sputniks was a result of their having been launched before the initial US satellite-tracking system was even ready. Naturally enough, its construction was immediately accelerated. And between 1958 and 1963 a trio of massive radar installations was built in Greenland (Thule), Alaska (Clear) and Britain (Fylingdales Moor), to form the Ballistic Missile Early-Warning System (BMEWS) (Englebardt, 1966: ch. 10). Other tracking telescopes and radars were added throughout the 1960s at stations around the world, and coordinated by the joint US–Canadian North American Air Defense Command (NORAD) into the Space Detection and Tracking System (SPADATS) (p. 239).[5]

The role of such systems was shown by the incident of the so-called 'Dark Satellite'. At the end of 1959 an unknown object was detected in orbit by a US Navy space surveillance system. The Pentagon was reproved for failing to take any serious notice of this satellite for some weeks, and the possible threat which it might represent to the United States was then discussed in secret at the highest level. It was finally identified as the 136 kg film-recovery capsule from the failed Discoverer 5 mission of the previous August, which had been ejected into a higher orbit instead of back down to Earth as intended.[6]

Even before Sputnik 1 had aroused Mr Christofilos's destructive response and the resulting Project Argus, senior technical officers in the US forces had begun to draw up plans for anti-satellite weapons. The first USAF study began in 1956 and General Gavin, head of US Army research and development, initiated another early anti-satellite project in June 1957 (Gavin, 1958: 16). However, it was another five years before the Army was able to put such a weapon into full development.

The first actual test of a missile for intercepting satellites took place on 13 October 1959, as an extension of Project Bold Orion, a USAF programme to develop a potential air-launched ballistic missile. Tracking facilities were primitive by today's standards, but it was later calculated that the missile had passed within 6.5 kilometres of its 'target', the Explorer IV satellite, close enough to destroy it with the radiation from a one megaton nuclear warhead.[7]

Another technically bold approach to the anti-satellite mission was the US Navy's 'Early Spring' study in the early 1960s. The idea was to use a non-nuclear warhead, launched on a Polaris missile, with an ability to 'hover' for up to 90 seconds at the required altitude until the target satellite was detected by its optical systems. Together with instructions relayed up from ground control, automatic homing systems on the asat itself would then have manœuvred it close enough to destroy the target with a swarm of steel pellets. This proposal was never more than a paper exercise. But the Navy did carry out exploratory tests for an air-launched asat in 1962 (Peebles, 1983: 81–2; Stares, 1985: 109–11).

By that date, a mere two years after the first workable 'pulsed laser' device had been built in the laboratories of the Hughes Aircraft Company, research had begun into possible applications of the new technologies of 'directed energy' for space weapons. Not surprisingly, in view of the low energies then available from such devices, one early programme, Project Black Eye, was aimed simply at disabling enemy satellites by blinding their sensors (*Aviation Week & Space Technology*, 25 June 1962). But already in March 1962 the USAF Chief of Staff, General Curtis LeMay, was prophesying that:

> beam-directed energy weapons would be able to transmit energy across space with the speed of light and bring about the technological disarmament of nuclear weapons . . . Whatever we do, the Soviets already have recognized the importance of these new developments and they are moving at full speed for a decisive capability in space. If they are successful, they can deny space to us. (*New York Times*, 29 March 1962)

Lasers did not start to be foreseeably powerful enough for weapons, however, until the late 1960s and early 1970s (see chapter 7).

The American work on possible space weapons that has been mentioned so far neither spent any great sums nor led to any actual deployments of anti-satellite or other systems. Before turning to US programmes which did both, the earliest known Soviet efforts in the same field should be surveyed.

To begin with, as we have seen, American fears of Russian nuclear bomb satellites were wholly of their own making. Soon, however, the Soviet leader Nikita Khrushchev decided to play on those anxieties. In August 1961 he referred to Soviet rockets that could be 'directed to any place on Earth'. In March 1962 he grew more explicit:

> [W]e can launch missiles not only over the North Pole, but in the opposite direction too . . . Global rockets can fly from the oceans or other directions where warning facilities cannot be installed. (In Peebles, 1983: 62)

Within a year the Commander of the Soviet Union's Strategic Rocket Forces, Marshal Biryusov, was claiming that:

> It has now . . . become possible at a command from Earth to launch rockets from satellites at any desired time and from any point of the satellite's trajectory. (In Peebles, 1983: 62)

Western analysts seem to have been less alarmed by the Marshal than by Khrushchev. Their fear was, not that the Soviet Union might develop a cumbersome orbiting bomb platform of the sort once proposed by von Braun, but that it was thinking of a method of delivering nuclear warheads known as a 'Fractional Orbit Bombardment System' (FOBS). With this technique, a launch rocket is fired at a lower, less efficient angle than normal ('depressed trajectory'). The warhead is released into an orbit as low as 200 kilometres, from which it must later be deflected downwards by firing a retro-rocket. At this point it would be only about 800 kilometres, or three minutes, from the target. Such a re-entry would usually be carried out before the warhead had actually completed one full orbit – hence the title bestowed on the system by the Pentagon. This system would be capable of attacking an enemy from any direction and would have unlimited range. It might be especially effective as the first wave of an attack, to destroy radars and other components of the kind of anti-missile defences that were being developed in the 1960s (see chapter 3).

The Americans, with their lead in submarine-launched missiles and with numerous nuclear weapons bases around the Soviet Union, were never attracted by the FOBS delivery method. They were accordingly willing to follow up an independent Canadian initiative for a ban on placing nuclear weapons in space, which was finally adopted by the UN General Assembly in October 1963, in

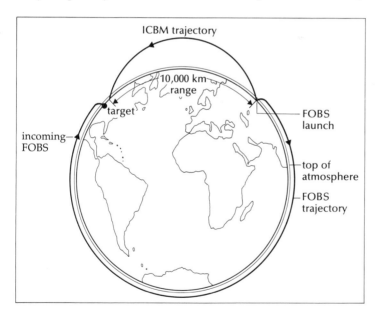

Figure 2.2   ICBM and FOBS Trajectories

the form of a Mexican proposal. However, the resolution did not forbid the *development* of systems capable of being used for that purpose. And it is hard to see how any such stronger ban could have been worded, without putting an end to all space programmes.

From late 1966 to the end of October 1967 the Soviet Union conducted a series of tests in which the payload did not pass over the United States, and in which it was brought down by retro-firing before completing one full orbit. This was the kind of testing that a FOBS would need. When the launch vehicle was paraded in Red Square it was said to be capable of both inter-continental and orbital flights. Though Soviet representatives later rebutted the American charge that they had been developing an orbital bombardment system, there seems little doubt in the matter. Thereafter similar Soviet tests were limited to about two a year, suggesting that the FOBS was now regarded as an operational weapon.[8]

To tell the rest of the Soviet FOBS story, operational flights were discontinued after 1971, perhaps because the ABM Treaty signed the following year (appendix 1) meant there would be virtually no US anti-missile defences, hence no need to by-pass them by such an expensive and unreliable method. The 1979 SALT 2 Treaty included a provision that the Soviet Union should destroy 12 of its 18 FOBS launch-pads and convert the remainder to other uses. Despite the refusal by the United States to ratify this treaty both governments agreed to observe its provisions, and the Soviet Union has complied with this one.

By contrast with the United States, the Soviet Union appears not to have had any full-scale asat development programme before the late 1960s, though the idea of such weapons did receive some more or less speculative Russian discussion in the aftermath of the Sputniks, and military manuals occasionally referred to the problem of defence against space threats in general. There were some Soviet diplomatic protests about US reconnaissance satellites at first, but these died away as their own programmes slowly became established from about 1963. Western conjectures that a special anti-satellite version of the 'Galosh'[9] anti-ballistic missile was deployed in the mid-1960s have never been confirmed, and seem to have been a case of 'mirror-imaging', in this case transposing the US Nike–Zeus asat (p. 22) into a subjectively apprehended Soviet counterpart.

Before the innocuous explanation of the 'Dark Satellite' incident had been discovered, Eisenhower's Science Adviser, George Kistiakowsky, confided to his diary the fear that it would be exploited by USAF space weapon enthusiasts in their lobbying for an asat programme (1976: 245–6). The project that concerned him had originated in an Air Force study as early as 1956, and was known as SAINT, from its original conception as a SAtellite INTerceptor and inspector. Though both the Eisenhower and the Kennedy administrations decided that SAINT should be restricted to the inspection role only, there was continued friction between the White House and the Pentagon over the issue, and some of the programme's Air Force backers continued to hope and unofficially to suggest that the satellite interception role could be put back into SAINT at a later stage.

As it was, the inspection requirements alone proved too difficult and too expensive to meet by this method. The initial programme called for a payload which could be launched with maximum promptness and reliability into an elliptical orbit with its high point close to that of a test target at 640 kilometres. The SAINT would then be manœuvred into an identical orbit to that of the target, and proceed to inspect it with several advanced sensors, including television, infra-red cameras, and nuclear radiation detectors, to discover whether it was armed with a nuclear weapon. The idea suffered both from the political objection that it undermined the claims made for the essentially peaceful nature of the US military space programme, and from the USAF's reluctance to fund SAINT properly if, as Kennedy insisted, it was to produce neither a weapon nor a manned space vehicle. The project was finally abandoned in December 1962. The disagreements to which it had given rise were partially resolved by giving the Air Force greater access to NASA's manned Gemini programme, under which further secret experiments with satellite inspection methods continued.

The political objection to anti-satellite work with objects put into orbit could be softened, however, by adopting an alternative technique of 'direct ascent'. Since the Eisenhower administration's decision, in 1960, to stop referring to long-distance missiles as 'space weapons', it was thought possible to get away with regarding missiles fired 'straight up' at satellites in low orbit as being outside the space programme proper. This approach was also technically more feasible for some missions if large nuclear warheads could be used. By September 1964, President Johnson could announce:

> We have now developed and tested two systems with the ability to intercept and destroy armed satellites circling the earth in space. I can tell you today that these systems are operationally ready, that these systems are on the alert to protect this nation and to protect the free world. (In Stares, 1985: 106)

Johnson's careful phrasing about 'the ability to . . . destroy armed satellites', as if it were not also an ability to destroy the unarmed variety, was a reference to projects under way in different services. The Army began work on an anti-satellite version of its anti-ballistic missile system, Nike-Zeus, in 1957. This was accelerated from about 1962 to develop a weapon with a range of 240 kilometres and a warhead of around one megaton. Tests were conducted in New Mexico and then at Kwajalein Atoll, at the western end of America's Pacific Missile Range, and by mid-1963 the system was declared operational. However, this seems to have amounted to little more than a single missile on stand-by, and the requirement that this status be maintained was dropped after 1964. In 1966 Defense Secretary McNamara decided to scrap the Nike-Zeus asat. However, the ABM system to which its follow-on missile was meanwhile giving rise, the Nike-X/Sentinel (later Safeguard) project, was assessed by the Pentagon's Director of Research and Engineering, Dr John Foster, as having:

> an anti-satellite capability against satellites passing within the field of fire of the deployed system. (S. Aeronautical and Space Sciences, 1969: 856)

Since in the end it was to be deployed for no more than a few weeks (p. 35), it can hardly be said to have made much of a contribution to the security or insecurity of either side.

Work on the Air Force's direct-ascent asat began in 1962 and was accelerated in response to a directive from President Kennedy a year later. This was part of the 'Contingency Plan for U.S. Reaction to Soviet Placing of a Nuclear Weapon in Space', drawn up by Raymond Garthoff in May 1963 (Stares, 1985: 88), with which the administration wanted to support its negotiating position at the United Nations. The emphasis was on as simple a system as possible, able to be fired at only two or three days' notice in a crisis, and using existing technology for early deployment.

The time allowed for rapid response was later cut by half again. The launcher selected was the Thor, numbers of which were now surplus as the missile bases in Britain were about to close. By November 1963 initial training and prototype deployments had been completed. Two missiles were kept ready for launch at Johnston Island in the Pacific, with four back-ups at Vandenberg Air Force Base (AFB) in California. The first three tests were successful, and despite a technical malfunction on the fourth the unit was declared operational in June 1964, with both Thor anti-satellite missiles now placed on alert. The missile had a 1.5 megaton warhead, and could intercept targets up to 2700 kilometres away, to an altitude of 370 kilometres. President Johnson's later announcement of the deployment, however, seems to have been a political response directed less at deterring the Soviet Union than at refuting Senator Goldwater's charges that the administration had done nothing about possible threats from space.

Program 437, as it was called, remained nominally operational till 1975, though only six of its sixteen launches during this time were 'combat tests' proper. Gradually the launch facilities and the missiles were turned over for use in experimental tests for possible successors, including ABM systems, for which non-nuclear homing devices were beginning to be considered. What happened was that evaluation of the high-altitude nuclear tests of 1958 and 1962 showed that nuclear explosions in space would not only damage one's own satellites as much as those of the enemy, but would threaten similarly indiscriminate disruption, by the electromagnetic pulse effect (box 2.1), to vital communication links and other electronic systems needed for war-fighting on Earth. This nuclear 'own-goal' bestowed an obsolescence on the anti-satellite version of the Thor rivalled only by its previous brief career as a militarily useless Intermediate Range Ballistic Missile (IRBM).

In 1970 the missiles and most of the personnel were withdrawn to Vandenberg AFB, with a far lower operational status, and reaction time extended to thirty days. This was the formal situation until the programme was officially closed down in 1975, but there was another side to the story. Johnston Island is no more than a flat speck of coral, some of it artificially constructed, 1300 kilometres south-west of Hawaii and nowhere near anywhere else. In August 1972 Hurricane Celeste forced the evacuation of all personnel and caused massive damage to the computers and other delicate and expensive machinery. Program 437, and with it the first phase of American anti-satellite efforts, was thus effectively ended by an 'act of God' which could only be accepted after the fact by inspectors from its computers' manufacturers, Univac, and by the US Air Force.

Box 2.1   Electro-Magnetic Pulse

The electro-magnetic pulse effect (EMP) is a complex and still imperfectly understood physical process. The gamma radiation from a nuclear explosion in the atmosphere leads to ionization of air molecules. This generates a brief surge of electro-magnetic radiation across many different frequencies, but above all in the long-wave radio band. Electrical circuits and indeed any conductors, such as radar aerials, respond to this with a surge of current that can be high enough to damage unprotected equipment. The phenomenon reaches a peak within a fraction of a millionth of a second, which makes it difficult to protect systems by installing automatic cut-outs. Systems using transistors and microchips, such as missile-guidance computers, are especially vulnerable.

At higher altitudes, say, above 20 kilometres, the gamma radiation can travel very great distances through the thin atmosphere. At 300 kilometres a single megaton-yield explosion would generate the effect over continental areas.

Nuclear explosions in space can also have a more localized EMP effect on satellites and other space objects. Both gamma rays and X-rays would cause 'system-generated' EMP in the material of such objects, again creating a surge of potentially damaging current.

Source: Glasstone & Dolan, 1979.

3
The First Anti-Missile Weapons

*If Nike-Zeus does not fill the bill, where, pray, is the system which
does? If there is no effective anti-missile program, why isn't there?*
S. T. Possony, 1963

The extreme difficulty, or as it seemed the impossibility, of using
counter-weapons for defence against long-range ballistic missiles,
was recognized from the outset. The great speculative pioneer in
the field of rocketry and space travel, the Russian Konstantin
Tsiolkovsky, remarked on this aspect of the matter in the early
years of our century. By the 1920s it was well understood amongst
the growing but still small number of pioneers in practical rocketry
in the Soviet Union, the United States and Germany. Another two
decades, and the task was actually addressed for the first time. At
the end of 1944 the British Air Ministry began to evaluate a
proposal to counter the world's first operational ballistic missiles,
the German V-2s, by exploding barrages of anti-aircraft shells in
their radar-predicted paths. Scientists disagreed about the feasibility of the idea, with the primitive radar and fire-control technologies of the day, and it was not officially adopted (Collier, 1976:
135–6). Just as this decision was reached, the last V-2 fell on
London. The German units firing the weapons had been forced to
retreat out of range by the Allied advance. This was the first
practical demonstration of an important military principle, that the
first line or 'layer' of any anti-missile defence consists in offensive
action against the relevant launch facilities.[1]

At the end of World War 2 the idea of uniting the most
formidable new weapons of the defeated and the victorious sides,
the long-range missile and the atomic bomb, had become an
obvious medium-term possibility.[2] This brought closer the long
anticipated prospect of the total vulnerability of all major powers
to each other's long-range missile forces. The problem arises

mainly from the enormous speed to which such a missile is accelerated by the pull of gravity, as it falls towards the surface of the planet after climbing to the top of its elliptical trajectory. In the case of modern intercontinental ballistic missiles (ICBMs), which reach altitudes of more than 1200 kilometres, that speed is approximately 7 km/sec as the warhead re-enters the upper atmosphere. It can still be moving at up to 4 km/sec, depending on its design, as it approaches the lowest feasible altitude for any ABM system to stop it. In the mid-1940s Truman and other leading members of the US government tried to put a bold face on the prospect of American defencelessness, with claims that anti-missile technologies were already at a promising stage of development. (Thus began the exchanges of bluff and counter-bluff with which the history of our subject is bedevilled.) Ten years later, however, long-range nuclear missiles were approaching deployment on both sides, with little but fears and promises yet to show by way of any direct shield against them.

As usual, the history of US programmes can be traced more clearly than that of their Soviet equivalents. However, since the first ABM deployments were on the Soviet side, we shall tell their part of the story first. Like the United States, the Soviet Union had imported numbers of German rocket engineers and scientists immediately after World War 2. In particular, some of these personnel had worked on the 'Wasserfall' anti-aircraft missile project, a member of the V-2 family with a planned operational ceiling of 10 kilometres and a lateral range intended to reach 32 kilometres, once a system enabling its human ground-controllers to track and guide it by means of a radio beam had been perfected. In an intensive and successful development programme, Soviet and German scientists launched their first captured V-2 fourteen months after their counterparts in the United States (McGovern, 1965: 223), and went on to enable the Soviet Union to deploy one of the world's earliest operational anti-aircraft missiles, the SA-1 'Guild' in NATO terminology, by about 1954. If its physical resemblance to Wasserfall was slight, the estimated range was much the same. Thereafter successive new Soviet anti-aircraft missiles were developed and very widely deployed, in keeping with the emphasis on defence of the national territory which is so dominant in Soviet military thinking.

As the process continued, Western analysts became concerned that improvement of surface-to-air missile systems (SAMs) for anti-aircraft defence would eventually endow them with a growing degree of anti-missile capability. This 'SAM-upgrade' route to anti-missile defence, as it is known in military jargon, avoids the high cost and risk of large-scale purpose-built ABM systems. It has certainly been explored in the Soviet Union, where one authoritative text was revised in 1967 to include the statement that 'modern air defense is built to be antiaircraft, antimissile and antispace united in a single system' (Sokolovskiy, 1975: 297). One early, brief, and limited deployment of surface-to-air missiles, in the Leningrad area, is thought in the West to have been an attempt at providing a limited capability against ballistic missiles as well as high-flying aircraft. The original system, deployed around 1963 with the SA-5 'Griffon' interceptor missile, was scrapped so quickly that it must have been found decidedly unsatisfactory. Though the Griffon was succeeded by a wider deployment of its successor, a three-stage solid-fuel missile with either conventional or nuclear warheads, known as the SA-5 'Gammon', this is now generally agreed to be no more than an anti-aircraft system. Its very slight anti-missile capabilities would be no greater than those of the American Nike-Hercules (p. 27), which has also been tested against missiles but has not been thought to constitute a serious anti-missile deployment.

A distinct Soviet programme for developing ABMs as such first became known to the West after its detection by an American U-2 spy-plane in April 1960 (Stevens, 1984: 191). The installations were in a remote interior location near the village of Sary Shagan, on the edge of Lake Balkash. A suitable missile range for ABM testing, and six very large early-warning 'Hen House' radars, dispersed around the edge of the Soviet Union, had been constructed in the late 1950s. Then in the early 1960s there were ABM test firings, sometimes associated with high-altitude nuclear explosions. These were probably aimed at discovering how well the system would function, with its many vulnerable electrical components, under the conditions of an actual nuclear war. By late 1961 Marshal Malinowsky felt able to declare to the 22nd Party Congress:

[T]he problem of destroying enemy missiles in flight has been successfully resolved. (*Pravda*, 25 October 1961)

Khrushchev repeated the point in his own way, with a boast that Soviet forces could now 'hit a fly in space' (*New York Times*, 17 July 1962). The fly-swat he was referring to, first shown at a Moscow parade in November 1964, was of no mean dimensions. A three-stage missile over 18 metres long, with a nuclear warhead of between one and three megatons and a range of 320 kilometres, it was much larger than the Minuteman missile whose warhead it was designed to intercept.

One view of the rationale behind such a development was given at the time in an article by a Soviet Major-General:

> [T]he creation of an effective anti-missile system enables the state to make its defences dependent chiefly on its own possibilities, and not only on mutual deterrence, that is, on the goodwill of the other side. (Talensky, 1964: 18)

This ABM-1 'Galosh', as it was known in the West, was deployed in the late 1960s in a ring some 65–100 kilometres around Moscow, with eight squadrons of eight missiles each. Several years later they were replaced by an improved version, the ABM-1B, said to be capable of stopping and starting the liquid-fuelled motor of its third stage. This might mean it had a limited capacity to 'wait on' near the top of its trajectory and accept new guidance instructions for re-targeting, thereby enhancing the battle-management aspects of the system. However, the warhead is not thought to have had any capacity for independent homing on its target.

The Galosh launchers were served by two large phased-array radars (LPARs–box 3.1) – 'Dog House' and 'Cat House' – for early target acquisition,[3] and by sixteen 'Try Add' mechanically steered target and missile tracking radars. The battle-management capability of the latter is limited to handling one or two interceptions at a time, using data received from the LPARs. Despite signs that twice as many ABMs had at one stage been proposed for the system, work on any additional sites ended in 1968 and the system became operational in about 1970 as described, with only sixty-four ABM-1 interceptors. In the ABM-1B modernization these were reduced to thirty-two, thought to be nuclear-armed, which remained operational until very recently. (Further development of the Moscow ABM complex is described in chapter 9.)

However, 1970 also saw the initial deployment of a new US weapon technology, missiles carrying several independently aimed warheads (MIRVs). Developed at first for other reasons,[4] this now offered the shortest route to overcoming ABMs, simply by overwhelming their limited interception capacities with very large numbers of warheads (p. 57). But even if such US missiles had not

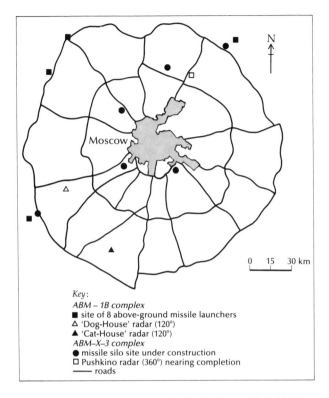

Key:
ABM – 1B complex
■ site of 8 above-ground missile launchers
△ 'Dog-House' radar (120°)
▲ 'Cat-House' radar (120°)
ABM–X–3 complex
● missile silo site under construction
□ Pushkino radar (360°) nearing completion
——— roads

Figure 3.1   Moscow Galosh Systems (ABM–1B & ABM–X–3)
Source: US Department of Defense, 1985c.

been deployed almost as soon as the first Galosh system was completed, experts are generally agreed that it could have had little military value. At all stages it has been extremely vulnerable to decoy methods, such as the dispersal of metallic 'chaff' alongside attacking warheads, and its main radars are wide open to destruction by any warheads that might 'leak' through. The actual interceptor missiles themselves have been 'soft' (unprotected against nuclear blast effects) as well, until the current modernization. And there were even some possible 'attack corridors', left uncovered by the Soviet battle-management radars, along which attacking missiles could reach Moscow (Stevens, 1984: 197).

Thus even if the Moscow area is assumed to be a particularly important target for US purposes, the Galosh system can have had very little effect on the strategic 'balance' between the superpow-

Box 3.1   Radar for Ballistic Missile Defence

Radar (RAdio Detection And Range-finding) was first applied to military operations in World War 2. It is crucial to any anti-aircraft or anti-missile defence. The basic technique consists in sending out pulses of radio microwave energy, with wavelengths between 1 metre and 1 centimetre, a little shorter than those used for FM broadcasting. The pulses are emitted by an antenna which can also act as an aerial, to collect the portion of radio energy reflected back to it by solid, especially metallic, objects. By repeating the process rapidly a radar image can be created, either on a screen or in some other form, which can be used to locate and even partly identify moving objects.

A mechanical radar antenna scans its detection field, or 'fan', by rotating on a fixed base. An alternative technique, also used in World War 2, is to steer the radar beam from a large array of fixed antennas, by adjusting the timing of their individual outputs within the array to create an artificial directionality. Radio emissions at the same wavelength from a simple row of antennas can be timed so that each emits a brief pulse in which the energy waves are 'peaking' at intervals which come later, by some minute fraction of a second, than in, say, the pulse coming from the one on its left. The pattern of emitted pulses will then resemble that which might have been emitted simultaneously (in phase) from a mobile array that had been rotated to face towards the right front. The apparent angle to the real array will depend on the phasing interval ('phase-length') chosen.

What matters for radar, however, is not the ability to create such directionality by phasing, but the ability to change it fast and often, by altering the phase-length. Although the principle of phased-array radar was understood and even applied before World War 2, it did not come into its own until the development, in the 1950s, of electronically operated phase-shifting devices to replace the earlier mechanically actuated ones. Whereas phased-array radars (PARs) using the older technology could at best cover their fans ten times a second, speeds ten times faster than that are regarded as slow for electronically steered PARs.

Modern PARs consist of hundreds or thousands of antennas arrayed in two dimensions, and change the phasing or
'direction' of their output by altering the frequency of the
electro-magnetic signals fed to the array. They are preferable
for anti-missile defence because they can be hardened
against nuclear explosions to a greater degree than mechanically steered antennas, and because they can handle very
large amounts of virtually simultaneous data. The key to
their potential for BMD 'battle management' lies in the
computers and software which enable them to analyse
incoming signals very rapidly, and thus to 'time-share' the
different tasks which must all be done at once. 'Synthetic
aperture' radar applications use moving platforms, such as
satellites, to create the effect of very large antennas and
achieve high resolution in the cross-range dimension.

The electronic capabilities of military radars are hard to
monitor externally, but the power and wavelength of radar
emissions can easily be detected, and provide some clues
about the role of any particular radar. Long wavelength systems, though large, are cheaper for a given range and more
suitable for initial search and detection applications. Shorter
wavelengths, which can use smaller antennas but need more
precise electronic controls, are useful for exact tracking of
objects such as missile warheads.

The size of a PAR is described by its 'power aperture' or
'potential', the product of its average power and the antenna
area. It requires at least a 16-fold increase of average power
to double the effective radar range of a given antenna. Range
can be 'bought' in a similar and often easier way, by enlarging the antenna area instead. This means that PARs for very
long-range detection are too large to be mobile and too
expensive to be built in great numbers. Since radar works by
being unable to 'see through' solid matter, detection fields
are also limited in their lower altitudes by the curvature of
the Earth, leaving more and more 'dead ground' beneath
them at longer ranges. Cruise missiles and ballistic missiles
fired on depressed trajectories seek to exploit this vulnerability.

Sources: Skolnik, 1981; Weiner, 1984.

ers. It is slightly more plausible to see it as in theory capable of thinning, though not completely blocking, the sort of strike on Moscow that could be delivered by a minor nuclear power such as Britain or China.

In the United States the first ABM system to be deployed was also a descendant of anti-aircraft missiles built after World War 2. It was developed by the Army through several years of competition with a rival, purely anti-missile approach promoted by the Air Force. The abandoned USAF Wizard project did however bequeath many design concepts to the Army system, especially in the crucial radar parts of its 'architecture'.

The Army's first anti-aircraft missile system, Nike-Ajax, was completed in the mid-1950s, at the same time as the Soviet SA-1, and carried a conventional shrapnel warhead. It was deployed in thousands in the United States, the Far East and Europe. These were replaced by Ajax's successor, Nike-Hercules, in the early 1960s. This came in two versions, with either a conventional or a small nuclear warhead. Though intended only for interception of aircraft, it was tested successfully against missiles (Gunston, 1979: 171).

In January 1958, in the crisis atmosphere produced in Washington by the Soviet Sputniks, the Army was given exclusive rights to the ballistic missile defence (BMD) mission. For the next three years, however, funds for actual production of prototypes of its new nuclear-armed anti-missile missile, Nike-Zeus, were not made available. Leading government scientists remained doubtful whether any such system could succeed. In view of today's debates about the Strategic Defense Initiative, the points which worried them are worth recalling. Beside the problems of 'saturation' by heavy attack and of radar vulnerability, already mentioned in respect of early Soviet ABMs, they were concerned that even if the system's radars, missiles and computers survived for any length of time, they simply could not handle the complexity of the ABM task for even the least demanding of probable attacks. Above all, the sceptics argued that the system could never be tested under conditions even approximate to those it would have to handle in actual war.

A prolonged political struggle ensued, with the Army running an advertising campaign to spell out exactly where the many portions of the total $410 million contract would be spent. The resulting babel in Congress reached a crescendo with the House

Majority Leader, John McCormack, exhorting the administration to:

> close the gap in our missile posture, muzzle the mad-dog missile threat of the Soviet Union, loose the Zeus through America's magnificent production line. (In Coffin, 1964: 168)

But Eisenhower managed to keep the block on production funding for Nike-Zeus until leaving office in January 1961.

The Army's ABM programme fared no better under Kennedy's Defense Secretary, Robert McNamara. The Zeus did succeed in a test interception against an ICBM in July 1962, but such a 'demonstration' was inadequate to prove the system's effectiveness as a whole. In 1963 it was decided it would be necessary to develop a novel two-tier or 'layered' system, using newly emerging radar, computer and missile-propulsion technologies to overcome the weaknesses of Nike-Zeus. A successor to the Zeus missile, the two-stage Spartan, would be the main defensive interceptor, for destroying enemy warheads just outside the atmosphere or on its fringes ('exo-atmospheric interception'). And an entirely new high-acceleration missile, the Sprint, would be used as a low-level ('endo-atmospheric') sweeper in the last few seconds, against anything that got past Spartan. To do this it would benefit from the improved discrimination between real warheads and the mostly lighter decoys, which would become possible in the last few seconds of their trajectory once atmospheric resistance had begun to retard the latter more than the former.

Both new ABMs were developed and tested satisfactorily against Minuteman and Polaris missiles in the late 1960s. Spartan had a slant range of 160 to 480 kilometres and a five-megaton warhead. Sprint, whose few seconds of flight have been described as 'a controlled explosion', had a slant range of a few tens of kilometres (at most) and an effective operational ceiling of 30 kilometres, with a smaller, kiloton-range nuclear warhead.

The key to the Nike-X system, as it was renamed, was its radars. There was to be a new large phased-array radar for 'perimeter acquisition' of potential targets 2700 kilometres away, so that analysis of their trajectories could begin sooner. This information would be passed to phased-array Missile Site Radars which would direct the anti-missile battle, tracking both targets and interceptors, and sending guidance instructions to the latter. A weakness in the system's overall design was the fact that Spartan would need to receive a signal from the ground in order to detonate. This was one of what may be called the 'self-damage limitation' elements of the system, in this case intended to ensure that only warheads which had successfully reached their targets would explode. Another was a device to prevent the detonation of Sprint warheads below 15 kilometres, to avoid unacceptable damage on the ground.

After considering two Nike-X deployment patterns, the Johnson administration announced in 1967 that an ABM system, now entitled Sentinel, would be deployed north of twenty-five US cities (including Honolulu). Its purpose would be to protect part of the country against light-weight attacks, in what was described as a 'thin' area defence. McNamara made it clear the system could not prevent the nuclear devastation of the United States by Soviet missiles, and argued that it was mainly intended to deter any future threats from China, which had just begun to develop ICBMs. Whether he entirely believed his own explanation can be doubted. One graphic account of the day McNamara first laid the new line before his staff concludes:

> When Warnke saw McNamara later that day, he asked, 'China bomb, Bob?'
> McNamara looked down, shuffled some papers around on his desk and muttered, 'What else am I going to blame it on?' (Kaplan, 1983: 347)

In fact the system still made no real sense, but after so many years of development the momentum for some kind of deployment had become irresistible, both from the aerospace contractors involved (Goulden & Singer, 1972), and in terms of internal politics. And there is little doubt that, rejecting McNamara's doubts about the possible destabilizing effects of large-scale ABM deployments, Sentinel's supporters expected to follow it up before very long with larger and more effective systems (Lapp, 1968: 29).

It turned out that people in some of the cities destined to enjoy the benefits of Sentinel's fragile shield, such as Boston and Chicago, were strongly antagonistic towards the idea. Perhaps it reminded them of things they preferred to forget. Perhaps they believed that in the event of war it would draw down an even heavier attack on their districts. When Nixon took office in 1969 the deployment was adjusted to move most of the sites further from population centres, the rationale was changed from the protection of cities to defence of US retaliatory missile forces against the threat of a Soviet 'disarming first strike', and the name was altered to Safeguard.

Persistent doubts on both sides about the effectiveness of the anti-missile technologies of the day, centred above all on the vulnerability of ABM radars, made it possible for them to sign the Anti-Ballistic Missile Treaty in May 1972 (appendix 1). This amounted to a joint decision not to disturb the strategic balance by deploying large-scale anti-missile defences, nor even to conduct any serious development programmes to investigate new approaches to the ABM mission, such as might one day suggest it was possible after all.

In the United States the Treaty had an interesting sequel. The only Safeguard deployment ever completed, at the Minuteman ICBM field near Grand Forks, North Dakota, remained operational for just one month in 1975 (Collins & Cordesman, 1978: 103),[5] after which its defensive missiles were stood down and their silos left to gather rainwater. The actual contribution of Nike-Zeus and its descendants to US strategic capabilities had thus been negligible.

4
Military Satellites and Asats

The Soviet incentive to fight in space cannot be reduced. U.S. military space activities, which are increasing in scope and significance, create that incentive.

Colin Gray, 1983

Military Satellites

During the 1960s and 1970s satellites became increasingly important for military roles. Despite the Soviet achievement with Sputnik 1, it was the Americans who rapidly took the lead. Whereas US photo-reconnaissance satellites began providing valuable information by August 1960, it took the Soviet Union almost three more years before their Cosmos reconnaissance satellites were in regular operation (Stares, 1985: 62, 71; Turnill, 1984: 59). The military value of the American satellite programme was recognized by the immediate decision of the incoming Kennedy administration, in January 1961, to impose a press black-out on all reconnaissance satellite launches (McDougall, 1985: 346–8), whilst media magnate Henry Luce was still throwing all the publicity weight of the Time–Life organization behind NASA's 'civilian' space exploration programmes (Wolfe, 1979: chapter 6).

Continuing improvements in American space technology led to reliance on fewer, longer-lived military satellites. For example, in 1966 a total of 66 US military satellites were successfully launched, compared to 22 in 1971 and 16 in 1974 (Stares, 1985: 160). Miniaturization and hence reduced weight meant that higher orbits could be used for communication, early warning and navigation. Reconnaissance satellites, on the other hand, could be reliably maintained in lower orbits to provide better-resolution photographs, by using on-board boosters to counter orbital decay (box 4.1).

Box 4.1   Artificial Satellites

Artificial satellites stay 'up' in orbit around Earth because their speed gives them a centrifugal force to offset the force of gravity pulling them down towards the surface of the planet. The speed required to do this varies according to the altitude of the satellite, because Earth's gravitational pull gets less at greater distances. At an altitude of 500 kilometres, the velocity needed for a circular orbit is 8 km/sec, and the time for one complete orbit would be about 90 minutes. This time, known as the satellite's 'period', naturally varies according to the altitude and shape of its orbit. At speeds greater than 8 km/sec, a satellite launched to 500 kilometres would enter an elliptical orbit, slowing down as it moved further from Earth towards its most distant point, known as the 'apogee', then moving steadily faster as it came back around the ellipse, until it reached its closest and fastest point, or 'perigee' (figure 4.1). (These two distances, commonly used to describe a satellite orbit unless it is so nearly circular that a single number will suffice, are altitudes above mean sea level, not distances from the planet's gravitational centre.)

If the satellite moved faster still, it would reach 'escape velocity'. This is the speed at which the centrifugal force becomes greater than the pull of planetary gravity. The object would then cease to be an Earth-satellite, and start moving away from Earth. At 500 kilometres, escape velocity is 10.8 km/sec.

Every satellite orbits within a plane that passes through Earth's gravitational centre. The angle formed between that plane and that of Earth's equator, measured on its north-bound pass over the equator, is known as the satellite's 'inclination'. Orbits with inclinations at or close to 90° are known as 'polar' orbits. 'Equatorial' orbits are those in or very close to the plane of the equator. The rest, between these two limits, are 'inclined' orbits. The combination of the satellite's own motion and that of the rotating planet beneath it produces a 'ground track' joining the successive points on the planetary surface which fall directly beneath the satellite. The surface area of the planet in line-of-sight or direct communication with any satellite is a function of its altitude and ground track. In the lowest feasible orbits, the area that can be 'seen' by satellite sensors is no more than that of one of Earth's larger cities.

With a perfectly spherical planet of even density, no air resistance, and no minute external gravitational pulls from neighbouring bodies (such as the Sun, the Moon, and the other planets), a satellite would stay in orbit forever. In the real world these factors upset the balance of forces which sustains the orbit, causing it to 'decay', so that the satellite eventually falls back to Earth. For practical purposes, satellites which go below 300 kilometres encounter air resistance serious enough to require the intermittent use of on-board boosters to maintain their orbits.

Sources: Carter, 1986; Marsh, P., 1985.

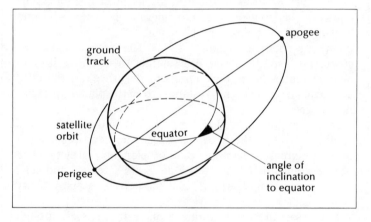

Figure 4.1   Satellite Orbit Characteristics

By the early 1980s, over 2000 satellites launched since Sputnik 1, or about 75 per cent of the total, had been used for military roles (Jasani, 1984: 5; 1982: 41). The most important of these have been reconnaissance, communications, and navigation, but meteorological and geodetic surveying satellites are also significant for military operations (table 4.1).

Reconnaissance

Reconnaisance satellites are classified under five headings – photographic, electronic intelligence, ocean surveillance, early-warning and nuclear explosion detection.

Photographic reconnaissance satellites are the most important of these during peacetime, and for monitoring conflicts around the world. Of all military satellites launched by China, the United States, and the Soviet Union, about 40 per cent have been used for photographic reconnaissance from low Earth orbit (LEO). Orbiting at altitudes as low as 200 kilometres, some of these photo-reconnaissance satellites are thought to be able to resolve details smaller than 30 centimetres (Jasani, 1982: 43, 46).

Box 4.2   Military Orbit Types

Low Earth orbit (LEO)   Between about 200 and 5000 kilometres. Polar or highly inclined orbits are favoured for general reconnaissance missions since they give planet-wide coverage. Periods are between 90 minutes and a few hours.

Semi-synchronous orbit   Circular orbit at 20,700 kilometres with a period of 12 hours. The term is sometimes extended to all orbits between LEO and this altitude.

Molniya orbit   Highly elliptical, approximately 500 × 40,000 kilometres, with a 12 hour period. Most stable at inclination of 63°. (At other inclinations gravitational anomalies resulting from irregularities in the shape and density of the planet cause the major axis of such an orbit – the line joining the perigee and apogee points – to rotate inconveniently.) Used by Soviet Union to provide satellites spending 11 hours out of 12 above the northern hemisphere.

Geostationary orbit (GEO)   Circular, *equatorial*, altitude 35,700 kilometres. With a period of 24 hours, such satellites appear to remain almost stationary above a fixed point on the equator. (In practice, they sometimes describe very small 'figure-of-eight' ground tracks about such a point.) Three or more evenly spaced geostationary satellites can cover most of the planet, except the polar regions, for purposes such as communications or missile early-warning.

Geosynchronous orbit   Circular, *inclined*, altitude 35,700 kilometres. Little used for military or any other purposes, this orbit has larger figure-of-eight groundtracks, according to its inclination. In military discussions the term 'geostationary' is tending to be replaced by 'geosynchronous', because the former is the limiting case of the latter, because even the slightest inclination causes a geostationary orbit to become 'strictly speaking' a geosynchronous one, and because *military* geostationary satellites may sometimes have a use for such a ground-track, though seldom for the much wider, true geosynchronous orbit. However, the distinction is important when discussing the interaction between civil and military uses of space. (N.B. this orbit is *not* shown in figure 4.2.)

Super-synchronous orbit   Orbits above GEO are little used so far, but offer many options for future military satellites taking refuge from ground-based or LEO asats. Certain points of equilibrium between solar, lunar and terrestrial gravitation are especially interesting. (N.B. this orbit is *not* shown in figure 4.2.)

Sources: Carter, 1986; Marsh, P., 1985.

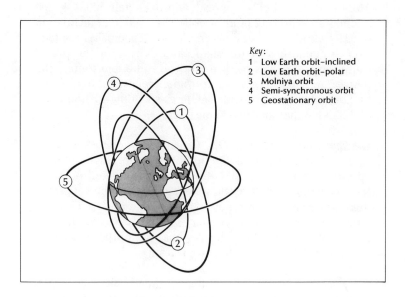

Key:
1 Low Earth orbit–inclined
2 Low Earth orbit–polar
3 Molniya orbit
4 Semi-synchronous orbit
5 Geostationary orbit

Figure 4.2   Satellite Orbits
Source: French, 1985. Adapted with permission.

As with other modern satellites the American LEO (box 4.2) photo-reconnaissance models have longer life-spans than their Soviet counterparts. The USAF's Key-Hole 9 (KH-9) or 'Big Bird' satellites are operational for several months and can return film capsules to Earth by ejecting them, usually over water near Hawaii. These are either caught by aircraft or picked up by a back-up ship, and can provide very high-quality pictures. Photographs can also be developed and scanned on board the satellite and the information relayed back to ground-stations immediately by radio signals. Although this technology does not provide quite such high resolution, it means that information is received sooner, which is a significant advantage in some circumstances. An almost immediate, or 'real time', transmission capability has been provided on the CIA's KH-11 satellites by cutting out the film stage altogether, so that 'digital imaging' takes place by means of sophisticated sensors within the optical system itself. The KH-11 has been described as equipped with:

> sensing systems and high-resolution cameras . . . to distinguish military from civilian personnel . . . [and] infra-red and multi-spectral sensing devices [which] can locate missiles, trains and launchers by day or night, and distinguish camouflage and artificial vegetation from real plants and trees. Its sideways looking [i.e. downwards – authors] radar can see through cloud cover. (Turnill, 1984: 244)

These satellites were placed in higher orbits than Big Bird, and could stay up for over two years.[1] The intelligence advantage thus gained by the United States is said to have been partly offset, however, by much improved Soviet camouflage at new military construction sites.

Soviet photo-reconnaissance satellites, by comparison, have life-spans of between about two weeks and, at most, two months. Until recently their exposed film was recovered only when the satellite was brought down, using a re-entry trajectory and parachutes. Capsule ejection technology has been achieved in recent models, since the late 1970s, but they still lag behind their US counterparts in sophistication. The latest generation have life-spans of about six weeks. Since they orbit below 200 kilometres at their lowest point, they need regular boosting from on-board rockets to maintain orbit, just like the US equivalents. One authority believes they may now have a digital film-scanning and transmission capability: 'Imagery is probably returned digitally from orbit, and film capsules are returned to Earth around

Table 4.1 – Current/Recent Military Satellites[a]

Satellite	Altitude (km)[b]		Orbit type[c]	Approx. number[d]
Photographic				
KH-9 (Big Bird)[e] [f]	130 ×	290	2	1
KH-11	300 ×	500	2	1
:*Cosmos-1347[g] [h]*	*173 ×*	*340*	*1*	*1*
Electronic intelligence				
Rhyolite	35 700		5	6–8
:with Big Bird 17	700		2	1
:with Big Bird 18	1 290		2	1
:*Cosmos-1340*	*626 ×*	*654*	*2*	*2*
:*Cosmos-699*	*440*		*2*	*1*
Ocean surveillance				
NOSS	1 000 ×	1 200	1	4
Rorsat	*250 ×*	*265*	*1*	*2*
Early Warning				
DSP	35 700		5	3
:*Cosmos-1367*	*581 × 39 624*		*3*	*9*
Navigation				
Transit/Nova	1 000 ×	1 200	2	7
Transit-like	*960 ×*	*1 000*	*1*	*10*
Navstar	20 200		4	5
Glonass	*19 000 × 19 200*		*4*	*10*
Meteorological				
AMS-5 D-2	720		2	2
Meteor 2	*855 ×*	*895*	*2*	*2–3*
Communications				
DSCS	35 700		5	10
SDS	390 × 33 800		3	3
FLTSATCOM	35 700		5	5
:*Cosmos-1452*	*758 ×*	*810*	*1*	*24*
:*Cosmos-1357-64*	*1 400 ×*	*1 500*	*1*	*40*

Notes:
[a] Altitudes and orbital types of principal current or recent *military* satellites only are shown. But no sharp distinction can be drawn between military and civilian satellites. Civilian systems such as the Soviet Molniya communications satellites or the US GOES weather satellites are also militarily important assets. Soviet Meteor 2 weather satellites are listed as military by some authorities (OTA 1985b: 37), civil by others (Turnill, 1984: 158–9).

Table 4.1 contd

[b] Single values in the second column represent nearly circular orbits, otherwise elliptical orbits are described by approximate perigee x apogee.
[c] See figure 4.2.
[d] The last column gives a *rough approximation only* for currently (early 1986) orbiting satellites of the same or a similar type.
[e] Turnill's distinction between Big Bird and KH-9 (1984: 243) is not supported by other authorities. e.g. Stares (1984: 45–6).
[f] US systems are entered in roman type.
[g] Soviet systems are entered in italic type.
It is sometimes more convenient to list individual examples of a type; such entries are preceded by a colon.

Sources: Aviation Week & Space Technology, 9 December 1985; Carter, 1986; Jasani, 1984; OTA, 1985b; Turnill, 1984.

the ninth and eighteenth days of the mission' (Perry, 1985: 84). The disparity between US and Soviet satellite lifetimes means that although between 1975 and 1982 the Soviet Union launched about 35 photo-reconnaissance satellites annually, compared with about three for the United States, actual coverage was roughly the same for both (Meyer, 1983: 206).

Electronic intelligence satellites are used to monitor military, diplomatic and other radio communications and similar activities, both those of other countries and, if wanted, those of a government's own citizens. They can and do detect such things as the radio signals from on-board instruments during missile tests, known as the test weapon's 'telemetry', or the patterns of microwave pulses emitted by the ship-defence radars of other navies. But their major function is to 'trawl' from space for the most revealing communications between radio users on the 'other side' that can be picked up. Even when heavily encoded these can sometimes yield valuable information to intelligence analysts, if only from the patterns in which they occur.

Ocean-reconnaissance satellites can carry sideways-looking radars to enable them to locate ships and take other maritime measurements in all weathers and at all times of day. Very precise satellite radar, using 'synthetic aperture' techniques (box 3.1), may shortly be able to measure the level of the sea so accurately as to detect the passage of a submarine beneath its surface in some areas. (Other sensors are also being developed for this purpose.)

Early-warning satellites are equipped with infra-red detectors which can provide either superpower with about 30 minutes' warning of an attack by the ICBMs of the other. The United States has three geostationary early-warning Defense Support Program (DSP) satellites. One watches the Soviet ICBM fields, and the other two the Pacific and Atlantic oceans for SLBM attacks. For geographical reasons, GEO is less attractive to the Soviet Union, and its early-warning satellites are in Molniya orbits. Complete coverage is obtained by a constellation of nine satellites, with shorter life-spans than their US counterparts. By ensuring the virtual impossibility of a surprise missile attack 'out of the blue', early-warning satellites may be regarded as playing a stabilizing role during peacetime.

Satellites for detecting nuclear explosions, however, have a dual character. During peacetime they can help to maintain the Limited Test Ban and Non-Proliferation Treaties by watching for above-ground nuclear tests. The United States launched six successive pairs of super-synchronous Vela satellites for this purpose between 1963 and 1970. Designed at first to operate for only six months, the first three pairs exceeded this by enormous margins, often working for nearly ten years. The second-series pairs were designed for a transmission life of 18 months. But the level of performance which they may actually have achieved can be illustrated with a famous but technically controversial example. On 22 September 1979 an American Vela Hotel satellite, which had by then been in orbit for almost nine and a half years, may have detected a small nuclear explosion in the atmosphere high above a remote area of the Indian Ocean, south-east of South Africa (SIPRI, 1980: xxxii; *The Guardian*, 22 May 1985).[2]

Today, US nuclear explosion detection equipment is carried on other platforms. Navstar satellites, principally dedicated to the Global Positioning System (GPS), also carry the Integrated Operational Nuclear Detection System (IONDS). Besides its capacity to monitor atmospheric nuclear tests, should any country irresponsibly renew them, IONDS has a potentially more active military role. In 1982 the USAF's Brigadier General Randolph explained what it could do for nuclear war-fighting:

We are able, with the GPS fully deployed, and with so-called IONDS, the NUDET detection capability onboard, to be able [sic] to detect nuclear detonations within 100 meters. So, therefore, when we try to destroy hard targets in the Soviet Union, we are able to demonstrate or to understand our success in destroying those hard targets, and, therefore, not have to go back and restrike those targets, and we can retarget in real time.

So, I think in the total sense, it is a war-fighting capability, it is in fact a revolutionary capability as far as navigation and bombing are concerned. (S. Armed Services, 1982a: 4624–5)

Communications

Communications satellites have a similar dual character. During peacetime rapid and efficient communications may help to prevent misunderstanding and to defuse potential crises. Most obviously, the famous Washington–Moscow 'Hot Line', which has sometimes proved useful in calming potentially dangerous situations, has been carried by satellites since 1978. However, the rapid and fairly secure communications capabilities of satellites would also be very useful for actually fighting a war, whether nuclear or conventional, provided only that they survived under wartime conditions.

More than half of all long-range US military communications are now routed via satellites, many in geostationary orbit.[3] Because of world-wide military commitments these satellite links are of great importance for the United States. Soviet communications satellites, on the other hand, are slightly less vital, both for geographical reasons and because of the historically 'very strict and centralized command and control structure of the Soviet Armed Forces' (Meyer, 1983: 208). Soviet tactical communications for the Eurasian theatres are maintained by a constellation of forty or more small satellites in near-circular LEO at about 1500 kilometres. Longer-range military communications may sometimes use the civil Molniya satellite network, from which that class of orbit takes its name, but are also thought to pass through a group of 'store-dump' LEO satellites at 800 kilometres. With this technology data can be stored on tape as the satellite comes over the sender's transmitter, and then played back later on as it passes over a receiving station in another part of the world (Turnill, 1984: 255).

Navigation

The third major military function of satellites is that of providing *navigational* information. The ability to obtain rapid and accurate position fixes is important to all military forces, but has particular significance for nuclear-missile submarines. Ballistic-missile accuracy depends on precise knowledge of the initial launch point, and SLBMs have in the past been inherently less accurate than land-based ICBMs. However, improvements in navigation techniques, of which satellites are by far the most important, may make the next generation of SLBMs, specifically the US Trident D-5 warhead, as accurate as ICBMs. Two techniques are used. The earlier one, still serviced by US and Soviet LEO satellite networks, was to measure the (Doppler) frequency shift of radio signals received as the satellites moved across the sky. A more accurate method is supplied by the Soviet Glonass system and the US Navstar GPS, in semi-synchronous orbits, of which the latter is still in its initial testing phase, with a limited deployment of pre-production models only. These systems work by comparing the minute differences in time of arrival between signals emitted simultaneously from numerous satellites at known positions within a 'birdcage' network of orbits, designed so that several are always above the horizon at any point on Earth.

Scientific

The *meteorological* information needed for planning military operations, including special operations and reconnaissance, is supplied by military and civilian weather satellites on both sides. *Geodetic* satellites, measuring the precise geography of the planet's surface, its gravitational and magnetic fields and other physical phenomena, also make their contribution, above all to the potential accuracy of long-range missiles. Their functions are, of necessity, confined to peacetime.

A few generalizations can be made about the military satellites used by the United States and the Soviet Union. They are probably more important to the former because of her geographic position, thousands of miles away from many of her perceived national security interests. Many US military satellites, including important ones for communications and early-warning, are in geostationary orbits. Here they remain well out of range of current anti-satellite weapons. The higher technical capabilities of US military satellites are reflected in their longer life-spans and greater reliability. In contrast Soviet satellites tend to be short-lived, and few have yet been placed in geostationary orbits. To compensate for the shorter life-spans of its satellites the Soviet Union needs to launch many more each year than the United States. Of every 100 military satellites, and more than that number are now launched every year, approximately 85 are Soviet and 15 American (Jasani, 1982: 41). However, only about half the Soviet ones would still be in orbit a year later (UCS, 1984: 195). As in many other spheres of military activity, the Soviet Union can only match the US lead in high technology by using greater numbers of slightly less capable systems.

Asats (Anti-Satellite weapons)

The Soviet Orbital-Pursuit Asat

Exactly why the Soviet Union chose to develop an asat is unclear, but part of the reason seems to have been a realization that satellites were becoming increasingly valuable for military operations, with the United States leading the way. The early American work on asats (chapter 2) also probably stimulated Soviet development.

Soviet asat testing was first observed in 1968, though initial research obviously started earlier, perhaps by 1963. A total of seven tests occurred up to 1971. The Soviet asat consists of a homing satellite, weighing about 2000 kg, launched on a modified SS-9 rocket, the 'F-LV' launch vehicle. In this first series the asat homed on to the target during its second orbit by means of active radar, and was judged 'successful' by Western intelligence analysts in five of its seven tests. It appeared to be designed to detonate a conventional explosive at its nearest approach to the target satellite, which would be destroyed by impact of the resulting debris. However, target destruction has not been demonstrated in tests, though the asats themselves have sometimes been detonated after the fly-past. (This is frequently done with military satellites, to prevent any possible debris falling to Earth and into the 'wrong' hands.)

The second series of Soviet asat tests, between 1976 and 1982, has been interpreted in the West as falling into three sub-groups. In the first, interception was attempted during the first orbit – two out of four tests were judged successful. In the second group, a new homing device, probably based on optical/infra-red technology, was tested – all six tests apparently failed. Lastly, three tests were carried out using the earlier technique of second-orbit interception with radar homing – two successes out of three.

Interception in the second orbit has the drawback that it takes about three hours, thus allowing time for the target to take evasive action. The F-LV liquid-fuelled boosters also take at least 90

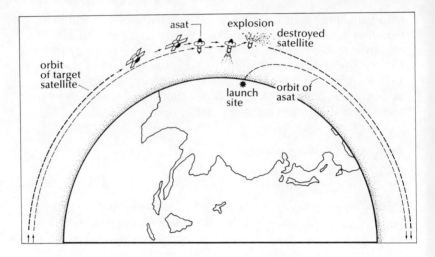

Figure 4.3   Soviet Asat Interception
Source: French, 1985. Reproduced with permission.

minutes to prepare for launch, and have been launched from only one military rocket base, at Tyuratam in Kazakhstan. A handful of launchers, perhaps as few as four, are available at this site, which is thought to possess two designated F-LV launch-pads, plus two further compatible ones. All the interceptors have so far been launched into orbits whose inclination matches that of their target. Except for the first two tests (which had inclinations between 62° and 63°) all the target satellites and interceptors have been orbited at about 66° (Stares, 1985: 262). Used in this way the Soviet homing asat could only intercept satellites whose ground tracks passed within a few hundred kilometres of the launch site. Very few US satellites in fact do so.

The inflexibility of the Soviet anti-satellite system suggests that even if it had an unlimited supply of asats and launchers, 'the Soviet Union would require well in excess of a week to destroy all the low-orbit US satellites that are within its ASAT's range' (UCS, 1984: 199). So far the highest altitude reached by a Soviet asat is 2300 kilometres,[4] far below GEO where important US communications and early-warning satellites are kept. For these reasons the judgement reached by USAF Chief of Staff General Lew Allen in 1979 still appears valid:

So I think our general opinion is that we give it [the Soviet asat] a very questionable operational capability for a few launches. In other words, it is a threat that we are worried about, but they have not had a test program that would cause us to believe it is a very credible threat. (S. Foreign Relations, 1979: 424)

Or as one expert put it, by comparison with the new US asat the Soviet system is something of an 'old blunderbuss' (*The Times*, 6 September 1985).

The US Direct – Ascent Asat

American interest in developing a non-nuclear homing device for anti-satellite roles dates back to the US Navy's Early Spring programme in the 1960s. The relevant technologies were under actual development long before the Soviet Union resumed its asat testing in 1976 (Stares, 1985: 203). In 1982 the Reagan administration explained its version of the rationale for such programmes, in a Fact Sheet on National Space Policy:

> The United States will proceed with development of an anti-satellite (ASAT) capability, with operational deployment as a goal. The primary purposes of a United States ASAT are to deter threats to space systems of the United States and its Allies and, within such limits imposed by international law, to deny any adversary the use of space-based systems that provide support to hostile military forces. (US WH, 1982)

The new US asat consists of a homing warhead mounted on a small two-stage rocket, to be launched from an F-15 fighter, which has an operational ceiling of about 20 kilometres. The homing device is known as the Miniature Homing Vehicle (MHV) and resembles the ABM successfully tested against a dummy Minuteman warhead on 10 June 1984 (p. 124). The MHV is about 30 centimetres by 30 centimetres in size, and weighs some 15 kg (figure 4.4).

Using tracking information from NORAD's Space Surveillance Network in Colorado,[5] the asat's internal computer first guides the plane, and then determines the release point and initial aiming of the two-stage missile towards the calculated intercept point. The two booster stages would then lift the MHV out of the atmosphere. Once

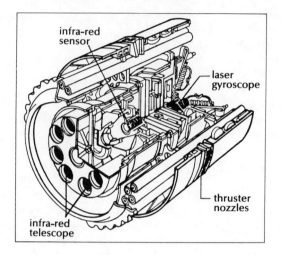

Figure 4.4   US Miniature Homing Vehicle Cut-Away
Source: Union of Concerned Scientists, 1984. Reproduced with permission.

they are jettisoned it would home onto its target, guided by its on-board infra-red sensors, and steered by a ring of small one-shot sideways gas thrusters. No explosive warhead is needed, since the target is destroyed by direct collision at high speed.

The F-15 asat is intended for interception of lower LEO satellites, such as those used for photo-reconnaissance and ocean surveillance. According to one report the vertical range of the system is almost 1500 kilometres (*Aviation Week & Space Technology*, 19 December 1983). The missile would only take a few minutes to reach such altitudes, and the complete operation time, from ordering the F-15s into the air to satellite destruction, could take less than half an hour. Preliminary system tests in 1984 culminated in the launch of a missile towards a fixed reference point, but not an actual target. In its first full-scale test on 13 September 1985, the system took about two hours from take-off to reach and destroy its target, a 'Solwind' satellite of the US Naval Research Laboratory, at an altitude of about 600 kilometres (*Aviation Week & Space Technology*, 23 September 1985; *The Guardian* 21 September 1985).[6]

The system is intended to be operational from 1987, with an eventual deployment of 112 weapons (UCS, 1984: 33). The fighters need only short runways. Current plans are for two squadrons, one at Langley AFB, Virginia, on the East coast, and one at McChord AFB, Washington, on the West. With in-flight refuelling such

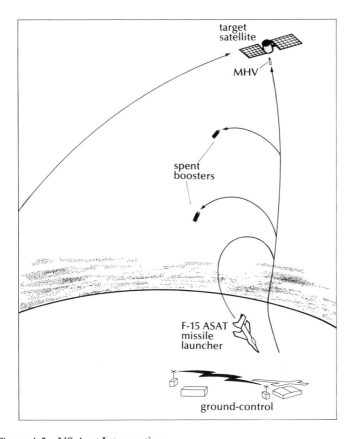

Figure 4.5   US Asat Interception
Source: French, 1985. Adapted with permission.

planes could provide a virtually global anti-satellite capability, since carrier-based aircraft could also carry the weapon. Overall, the system is intended to provide a flexible and rapid means of destroying low-orbit Soviet satellites.

Asats capable of destroying some low-orbit satellites are thus potentially available to both superpowers. The US system, not yet deployed, will be far more flexible than the limited Soviet system. Moreover, because many key Soviet satellites tend to be in orbits that are wholly or partly lower than those of their US counterparts, these second-generation low-orbit asats are probably more threatening to the Soviet Union than to the United States.

Future Asats

A higher-altitude version of the US F-15 asat has been considered, either as an air- or else as a ground-launched system. The Pentagon has also claimed that the Soviet F-LV asat could be adapted to the far more powerful Salyut (D-LV) or Soyuz (A-LV) launchers, though neither has ever been so tested. But the operational weaknesses of such direct-ascent asats – their vulnerability and slowness to reach their targets – would be increased still further against targets at higher altitudes. In the Soviet case, the long operational reaction time would be an additional disadvantage.

Space mines and beam weapons may now be considered by both the United States and the Soviet Union to be more feasible future candidates for the anti-satellite role, because they might be fast enough to beat evasive action by even a highly manœuvrable target. Space mines would simply be explosive satellites like the current Soviet asat, but would have been pre-placed in orbits near their targets, ready to be detonated on command. Such a deployment might be difficult during peacetime, because it could alert the other side, perhaps even leading to war. For example, there is an international convention by which satellites in geostationary orbit are separated by a minimum angle of 2°, to avoid signal interference. It might, however, be possible to disguise space mines as unarmed satellites, but give them the capacity to close rapidly on their targets. As yet no such devices seem to have been tested.

Space mines carrying ordinary nuclear explosives would be indiscriminate weapons, only likely to be used if destroying the enemy satellites was more important than preserving one's own. A nuclear bomb-pumped X-ray laser or similar device, however, could disable satellites from a considerable range instead of having to approach them. Obviously it would be designed to damage enemy satellites only, by channelling the power of its nuclear detonation into directed beams (chapter 7).

All directed-energy or beam weapons (DEW), including lasers and particle beams, could be used effectively as asats if they ever became technically feasible. Because their effects travel so fast, at or near the speed of light, there would be no warning or time for evasion. However, developing anti-satellite beam weapons is by no means easy (Hecht, 1984: chapter 12). If based on the ground or on aircraft, it would be hard to penetrate the atmosphere with a sufficiently damaging beam. Orbital anti-satellite battle stations might be used instead, but would of course themselves be vulnerable to attack.

To reduce this vulnerability satellites might be armed so that they could defend themselves against others. As Gray has noted:

To be able to defend satellites actively, and attack the satellites of the enemy, is a distinction without a difference. (1983: 49)

The main question about active satellite defences (DSATs) is whether it would be cost-effective to add such devices to satellites, as well as the sort of protective shielding now coming into use. Making satellites more sophisticated and heavier necessarily means producing fewer for any given budget. A better approach might be to use larger numbers of cheaper satellites, each designed for a limited number of tasks, which might include attacking or defending other satellites. The greater annual launch capabilities of the Soviet Union might be a relative advantage in that respect.

The prospects for fast-acting and effective anti-satellite beam weapons may be dim, but they are brighter than those for effective missile defences (chapters 7 and 8). The fact is that satellites are harder to protect and easier to hit than nuclear warheads. Future developments in anti-satellite technology threaten to make space a potential battlefield, and even current asats make it difficult to guarantee the survival of low-orbit satellites. Thus they undermine the present stabilizing role of satellites as 'national technical means of verification' for monitoring arms control agreements.

5
From 'Deterrence' to 'Star Wars'

Mad world! Mad kings! Mad composition!
Shakespeare, *King John*

The joint renunciation of anti-missile defences by the United States and the Soviet Union, in their ABM Treaty of May 1972, was based on the belief that mutual security could be guaranteed by the 'deterrent' character of each other's offensive missile forces. Neither side could be sure of eliminating all the other's missiles, either by the offensive route, with a 'disarming first strike', or by the defensive one, with an effective anti-missile shield. Both sides would thus be 'deterred' from ever starting a war, the theory went, since the certainty of an annihilatory 'second-strike' response by the surviving nuclear forces of the opponent would make an attack tantamount to national suicide. This relationship between opposing nuclear forces became known as 'Mutual Assured Destruction', or MAD.

The 1972 ABM Treaty was not the first time such ideas about reliably avoiding nuclear wars through 'deterrence' had affected decisions about nuclear forces. Ever since nuclear bomber crews had been sent to camp out round the clock beside their planes in the 1950s, the relationship between deterring nuclear war and the survivability of an effective, because unstoppable, 'retaliatory capability' had dominated Western nuclear strategic thought. It seems to have played a large part in Soviet strategic planning also. Reinforced concrete 'silos' for land-based missiles and untraceable missile submarines were all part of the same approach. But the ABM Treaty, with its declaration of the potentially destabilizing character of large-scale anti-missile defences, was and remains the most notable formal endorsement of this thinking by the two superpowers.

For the theory to have had any chance of working, however, it

54

would have been necessary *to apply it to both ways of eliminating an opponent's nuclear capability*. Putting a brake on major anti-missile defences could not ensure 'stable deterrence' unless offensive (counter-force) anti-missile capabilities were, at least, strictly controlled at a non-threatening level. Without this, both sides would continue to fear the sort of pre-emptive attack – sometimes described as 'getting one's retaliation in first' – which masquerades, in the old saw, as 'the best form of defence', but at the same time both would seek to acquire forces capable of delivering it. In the event, the development and deployment of increasing numbers of new and more threatening counter-force weapons continued apace – an arms race that ran directly counter to mutual second-strike deterrence. The historical record thus casts considerable doubt on whether the theory of mutual deterrence was ever wholly accepted by the military and political authorities of either side.

In fact, at the operational level the ABM Treaty had little effect on either superpower's nuclear strategy. US planners had included military targets for as long as they had been able to find them – certainly since the start of the 1950s. In 1962 the newly coordinated targeting plan (known as the Single Integrated Operational Plan, or SIOP) was revised to allow the option of hitting only military targets whilst deliberately avoiding cities (Ball, 1983). Secretary of Defense Robert McNamara soon started to back away from this nuclear war-fighting strategy in public, stressing instead an assured destruction objective. But the targeting policy remained unchanged. Henry Rowen, who helped to produce the 1962 SIOP, explained:

> The primary purpose of the Assured Destruction capabilities doctrine was to provide a metric for deciding how much force was enough: it provided a basis for denying Service and Congressional claims for more money for strategic forces . . . (However, it was never proposed by McNamara or his staff that nuclear weapons actually be *used* in this way.) (Rowen, 1979: 146)

Indeed the whole idea of MAD as a military strategy may be misleading. As the then Chairman of the Joint Chiefs of Staff, General David Jones, explained in 1979:[1]

> It is not the strategy we are implementing today . . . I do not subscribe to the idea that we ever had it as our basic strategy. I have

been involved with strategic forces since the early 1950s. We have always targeted military targets. (S. Armed Services, 1979: 169)

The close connection between the potential uses of military force in actual war and its application to the conduct of 'peacetime' international relations suggests not only that MAD was (obviously) unacceptable as a guideline for nuclear war planners, but also that it may never have been US or Soviet nuclear 'grand strategy' either, in the sense of a conscious and *settled* acceptance of mutual vulnerability as a valued policy goal. But in another sense it is and has been for many years a reality of the nuclear world. This was explained by one US scholar in the international relations field, commenting on the ABM Treaty, as follows:

> Had there been an apparent technological solution to these dilemmas, it is doubtful that the basic concept of mutual deterrence and its implications for strategic arms control could have gained sway with any government . . . The doctrine of mutual deterrence did not so much state a strategic objective as describe a technological fact. (Frye, 1974: 69)

The signing of the ABM Treaty was a recognition by US and Soviet leaders that this grim fact of modern life could not be overcome by technical developments in defensive weapons. But its effect on long-term military planning in the United States was slight. Just as the Treaty was signed a US government study group, known as the Foster Panel, produced a proposal for changes in US nuclear targeting (Ball, 1974). This formed the basis of Secretary of Defense James Schlesinger's 'limited nuclear options' posture, which further emphasized the utility of a selective counter-force capability.

Soviet nuclear preparations were similarly unaffected by the MAD implications of the ABM Treaty. Of course it was recognized that the virtually inevitable devastation in the event of nuclear war made mutual destruction the most likely outcome. However, Russian military planners have seemingly been as intent as their American counterparts on limiting the mutuality of the arrangement, to their own relative advantage, if only by passive and inadequate civil defence measures.

The 'declaratory' role often reserved for the MAD concept is evident from the history of American ICBM technology in the 1960s.[2] There are two basic elements to the sort of anti-missile capability, sometimes euphemistically cloaked as 'damage limitation', which is based on offensive missiles rather than defensive systems. First, the attacker's warheads must be delivered close enough to all or very nearly all of the defender's ICBMs to destroy them despite any protective 'hardening' of their silos. Second, a long-range missile is a complex engineering system with less than 100 per cent reliability. Even if it performs perfectly, its 'accuracy' against a hardened military target is only a (fairly high) statistical probability of destroying it. So an attacker would always need to send more than one warhead against each target.

There are also many other targets, besides ICBMs, which are thought to require 'prompt' destruction by means of nuclear warheads. In the early 1960s Lawrence Livermore Laboratory physicists Richard and Albert Latter proposed a simple method of meeting this 'requirement'. The United States would need to possess an arsenal of many more warheads than the Soviet Union would ever have missiles. With support from Edward Teller, John Foster, Harold Brown and other leading 'Wizards of Armageddon', they polished up their arguments, identified the necessary technology in the USAF's Titan IIIA 'Transtage' programme for launching multiple satellites with a single booster, and convinced Defense Secretary McNamara to authorize, in December 1964, the programme for the first 'Multiple Independently-targeted Re-entry Vehicles' (MIRVs) (Kaplan, 1983: 360–4; Durch, 1984: 24).

It may be wondered with good reason what such an initiative can possibly have had to do with the Johnson administration's proclaimed acceptance of the principle of stable deterrence by means of mutual retaliatory nuclear capabilities on both sides. The answer is that McNamara himself embodied the contradiction between, firstly, a public acceptance of mutual vulnerability and mutual deterrence, and, secondly, a constant striving to discover the 'technological fix' that might get deterrence back to its bygone one-way form. The second of these two impulses was directed towards reviving a US capability to impose restraints, by means of 'nuclear superiority', both on the Soviet Union's ability to threaten major conventional or nuclear war, and on her ability to conduct herself in peacetime international affairs, with all their frequent conflicts of interest, on 'equal terms' with the United States. McNamara also had to mediate between military demands both for ABMs, which he opposed, and for MIRVs, which he saw as a useful argument against the ABM lobby. He was thus unable, even had he wished, to develop the concept of mutual deterrence consistently and effectively.

Those who originally promoted the idea of MIRV were not so hampered. They had never accepted MAD and were never going to do so: '[Richard] Latter . . . was a perpetual promoter of a U.S. buildup to defend against the Soviet threat' (Kaplan, 1983: 361). Nor was his colleague Edward Teller any devotee of the notion that a stable balance of terror could be achieved by a rough equivalence of vulnerability and of destructive capability, often known as 'parity', on each side:

> [A]s to the question of parity, in a very rapidly developing field, where a lot of ingenuity can be involved, I don't see any reason to expect that parity will be accomplished. I think that the harder worker will win out, and so far the Russians have worked much harder . . . I believe they are already ahead of us in knowledge, and they are, as far as I can see, moving faster. I am not worried about parity. I am worried about the United States becoming a second-class power, and next to Russia there won't be a power left if it is a second-class power. (S. Foreign Relations, 1963: 448)

Teller's remarks had two main implications. First, that there was still meaning to the notion of one of the two nuclear superpowers being decisively more powerful than the other, and even than all other states together, despite the advent of the immense destructive power of nuclear explosives. (An hour or so earlier during that long day's testimony, he had commented that 'There is no such thing as the ultimate weapon' (435).) And second, that the only way to prevent the Soviet Union holding such a position was for the United States to do so, and to continue doing so for as long as necessary.

The contradiction between this line of strategic doctrine and the alternative, 'balance of terror' approach was accommodated in practice by a combination of three elements within American nuclear policy during the 1960s. First, the mutual vulnerability of the two superpowers was recognized as a brute fact, resulting from each side's current lack of the technical capability to neutralize the other's weapons by either the defensive or the offensive option. Second, this fact was made the best of at the political level, above all by McNamara himself, with pronouncements about the benefits for security produced by an era of 'stable deterrence', or MAD. But finally, within the community of nuclear weapons designers and war planners, for whose overall direction McNamara and his aides were also responsible, the situation of roughly equivalent and 'mutual' vulnerability was simply never accepted as permanent.

The rejection by conservative American strategists of any policy which accepted MAD, and hence of the ABM Treaty which embodied that acceptance, was made clear in a study published by the Hudson Institute just before the Treaty was signed (Boylan, Brennan & Kahn, 1972). Its authors explained their view that effective, one-way deterrence of the Soviet Union by the United States depended on the latter securing an advantage in 'relative war outcomes' by means of a reliable damage-limitation capability, provided either by offensive counter-force or by defensive anti-missile systems. In doing so, they referred approvingly to the utility of a comparative US 'deterrent strength' for securing favourable outcomes to non-military or non-nuclear East-West confrontations (an idea they attributed to Harold Brown).

For US ABM systems to have the effect such authors desired, their powers of 'damage limitation' would have had to be considerable. However, Brennan had argued for the desirability of even a 'thin' and fairly ineffective ABM deployment (1969b), if it would keep the United States in the anti-missile business whilst more effective systems were devised, and the politics of their approval was worked through.

This arrangement, whereby 'declaratory' talk about mutual deterrence and equal vulnerability accompanies military preparations for achieving a relative one-way 'deterrent strength', is a double standard that has corrupted much American thinking about 'nuclear deterrence' for the past quarter century. It can be aptly illustrated from a short talk on the principles underlying the United States defence budget, given at the American Enterprise Institute in Washington just five months after the ABM Treaty had been signed. The speaker was a San Francisco lawyer and Republican politician, who had come to Washington in 1970 to serve as President Nixon's deputy, then full, Director of Budget, a certain Mr Caspar Weinberger.[3]

Mr Weinberger began his talk by expressing his complete concurrence with 'Hobbes's observation that states naturally seek to "enlarge their Dominions . . . [and] endeavour as much as they can, to subdue, or weaken their neighbours, by open force, and secret arts, for want of other Caution."' (Weinberger, 1972: 3). Wasting no time on the inconvenient implication that, if this were true of all governments, it would have to be true of his own, the speaker went on to quote and to adopt Hobbes's definition of deterrence. This has a marked one-way character and, in preferring comparative military strength to any absolutely anni-hilatory capability, comes closer to conservative American think-ing about the need for nuclear weapons to be somehow 'usable', than to McNamara's cold calculations about the weight of nuclear explosives required to destroy the Soviet Union:

> The multitude sufficient to confide in for our security, is not determined by any certain number, but by comparison with the enemy we fear; and is then sufficient, when the odds of the enemy is not of so visible and conspicuous moment, to determine the event of war, as to move him to attempt. (Hobbes, 1947: 110)

Mr Weinberger recognized the budget savings that had resulted from the decision to limit ABM deployments, embodied in the recent Treaty. And he duly paid his respects to the 'unpleasant fact' version of MAD:

> In light of a rapid build-up in Soviet strategic forces in the later 1960s, the Nixon administration has reassessed our strategic force needs. It is recognized that both the United States and the Soviet Union possess an assured destruction capability. Further escalation in the strategic arms race would prove both expensive and ineffective in altering the strategic balance. Over the past few years, the Nixon administration has developed a doctrine of 'strategic sufficiency' with emphasis on maintaining a stable, credible balance. (1972: 6–7)

But that 'stable, credible balance' based on 'strategic suffi-ciency' turned out, in Mr Weinberger's understanding at least, to consist in nothing less than superiority after all:

> Although strategic missile forces have not changed in size since the mid-sixties, qualitative improvements such as the MIRV . . . program have been introduced. This program will more than double the overall effectiveness of our strategic offensive forces and insure a commanding lead over the Soviet Union in the number of deliverable warheads. (1972: 7)

Mr Weinberger recently gave an even clearer exposition of his belief that, if MAD ever looks like becoming too well established a feature of the East-West relationship, it is necessary to redouble US efforts to keep ahead of the Russians:

[W]hy try to fix [alter] something that has preserved the peace for forty years? Because in fact, it has not. MAD is a recent condition. Most of the past forty years were safeguarded *first* by American nuclear monopoly, and *then* by significant American nuclear superiority. Only recently has the Soviet Union acquired an assured second strike capability against the United States. (1985: 5)

It little matters whether the first Nixon administration (1969–73), during which Mr Weinberger personally acknowledged that MAD obtained, can be seen from 1985 as having happened 'only recently'. What is interesting is the light shed by the later speech upon the earlier, across more than a decade of 'further escalation in the arms race' through attempts at expanding the 'effectiveness of our strategic offensive forces'. For Mr Weinberger has now made crystal clear what that build-up in US missiles, from MIRVed Minuteman to Pershing 2, has always been about. It has been about trying, and for a long time in his view managing, to prevent the Soviet Union acquiring 'an assured second strike capability.' And what that means is very simple. The Soviet Union can only be denied a *second-strike* capability if the United States has a real *first-strike* capability. In short, US forces would only have the 'strategic sufficiency' for Mr Weinberger's version of a 'stable, credible *balance*', if they were capable of such an effective first strike against Soviet missile silos that the latter could be destroyed and the Soviet Union thereby virtually disarmed.[4]

At the start of the 1970s this combination of a declaratory doctrine, for political consumption, of recognizing the 'assured destruction capability' of both sides, with covert 'action guidelines' for acquiring enough thousands of nuclear warheads to 'insure a commanding lead over the Soviet Union', seemed to be holding up quite well.[2] But then the whole MAD deal between the hawks and the doves, the armourers and the 'arms controllers', came unstuck from both sides, West and East. On the one hand, members of the US Congress began taking the Pentagon's public-relations material a bit too literally. They argued that, if assured vulnerability was not just a reliable form of security but also a more or less permanent objective fact, such costly and competitive efforts should not be necessary to maintain 'deterrence'. New strategic weapons programmes started to run into delays, whilst expenditures on social programmes continued to rise.

And for their part, the Russians proved shockingly recalcitrant. They were supposed to be 'commanded' into permanent strategic

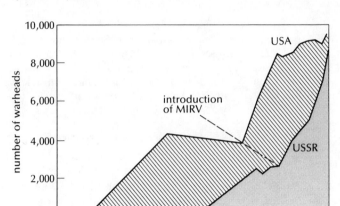

Figure 5.1   US and Soviet Arsenals 1950–85
Source: Openshaw et al., 1983. Reproduced with permission.

inferiority by the US 'lead' in warhead numbers. Instead, ten years after McNamara's fateful 1964 decision to proceed with MIRV, they started to catch up.

Figure 5.1 tells the tale of how and when the US defence establishment began to find itself hoist with its own MIRV petard. By the late 1970s the outnumbering of targetable missile silos by their potential attackers, the enemy's warheads, had produced a bizarre and increasingly dangerous situation. Both sides either had or would shortly have broadly equivalent force levels, and were busily seeking a formal agreement to accept just that, through ongoing Strategic Arms Limitation Talks (SALT). Indeed, it had been recognized since the early 1950s that no other basis for an overall arms control regime could possibly work.

Nevertheless, the advent of MIRVs made it increasingly possible for each side to see the other's nuclear arsenal as superior to its own, even if missile numbers were roughly equal. This was because MIRVs changed the 'exchange ratio':

As long as there was only one warhead on each missile, there would never be any advantage for a missile to attack another. The exchange rato between missiles destroyed and missiles used in an attack would always be less than one, an unfavorable position for the attacker because no missile with a single warhead could have complete assurance of destroying another missile in its silo. When a missile carries several warheads, however, this disadvantage for the attacker could in theory be reversed and turned into a gain. The exchange ratio could be greater than one, and thus favorable to the attacker . . . Thus now there could be a real gain by launching a first strike against the opposing side's land-based ICBM force. (Scoville, 1981: 65–6)

Once MIRVs had arrived, nuclear force planners looking at their own side of the 'balance' tended to see their *ICBMs* as outnumbered military targets whose survivability it was still abolutely necessary to preserve. Looking at the other side, however, they saw mainly the rapidly proliferating offensive *warheads*, which seemed, at several for each defending silo, to place that 'survivable deterrent' increasingly in jeopardy.[5] (It was found possible, somehow, to regard the large numbers of effectively invulnerable missiles on each side's submarines (SLBMs) as more or less irrelevant.)

The upshot was that each side both denounced and doggedly upheld a situation of growing 'arms race instability', in which neither seemed ready to accept the current nuclear 'balance' or any foreseeable version of it. But by the late 1970s it had become clear that the race for a real, or symbolic, 'superiority', in the form of an offensive first-strike capability, had actually resulted in nothing more than an ambivalent and increasingly dangerous stalemate.

To many Americans who had accepted, with however little thought, the desirability of that elusive and unholy grail, unilateral deterrence of Soviet military power by means of some decisive strategic superiority or 'deterrent strength', it began to seem even more urgent than in the past to set about expanding US strategic capabilities. Formerly 'unfashionable' US strategists, who had been arguing for decades that no lasting 'favourable balance' could ever be achieved without some direct means of blocking the opponent's offensive missile threat, began to receive a wider hearing. Perhaps technological advances now offered the basis for decisively removing the opponent's deterrent capability, whilst still keeping one's own offensive forces as what would then be an invulnerable and irresistible, but strictly unilateral, deterrent.

In the second half of 1980 many lobbying groups and influential strategic hawks were busy preparing to ensure that this BMD-based 'theory of deterrence' would be the one adopted by the increasingly probable, and then certain, future Reagan administration. On the academic side, studies poured forth from the Institute for Foreign Policy Analysis and the Hudson Institute. Lobbying organizations such as the Heritage Foundation and the American Security Council overhauled their grasp of the case for strategic defences, and their approaches to presenting it. One particularly well-connected, powerful but discreet player to emerge on the Washington scene was the 'Madison Group' of ultraconservative Congressional aides. They had started to come together the year before, at the initiative of Senator Helms's assistant John Carbaugh, to organize the successful legislative resistance to ratifying the proposed SALT 2 Treaty. The Group's failure to win a Senate battle for immediate full-scale development funding for chemical laser weapons in 1980 seemed only to make them more determined than ever to succeed under the new dispensation. And their negative attitude to arms control was readily transposed from SALT 2 to the ABM Treaty. That winter, after Reagan had been elected, the Group continued to meet for its weekly lunches, despite being driven out of the Madison Hotel for a short time by the glare of unwanted publicity.[6]

As Inauguration Day drew near in January 1981, the political sun seemed sure to rise again for US strategic defence programmes, probably using novel technologies and deployed in space. If it did not, it would not be for want of faithful prophets in attendance on the occasion, many of whom had hoarded the essential elements of this strategic 'vision' through ten long years in the deserts of conservative strategic heterodoxy.

Thus the President's signature was hardly dry beneath his Oath of Office before the Hudson Institute's Colin Gray and Keith Payne had completed a comprehensive study for Acting Assistant Secretary James Wade at the Pentagon, on the advantages of a new, 'denial-of-victory' strategic posture. Strategic defences would be essential for this, to back up the US 'commitment to initiate nuclear escalation' together with a negotiating stance in favour of 'deep cuts' in offensive weapons. To show this meant a radical break with the strategic thinking of the Carter administration, and its predecessors, they quoted from a recent Annual Report by Carter's Defense Secretary, Harold Brown, and then rejected it explicitly:

'In the interests of stability, we avoid the capability of eliminating the other side's deterrent, insofar as we might be able to do so. In short, we must be quite willing – as we have been for some time – to accept the principle of mutual deterrence and design our defense posture in light of that principle [US DoD, 1979: 102].'

A force posture appropriate for a DOV [denial of victory] deterrent would specifically emphasize a capability to destroy 'the other side's deterrent', and protect U.S. societal assets. (Gray & Payne, 1981: 1–78)

The authors went on to point out that though 'offensive counterforce capabilities obviously would be an integral part of U.S. damage limitation capabilities . . . American leaders could never be certain' of having 'the advantage of "taking the initiative"', nor of their offensive forces being able to knock out all Soviet ICBMs even if they did (2–1). Ballistic missile defence *of the American people* would therefore be essential if the Soviet Union was to be denied 'victory', whilst the United States could still 'limit damage to its homeland significantly' (2–3). The overall strategic objective being that:

> The Soviet Union should anticipate disastrous domestic political consequences in any excursion into an extremely destructive and unwinnable war – while the U.S. (particularly with this Soviet vulnerability in mind) could with relative confidence in its political integrity engage in a war that while perhaps militarily unpromising in the short term, would envisage the eventual attainment of the desired political objective. (2–4)

Or as Payne rephrased it elsewhere:

> An adequate theory of deterrence must also recognize the relationship between homeland defense and freedom of offensive action . . .
>
> The idea that U.S. – Soviet deterrent capabilities should be equal is inconsistent with American deterrent responsibilities . . .
>
> Whether such a force posture comes under the rubric of parity or strategic superiority is largely beside the point. (1981: 755, 759, 763)

6
The President's Programme?

This is the President's program. We can't tell the President he's got a nutty idea.

US government official, 1984

Soon after President Reagan's 'Star Wars' speech of 23 March 1983 there were reports of a lack of prior consultation on the passages about anti-missile defence between the White House and relevant senior Pentagon science officials (*New York Times*, 25 March 1983). John Gardner, Director of Defensive Systems, and Robert Cooper, Director of the Defense Advanced Research Projects Agency (DARPA) were mentioned. This information led many who knew no better, and some whose prejudices meant they hardly cared, to suppose that the entire strategic policy of putting greater emphasis on anti-missile systems, and greater effort into developing them, had somehow been dreamed up overnight by a naïve and irresponsible US Commander-in-Chief. The true story was rather different.

Three years earlier, in 1980, those who drew up the position on national security for the Reagan election campaign, and the Republican Party platform which went with it, had included a firm public commitment to new efforts in the ABM field. For some years, indeed since before Reagan's first try for his party's Presidential nomination in 1976, this group had taken as its major political focus the allegedly growing Soviet threat to the United States. For reasons already explained (p. 62) Soviet strategic capabilities were perceived as at or close to 'superiority'. In the same year, the last of President Ford's lame-duck administration, they managed to gain official recognition for their view of the situation. Under the patronage of George Bush, then the new director of the Central Intelligence Agency, they secured an upward revision of the US government's estimate of Soviet strategic forces (Scheer, 1982: chapter 5; Sanders, 1983: chapter 6).

During the second half of the 1970s this political view about America's 'present danger' formed natural alliances with the long-frustrated managers of ABM development programmes in both private industry and the armed services. In 1980 they prepared to bring the message to the American electorate for a second time. And as always, they saw the idea of anti-missile defences as an important part of a package of measures aimed at restoring a dangerously eroded American security. So the campaign's policy paper, 'Strategic Guidance', prepared under the direction of Reagan's senior adviser on military policy, Richard V. Allen, listed ABMs alongside civil defence programmes as a priority option, on the principle that:

> The United States will not accept permanent abstention from the right to protect its citizens from the effects of nuclear attack. (In Burt, 1980: 87–8)

Since just such a 'permanent abstention' was central to the arrangement for security through mutual vulnerability which had been pledged between the United States and the Soviet Union by their 1972 ABM Treaty, this was a radical position to adopt. But it was plainly shared by other prominent members of candidate Reagan's team, such as W. R. Van Cleave and W. Scott Thompson, who had jointly argued for it in a recent book (1979).

There had however always been scope for differences of interpretation and emphasis amongst conservative strategic analysts and the politicians they advised. Ever since the late 1960s (pp. 34–5) there had been the question of whether ABM systems should or could be deployed principally for limited defence of military sites ('hard site', 'point' or 'force' defence) or for the more ambitious task of giving comprehensive protection to society as a whole ('area', 'population' or 'city' defence). Within the Reagan camp, Van Cleave and Thompson came down firmly on the former side. There was also the delicate question, in view of the negative response to Sentinel from public opinion in 1968, of just how much overt emphasis to place on this aspect of policy during the 1980 campaign, and how to proceed with and present it once in office. Such matters are known to be ones on which Reagan likes to make up his own mind (e.g. Barrett, 1983: 310–4).

This was evident in the future President's handling of the ABM question long before March 1983. Natural politician that he is, Reagan's instinct may have been not to push such ideas too hard in the election campaign. Sometimes, however, he seems to have been almost bursting to tell. Thus he may simply have failed to get through to the *Los Angeles Times's* Robert Scheer, with his folksy, gee-shucks remarks about the North American Air Defense Command (NORAD):

> They actually are tracking several thousand objects in space, meaning satellites of ours and everyone else's, even down to the point that they are tracking a glove lost by an astronaut that is still circling the earth up there. I think the thing that struck me was the irony that here, with this great technology of ours, we can do all this yet we cannot stop any of the weapons that are coming at us. I don't think there's been a time in history when there wasn't a defense against some kind of thrust, even back in the old-fashioned days when we had coast artillery that would stop invading ships if they came. (Scheer, 1982: 104)

Throughout the conversation Reagan put out signals of his interest in 'active' defences, both with these phrases about stopping weapons and about space-tracking, and with less explicit hints. But the interview turned to 'passive' civil defence programmes and other matters. Only later, when writing up his material as a book, did Scheer remark that:

> What [his] statement reflected was a longing for the ultimate antiballistic missile defense. (1982: 105)

The demands of some American conservatives, for a break with the strategic deadlock of MAD, have been linked with aspirations for 'exotic' weapons technologies through which to do so since almost before the resented stalemate of mutual deterrence had by any reckoning been constructed. General LeMay's hopes for the early application of directed-energy weapons as anti-missile defences for this purpose have already been noted (p. 18). It is not possible to say whether Reagan's own interest in advanced weapons technology had similar origins. But it did lead to his being the first governor of California to visit the Lawrence Livermore Laboratory, in 1967, at the invitation of its leading light, Dr Edward Teller. Already by that date Teller had been ardently, almost obsessively campaigning, for more than ten years, for those schemes to discover 'good' defensive uses of nuclear weapons to counter the familiar 'bad' offensive ones which have brought him international renown, or notoriety, according to one's point of view.[1]

So Reagan had acquired an interest in the possibility of radical anti-missile systems, with novel technologies, well before coming to national office. In his first few weeks at the White House he began to follow it up. Early in 1981 there were press reports of the influence on the President of such ABM proponents as Senators Wallop, Tower, and (former astronaut) Schmitt:[2]

> Schmitt said President Reagan fully recognizes that this technological revolution is providing new strategic policy options that 'will in the not-too-distant future make weapons of mass destruction obsolete, to provide a strategic political option based on the principle of protection of human beings rather than their mass destruction.'
>
> President Reagan has expressed an interest in having laser weapons developed in meeting with both Schmitt and Tower, according to Wallop. The reason the President gave for favoring space laser weapons is ballistic missile defense. (*Aviation Week & Space Technology*, 25 May 1981)

At first, however, the administration's attention was taken up with more 'traditional' ABM systems. Towards the end of the Carter years, the US Army's long-frustrated Ballistic Missile Defense Organization had formed a loose and purely temporary alliance with the USAF over the vexed problem of selecting a 'basing mode' for the MX missile. The US military establishment had grown steadily more concerned, throughout the 1970s, that the accuracy and increasing numbers of second-generation Soviet ICBM warheads might give them a first-strike capability against American land-based missiles in fixed silos. Some sort of concealment or mobility for the latter was felt to be the only solution, and the MX had been designed as a partly mobile ICBM to provide it. But years of Congressional hearings and public debate under Carter had failed to produce a plan for basing the new ICBMs that could be accepted by Americans living in the relevant areas, and could also have some chance of 'working' in terms of the perceived Soviet counter-force threat.

In particular, the Multiple Protective Shelters (MPS) proposal, to cover vast areas of Utah and Nevada with 'race-tracks' around which both real and decoy MXs could be moved on trucks from shelter to shelter, had met with a storm of political and technical criticism. In the autumn of 1981 it fell to Reagan to pronounce the final veto on the original, purely passive-concealment version of MPS, on the grounds that: 'no matter how many shelters we might build, the Soviets can build more missiles, more quickly, and just as cheaply' (in Smith, R. J., 1982). But his commitment to building up US 'strength' meant he could not abandon the MX as such. Some less unpopular and implausible form of deployment would have to be found.

In Carter's last year, with the MX already in trouble, the Army's BMD Organization began lobbying for a proposal to solve the MX basing problem by adding a limited 'point defence' ABM system known as LoADS (Low Altitude Defense System). This was the most immediate and plausible ABM deployment available to the Reagan administration on taking office. But despite the administration's embarrassment over the MX basing fiasco, LoADS too was rejected in the autumn of 1981. The reason was interesting.

The idea which linked both the passive MPS scheme for MX, and the proposal for 'point defence' ABMs to protect its missile shelters, was basically simple. If a system could be built which would either passively absorb or actively block enough incoming missiles to leave the defender with enough surviving retaliatory forces either to destroy the enemy's society, or to make a similar counter-force strike in return, then the enemy's first strike would never be made, because its outcome had become too uncertain. (This military rationale is assessed in chapter 10.) Advocates of such deployments often judge them good enough to 'work' if the proportion of ICBMs required to 'survive' could be achieved by rendering ineffective 20 to 50 per cent of attacking warheads in a 'normal' strike, or if they could only be overwhelmed, in a 'saturation attack', provided the attackers were obliged to use up most of their nuclear armoury for the purpose (Moore, 1981; Starsman, 1981). President Carter had rejected this rationale completely, as his Air Force Chief of Staff, General Lew Allen, explained:

> The basis of the President's [Carter's] decision, the thing we were not able to convince him of, is that he believed the MPS system was fundamentally vulnerable to saturation. I believe that was the wrong way to look at it . . . If the system required the Soviets to disarm themselves in that leg of the triad [i.e. land-based ICBMs] to which they had clearly ascribed the highest value, then that system did the job we wanted it to do. That is, it changed the balance in

such a way that the Soviets could no longer face us with this great superiority of ICBMs. (In Smith, R. J., 1982)

As one commentator remarked, General Allen seemed less desirous of complete protection for MX missiles than of:

> a sponge to soak up accurate Soviet warheads. This is, to put it mildly, a highly unpopular view in Utah and Nevada (Smith, R. J., 1982).

But even with LoADS added to the passive MPS approach, General Allen found the Reagan administration as hard to convince as its predecessor. As Defense Secretary Weinberger expressed it:

> What we have now is not good enough. It works . . . perhaps with 50% of the incoming missiles. And this is not a situation in which a 50% average is very good. (In Smith, R. J., 1982)

The episode suggests that in its first year the administration was still strongly committed to throwing off the chains of MAD which they had resented for so long. Coupled with that was their appreciation that partial defence of missile silos could not end MAD but must only *reinforce* it (p.169). As long as the ABM issue remained focused on options in point defence, those inside and outside the administration who opposed the principles of the ABM Treaty, and who were growing increasingly anxious for early anti-missile deployments, could not line up sufficient political enthusiasm for their position, above all from the President. Put simply, the Reagan objection to point defence was that it was not, nor could it be publicly represented as, area defence and protection for the people. On the contrary, it was directed to preserving the viability of retaliatory nuclear forces, and thus amounted to a conscious renunciation of the feasibility, and hence the goal, of *ever* making 'weapons of mass destruction obsolete'.

1982 seems to have been given over to intensive lobbying by various supporters of anti-missile defence, aimed at political recoupment in the wake of the temporary reversal they had suffered in October 1981. The Army's BMD group went back to the drawing boards to come up with a new point-defence proposal, using the same 'Sentry' missile for terminal interceptions, but at slightly higher altitudes. This was known as 'Baseline Terminal

Defense'. Teller's first private visit to the Reagan White House
came later that year. A contemporary article by him in the
President's favourite magazine suggests he may have restated the
case for such programmes:

> [E]xtremely important research is being conducted on systems to
> defend against incoming nuclear missiles . . . exploding a very small
> nuclear bomb near an attack missile as it enters the upper-to-middle
> atmosphere . . . could totally *disarm* . . . without *detonating* it.
> (Teller, 1982: 140–1)

Doubtless Teller would also have been asked by the President
about the work being done at Lawrence Livermore on a possible
X-ray laser weapon powered by a nuclear explosion (p. 96), and
would have replied enthusiastically. But he would have had little
need to sell this idea to his long-converted auditor.

During the same period yet another ABM lobby went into
action. This was the Heritage Foundation's 'High Frontier' group,
led by General Daniel Graham. Whereas the Army were pushing
for early deployment of some sort of terminal-phase (p. 140)
point-defence system with at least an 'initial operating capability'
(IOC), and Teller and his associates were promoting very
long-term schemes for total population defence, requiring massive
and prolonged investment in basic research (p. 267), High Frontier
argued for achieving the latter goal, but by the former method. In
short, they claimed it was possible to use existing or near-term
techniques for missile interception to build a highly effective
space-based *area* defence over the next few years. In the end, they
failed to convince either Pentagon or President that their network
of 432 satellites, each armed with 40 to 45 homing interceptors for destroying Soviet ICBMs and warheads by direct collision,
either could or should be built. For one thing, it turned out that
the interceptor missiles were so 'slow' that they would need to
have been fired almost a minute before their target ICBMs. High
Frontier never explained how the required clairvoyance was to be
reliably achieved.

However, High Frontier was always after bigger game than any
particular ABM deployment. As one of its consultants put it
(Fossedal, 1982), missile defences in space would provide 'an
opportunity . . . to fast-thaw the nuclear-freeze movement'. If the
specific High Frontier proposal got into difficulties, perhaps a
more general campaign, mounted from some prudently distinct

organizational base, might capture the liberal centre of US and allied public opinion? Such at least was the hope of another High Frontier consultant, John Bosma, in an in-house discussion document circulated, on Heritage Foundation paper, in about April 1984 (Bosma, 1984).[3]

The paper has been widely excoriated (Mische, 1985; Thompson, 1985: 94–6). It was indeed an outrageous and unintelligent rigmarole, which fantasized about promoting anti-missile defences to European public opinion as a form of support for the 1977 Geneva Protocols codifying 'the international laws of war', whilst ignoring the awkward fact that the Carter administration had only signed them with the reservation that they were 'not intended to have any effect on and do not regulate or prohibit the use of nuclear weapons.' In a similar vein, Bosma proposed exhuming a twenty-year-old Soviet disarmament proposal, which had once endorsed the idea of building up strategic defences, merely so as to embarrass the Soviet Union and its supposedly numerous sympathizers in Western Europe.

But the oddest thing about the document is the way Mr Bosma assumed without comment that all the work for 'keeping BMD alive in 1984 and to make it impossible to stop by 1989' was still to do, as if the President's speech and the resulting SDI programme had never happened. When interviewed during the 1984 election campaign, High Frontier's Director, General Graham, took a rather more optimistic view:

> We [High Frontier] went public in March 1982 and by March 1983, we had the President of the United States saying this is the way to go. (*Washington Times*, 17 May 1984)

The confused and implausible goal of both hijacking peace-movement arguments about the immorality and insecurity of offensive nuclear deterrence, and also bringing the arms control process to an early end, was a general feature of the Heritage approach to Reagan's re-election campaign. And the President gave such ideas at least his broad approval, by endorsing the cover of Heritage's election platform book (Butler et al., 1984) with the testimony that 'one of the people it's been most useful to and used by is me.'

So far as the 23 March 1983 speech itself is concerned, Reagan seems to have been finally persuaded to throw his full Presidential weight behind a drive for missile defences at a meeting of his

Chiefs of Staff a few weeks earlier. The discussion had been ably manœuvred by the then Deputy National Security Adviser, Robert McFarlane, with assistance from the Chief of Naval Operations, Admiral Watkins, to produce recommendations for what could be made to seem like an SDI type of approach.[4] The President referred to that meeting in his speech as follows:

> In recent months . . . my advisers, including in particular the Joint Chiefs of Staff, have underscored the necessity to break out of a future that relies solely upon offensive retaliation for our security. (Reagan, 1983)

Some people's perception of that 'necessity' may have been less visionary than the President's. For instance, one influential associate of the Madison Group and the Heritage Foundation is reported to have been Mr Maxwell W. Hunter, a senior Lockheed executive. Immediately after the President's speech, Lockheed shares rose 12 per cent in value (*Business Week*, 18 August 1980 and 25 April 1983).

But what exactly did the President propose? It is worth getting that as clear as the President's own words and a brief commentary can make it, before the technical and strategic assessments of the next two chapters. The speech began with a reference to what was to come later on:

> I have reached a decision which offers a new hope for our children in the 21st century – a decision I will tell you about in a few minutes . . .

But its first and larger part was devoted to an attempt to raise popular support for the President's second, much increased defence budget. Some aspects of the basic Reagan approach appear in this text, and are worth reviewing. The President spoke of:

> a budget which . . . is part of a careful, long-term plan to make America strong again after too many years of neglect and mistakes.

Just as Mr Weinberger had argued eleven years earlier (1972: 4), there was only one way to draw up such a budget:

> We start by considering what must be done to maintain peace, and review all the possible threats against our security. Then a strategy

for strengthening peace and defending against those threats must be agreed upon. And finally our defense establishment must be evaluated to see what is necessary to protect against any or all the potential threats . . . There is no logical way you can say, let's spend X billion dollars less. You can only say, which part of our defense measures do we believe we can do without and still have security against all contingencies?

The Reagan view of US security identifies peace with the removal or at least the effective negation of *all* 'threats' to the United States, and can be satisfied with nothing less than a total security, 'against all contingencies'. The list of unacceptable threats is drawn very wide:

'Deterrence' means simply this: making sure any adversary who thinks about attacking the United States, or our allies, *or our vital interests*, concludes that the risks to him outweigh any potential gains. Once he understands that, he won't attack. We maintain the peace through our strength; weakness only invites aggression. (Emphasis added)

Once peace has been defined as an arrangement by which the United States is not to be deprived of anything it sees as really important, it becomes rather obvious that it cannot be achieved except by being stronger than every other country. Conversely, if the opposite of 'peace', namely 'aggression', is applied to all circumstances in which the United States has to put up with things she does not want to happen, this is more likely to come about if she is relatively 'weak'. But what happens if the 'strength' of two opposed rivals is ever roughly equal?

The question has already been touched upon in chapter 5. In a situation of 'rough parity', especially of more or less unstoppable and totally devastating offensive nuclear forces, each government can probably prevent the other from making any major changes in the world that it sees as too great a 'loss' to itself. But there is a price to pay for this negative security, namely that each is also prevented from making any major changes to its own 'advantage'. *It depends entirely on each government's subjective expectations whether this situation of parity is seen as an equitable if unpleasant necessity, or as an unacceptable form of 'weakness', to be remedied as rapidly as possible.*

The Reagan administration has a higher than usual expectation that the United States should be able to have at least most of her own way in the world. The President had, after all, come to power with an election campaign based on such rhetorical questions as: 'Since when has it been wrong for America to be first in military strength? How is military superiority dangerous?' (in Burt, 1980: 87). Such an approach to security is very likely to resent the other side's rough parity as a threatening 'superiority'. If the United States cannot risk making certain moves in the world because of the perceived size of Soviet forces, then she is weaker than she needs to be. And that translates directly into a delusion that she is weaker than the Soviet Union. After all, are not the latter's forces, allies or influence preventing the United States from getting her way over certain matters? So the Soviet Union must be getting *her* way, at US expense. 'Aggression' has therefore occurred. And its only conceivable cause, American weakness, American inferiority to the Soviet Union, must be present.

Throughout this thought-process no attention at all is paid to the manifold ways in which the story could be told the other way about, from the Soviet viewpoint. That would be to acknowledge, inconveniently, how much exactly similar 'strength' the United States possesses in the global confrontation.

It should also be noted that situations of antagonistic parity between two great powers or power blocs are the classic historical occasions for the most intense and dangerous arms races, in which 'arms race instability' (an uncontrolled arms race tending to disrupt good relations between the parties) may interact with 'crisis instability' (the risk of war through misperception or misjudgement during an acute clash of interests or 'incident' involving opposing states). In such a competition, each side struggles to restore its own security under a definition which ensures that it must tend to deprive the other of what the latter regards as the minimum essential for *its* security. From within the narrow logic of purely *military* necessity, the alternative, a 'stalemate' of mutual deterrence, could never be accepted. Only a *political* easing of the acute differences between the adversaries might make that possible.

Back to the speech:

> This strategy of deterrence has not changed. It still works. But what it takes to maintain deterrence has changed. It took one kind of military force to deter an attack when we had far more nuclear weapons than any other power; it takes another kind now that the Soviets . . . have enough accurate and powerful nuclear weapons to

destroy virtually all our missiles on the ground . . . what must be recognized is that our security is based on being prepared to meet all threats.

The message of absolute security was developing into one about maintaining 'deterrence', to wit, a US monopoly on sufficient power to scare off any nation that might disagree with her manner of pursuing 'vital interests' anywhere in the world, despite any changes in the 'threat' that might develop over time. Note, in this passage, the disingenuous transformation of some 2300 warheads on US ICBMs, under presumed threat from a Soviet disarming counter-force first strike, into 'virtually all our missiles', thereby puffing their dubiously vulnerable numbers with the illegitimate addition of the 4500 SLBM warheads that might reasonably be expected to survive at sea (OTA, 1985a: 105). The deception is surely deliberate, for even the most ardent of American 'hawks' cannot invent a plausible first-strike threat against American SLBMs at sea. This was an appropriate overture to the picture of 'increasingly obsolete', indeed almost useless American missiles bravely facing the Soviet 'massive arsenal of new nuclear weapons', which followed. Then the central theme was restated, of the intolerable nature of the Soviet Union's military power, if it hampered in any way that of the United States:

> As the Soviets have increased their military power, they have been emboldened to extend that power. They are spreading their military influence in ways that can directly challenge our vital interests and those of our allies.

Next came illustrated examples of two Soviet bases in Cuba, one for electronic intelligence-gathering, and the other for air reconnaissance, and of Soviet military equipment at a Nicaraguan air-field. Then a paragraph devoted to 'the Soviet-Cuban militarization of Grenada'. Doubtless his wish to avoid 'compromising our most sensitive intelligence sources and methods' prevented the President from actually naming any of the numerous American installations used for similar purposes in Japan, West Germany, Turkey, China, Britain and a few other countries. Otherwise his audience might have been able to work out the probable reason *why* the 'Soviet intelligence collection facility' at Lourdes in Cuba is 'the largest of its kind in the world', if indeed it is. Namely, that the Soviet Union does not enjoy the luxury of being able to spread her ground-based intelligence activities around so many diverse locations as her rival.

Four-fifths of his address having been spent on such matters, the President finally strained for something more inspiring:

> I have become more and more deeply convinced that the human spirit must be capable of rising above dealing with other nations and human beings by threatening their existence . . . One of the most important contributions we can make is . . . to lower the level of all arms, and particularly nuclear arms . . . If the Soviet Union will join with us in our effort to achieve major arms reduction we will have succeeded in stabilizing the nuclear balance.

In other words, if the Soviet Union would just accept the United States' proposals, the strategic confrontation could be managed to US advantage. But:

> Nevertheless, it will still be necessary to rely on the specter of retaliation – on mutual threat, and that is a sad commentary on the human condition. Would it not be better to save lives rather than to avenge them? . . . I believe there is a way. Let me share with you a vision of the future that offers hope. It is that we embark on a program to counter the awesome Soviet missile threat with measures that are defensive . . . What if free people could live secure in the knowledge that their security did not rest upon the threat of instant U.S. retaliation to deter a Soviet attack; that we could intercept and destroy strategic ballistic missiles before they reached our own soil or that of our allies?

What indeed? But the President was not being very specific. People were hardly going to live 'secure in the knowledge . . . that we could intercept and destroy' attacking missiles unless two conditions could be met. First, they would have to be absolutely confident that the ballistic missile defences provided were really as good as they were painted. And second, they would want protection in the form of a complete and not a partial shield, because the latter would make no real difference to their present situation.[5] But the President was being coy. Would there be a capacity 'to intercept and destroy' *all* or *most* enemy missiles? How 'thick' an anti-missile defence had he in mind? By gazing beyond existing anti-missile technologies, his next remarks suggested he took a pretty large view of what was required:

I know this [but what?] is a formidable technical task, one that may not be accomplished before the end of this century . . . It will take years, probably decades, of effort on many fronts.

At this point the speech turned aside to address two other possible objections. First of all, it was emphasized that throughout those long decades the strategy of deterrence by threat of retaliation ('flexible response'), on behalf of America's allies as well as herself, would have to be 'ensured by modernizing our strategic forces' – which had in fact been the main theme of the evening. Next, the dangers that might thus be created were obliquely referred to:

I clearly recognize that defensive systems have limitations and raise certain problems and ambiguities. If paired with offensive systems, they can be viewed as fostering an aggressive policy, and no-one wants that.

It was not explained, of course, how not wanting it would suffice to stop it happening. (This and other possible strategic flaws in the SDI are discussed in chapter 10.)

The speech closed with a statement of what should happen next, and who should make it do so:

I call upon the scientific community who gave us nuclear weapons to turn their great talents to the cause of mankind and world peace; to give us the means of rendering these nuclear weapons impotent and obsolete.

   Tonight, consistent with our obligations under the ABM Treaty and recognizing the need for close consultation with our allies, I am taking an important first step. I am directing a comprehensive and intensive effort to define a long-term research and development program to begin to achieve our ultimate goal of eliminating the threat posed by strategic nuclear missiles. . .We seek neither military superiority nor political advantage. Our only purpose – one all people share – is to search for ways to reduce the danger of nuclear war.

The route to security was clearly being signposted as passing through Silicon Valley.

The President upheld in his peroration the philosophy that what had to be done with any threats was always to 'eliminate' and never to live with them, and that the way to success was only through some comprehensive military-technological 'fix'. But in fact all that was actually being undertaken, on 23 March 1983, was 'directing an effort to define a . . . program to begin to achieve our ultimate goal.' To any thoroughly attentive listener, no matter how enthusiastic, this particular fix must have seemed almost as far off after the speech as it had been for the previous thirty years, and as many analysts believe it remains today.

Nevertheless, the goal had finally been declared with that fraudulent but unforgettable phrase about rendering nuclear missiles 'impotent and obsolete' by making it impossible for them ever to reach 'our shore or that of our allies'. Somehow this was to be achieved, or to start to begin to commence to be achieved, within the constraints of a treaty which had registered a clear and unlimited commitment, on both sides, not to engage in such quests for the one-sided disarmament of the opponent by military-technological means, because of the evident hazards that would result. And somehow there would be recognized the need for consultation with allies to whom the President's words had come as a total and almost without exception unwelcome surprise.[6] Small wonder if some of the President's less trusting auditors were left wondering whether 'to reduce the danger of nuclear war' had meant, not so much to lower the chance of one occurring, as to lessen the danger from such a war to the United States, should one ever have to be fought, thereby perhaps making it *more* rather than less 'fightable', for at least one country.

In the light of the ABM debates within the Reagan administration before 1983, it is at least possible that the main function of the SDI will appear in the long term to have been that of an excellent public relations front, focusing attention in the late 1980s onto the issue of population defence, whilst an array of force-defence, anti-satellite and space-to-earth weapons programmes were able to gather fresh technological and bureaucratic momentum, using the SDI as a spring-board from which to take the strategic offensive arms race ever onward and upward in the latter part of the century.

This hypothesis has of course been officially contradicted at the highest level.[7] But it does not seek to address the current intentions of the Reagan administration, so much as their likely upshot. It is possibly simplistic. It is certainly fallible. But it may be worth bearing in mind, for all that, when reading the assessments of the SDI and other space weapon programmes which follow.[8]

Part 2
Means

7
Weapons for Ballistic Missile Defence

The Secretary [McNamara] *has also underestimated the importance of actual operational tests* [of ABM systems]. *He is willing to look at the components. He doesn't care apparently whether all his components are put together and the working of the whole thing is investigated. It is like testing each part of the car but then not to give it a road test before you give it to the customer.*

Edward Teller, 1963

The Challenge – Strategic Nuclear Attack

The problems for any active defence against long-range ballistic missiles during their flight are very severe. To appreciate them properly, it is necessary to understand the basic flight characteristics of an intercontinental ballistic missile (ICBM).

Four phases can be distinguished during an ICBM's approximately thirty-minute flight from the Soviet Union to the United States, or vice versa (figure 7.1). The initial powered part of the flight, during which the missile's rocket boosters burn, is known as the *boost phase*. In the current generation of ICBMs this boost phase lasts for three to five minutes, ending at an altitude of 200 to 400 kilometres, well outside the atmosphere (box 7.1). For example, the new American MX ICBM has three stages. The first of these finishes burning at an altitude of 22 kilometres, 55 seconds after launch. The second stage then takes over, also burning for 55 seconds, and finishing at 82 kilometres. The final, third stage burns for 60 seconds, taking the remainder of the missile to about 200 kilometres (Carter, 1984a: 10).

At present the Soviet Union deploys about 1400 ICBMs, all of which the United States would have to target for boost-phase interception. Most modern ICBMs carry several warheads (MIRVs), which can be released towards different targets within some limited area or 'footprint' on the ground.

83

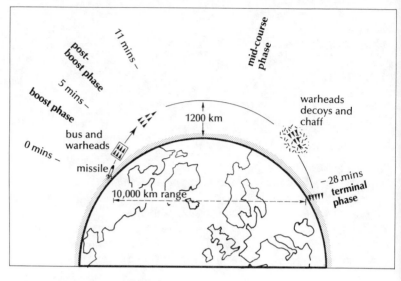

Figure 7.1   ICBM Flightpath
Source: Scientists Against Nuclear Arms

Box 7.1   *The Earth's Atmosphere*

The layers of gases around the Earth known as the atmosphere do not end abruptly at a particular height. They simply get thinner and thinner. So the definition of 'outside the atmosphere' depends on what phenomenon is of interest. At about 100 kilometres altitude air density is high enough to slow down objects with high aerodynamic drag, such as satellites. Specially shaped re-entry vehicles or decoys can descend somewhat lower before atmospheric drag becomes significant. Atmospheric interaction with weapons also varies for the different technologies under consideration. Even the very rarified atmosphere at 150 kilometres would render neutral particle beams useless, whereas X-ray lasers could be used above about 100 kilometres. Other lasers might be used at lower altitudes without suffering serious attenuation, but the disruptive effects of atmospheric turbulence would have to be compensated for, in order to permit long-distance propagation or focusing.

Once the rocket boosters had finished burning, each missile warheads-platform, or 'bus', would begin to release warheads onto separate trajectories, in what is known as the *post-boost phase*. To do this the bus is able to manœuvre a little in space by means of small jets. Apart from this the missile would be essentially unpowered after the boost phase, travelling on a flightpath determined by its initial velocity and the gravitational field it passes through. Most ICBMs can also release a variety of penetration aids, for example metallic chaff and hundreds of decoy balloons. The net result is that during the post-boost phase (which currently lasts about five minutes) each ICBM would give rise to a large, confusing swarm of objects, known as the 'threat cloud'.

Soviet ICBMs are thought to carry almost 7000 warheads altogether. Along with penetration aids, there might therefore be some hundreds of thousands of potentially dangerous objects travelling through space towards the United States (UCS, 1984: 130). This *mid-course phase*, as it is called, lasts for about fifteen to twenty minutes. Because of the absence of friction all objects from the same missile, no matter how light, would travel at the same speed.

Only after the threat cloud had entered the edge of Earth's atmosphere, after descending to an altitude of about 60 kilometres, would aerodynamic drag begin to strip away the chaff and lighter decoys. The *terminal phase*, from atmospheric re-entry until impact, lasts between 30 and 100 seconds, depending on the design of re-entry vehicle and the angle of entry. Typically a warhead from an ICBM would enter the atmosphere at an angle of about 23° to the horizontal, travelling at a speed of about 7 km/sec (or twenty times the speed of sound), and would have slowed to just under half that speed by impact (Bunn, 1984: 1). Some sophisticated, heavy decoys may accompany the real warheads through the upper part of the atmosphere.

The typical ICBM flight-path just described is designed to deliver the maximum payload with the least effort, and is thus known as the minimum energy trajectory. By sacrificing a little payload it is feasible to launch ICBMs at steeper or shallower angles, into 'lofted' or 'depressed' trajectories respectively.

ICBMs are just one possible way to deliver nuclear warheads from one nation to another. Whereas they make up the major part of the current Soviet arsenal, the United States deploys only about one quarter of its strategic nuclear warheads on ICBMs (table 7.1). Submarine-launched ballistic missiles (SLBMs) would have

Table 7.1   US and Soviet Strategic Nuclear Arsenals, Mid-1985

| | Launchers | | Warheads | |
	United States	Soviet Union	United States	Soviet Union
ICBMs	1 023	1 398	2 123	6 540
SLBMs	672	943	5 728	3 043
Bombers	263	173	2 496	386
Totals	1 958	2 514	10 347	9 969

Source: *Arms Control Reporter*, June 1985.

much shorter flight times if fired from nearby waters. For example, Soviet SLBMs, if fired from the Atlantic, could hit Washington in as little as five minutes. Other problems are posed for the defence by bombers and cruise missiles, the air-breathing delivery systems.

Layered Defence

The analysis in this chapter is largely illustrated by reference to the case of a prospective US defence against Soviet ballistic missiles. Technically speaking, the reverse case is of course equally relevant. However, President Reagan's March 1983 initiative, now being implemented by the SDIO (appendix 2), inevitably produces a degree of asymmetry in the remainder of the book, as the discussion moves increasingly from the past and present into the future, albeit often conditional, tense.

Although various, sometimes apparently inconsistent objectives have been put forward for the Strategic Defense Initiative, its originally preferred approach to the BMD problem was fairly clear. Instead of waiting to attack the enemy missiles late in their flight, with weapons based in the United States, the emphasis would be on trying to intercept most of them within a few minutes of launch, as they rose above the Soviet Union. This would be the crucial element of a layered defence, and because of the curvature of Earth's surface at least some parts of the defensive systems would necessarily be based in space.

During boost phase an ICBM constitutes a large target, getting smaller as successive booster stages drop off. It is relatively easy to detect and track because of the intense infra-red radiation from the exhaust plumes of its rocket boosters. The booster stages are in effect large fuel tanks undergoing massive acceleration against Earth's gravitational pull. This places their skin under considerable stress, making them relatively fragile targets. The most important advantage of boost-phase interception, however, is that it helps to negate the offensive advantage gained by putting multiple warheads on ICBMs. Unless the boost-phase 'layer' of BMD were highly effective, later layers would almost certainly be overwhelmed by the numbers of warheads and decoys. James Fletcher, who headed the Defensive Technology Study carried out in 1983 (chapter 6, note 8), has emphasized this:

> Without the ability to intercept missiles in their boost phase, a highly reliable, low-leakage defense would be exceedingly difficult to achieve. (Fletcher, 1984: 23)

The next layer of the defence could operate against the bus, as it dispensed its warheads during the post-boost phase. However, the bus and warheads carry much more protective hardening than the boosters and would be harder to spot. The bus is manœuvred by small jets which either burn liquid fuel or else expel an inert gas. In the first case the infra-red signal would be weak compared to that from the missile's main boosters, and would only be intermittent, as the bus positioned itself to dispense the warheads onto their different trajectories. In the second, only the still weaker long-wave infra-red signal from the bus's own physical temperature would remain (box 8.1), an even more difficult effect for the defence to make use of.

Those warheads and their accompanying decoys which still survived could then be attacked during the mid-course and terminal phases. Proponents of the SDI envisage a scheme in which Soviet missiles would have to run a gauntlet of US defences. With several defensive layers arranged in sequence the total efficiency of the entire defence could, in theory, be high even if each layer was only moderately effective. For example, four defensive layers which were each 70 per cent effective would give an overall system effectiveness of over 99 per cent, on the assumption that each layer operates in complete independence of the other three (Fletcher, 1984: 17). If one or more layers had common components, however, such as battle-management computers and software, or even just common design assumptions, their shared vulnerability or liability to prove somehow deficient would remove much of the advantage expected from layering, since interdependent probabilities cannot be multiplied as simply or as confidently as strictly independent ones. (This possibility is known as the risk of a 'common mode failure'.)

Recent 'architecture studies' within the SDI have produced

designs for up to six or seven layers, using different combinations of the hypothetical types of weapon described in this chapter (*New York Times*, 3 November 1985). However, no technical basis exists for any particular assumption regarding the effectiveness of any postulated layer. If each layer is supposed, for simplicity's sake, to have the same level of 'missile kill' capacity '*f*' (a number between 0 and 1), then for any postulated number '*NL*' of attacking warheads launched the number 'NP' of penetrating warheads likely to survive a defensive 'architecture' of '*n*' layers is given by the equation:

$$NP = NL(1 - f)^n$$

The 'efficiency' of the overall system design can thus be easily manipulated, on paper, by adjusting the values for n and f at will.

In reality, however, the actual performance of each successive layer would depend on the scale and nature of the attack, and on the performance of the other layers. Should the boost-phase layer fail to meet its specified efficiency, for whatever reasons, then later layers would probably be so overwhelmed that their efficiency would also be reduced and the entire system would rapidly 'degrade'.

Nevertheless, the age-old concept of defence-in-depth seems to be the best hope for any ballistic missile defence. Leaving aside the strategic and political desirability of such a system there remains the issue of its technical feasibility. An understanding of the technical issues is clearly important. Today, exactly as they did in the late 1960s (Brennan 1969a: 33), advocates of strategic defences claim that technological progress at last offers the potential for building such systems, and argue that the technical possibilities should be thoroughly investigated. However, it should perhaps be emphasized at this point that the question of overall feasibility for large-scale BMD cannot be resolved in terms of the problems to be overcome by, or the 'demonstrated' successes of, component technologies in themselves. Nor are the weapons concepts examined here intrinsically defensive. They could equally be used for attack, for instance against the defensive systems of the other side. Over the next two to three decades some of them may possibly become available to *both* sides of the East-West arms race. Doubts about the military feasibility of BMD flowing from such considerations (chapter 10) are not the result of any lack of respect for the abilities of scientists and engineers – if anything, the reverse is the case.

Weapons

Lasers

Lasers, in a variety of forms, are a key technology for the SDI (box 7.2). Their potential applications include high-speed optical computing, laser sensors, target designation, interactive discrimination and most importantly directed-energy weapons (DEW). Laser light combines high *intensity* with high *directionality*, so achieving high *brightness*, a measure of the capacity of a light source to concentrate optical radiation into a narrow beam. This property, coupled with the fact that laser beams travel at the speed of light (about 300,000 km/sec), makes their use for the interception of ballistic missiles over very long distances technically attractive, at least in principle.

Box 7.2   What is a Laser?

A laser is a special kind of light source, of which different types can generate electro-magnetic radiation at different wavelengths, in the visible, infra-red, ultraviolet and other parts of the spectrum (table 7.2). All lasers are applications of the process of 'light amplification', with the term 'light' used loosely to span the electro-magnetic spectrum from the very long wavelength millimetre-wave region, through the far and near infra-red, to the shorter wavelengths of the visible, ultraviolet and X-ray regions. The *light amplification* process is produced by using an external energy source to selectively 'excite' an optical medium or material. For example, the medium can be a gas, which may be energized by an electric discharge. Or it may be excited by a particular chemical reaction. What matters is the selective injection of energy which excites the individual atoms or molecules into a higher 'energy state'. If a bundle of light energy (a 'photon') of the right wavelength interacts with such an atom or molecule, the latter may be *stimulated* to *emit* a second photon identical to the first. Each of the two photons may now interact with further atoms or molecules in the excited state, stimulating the emission of further photons in a chain reaction. Light Amplification proceeds by Stimulated Emission of Radiation – in a word, a laser.

Laser action ('lasing') depends on stimulating the coordinated emission of photons throughout some medium, in appropriate atomic transitions from a higher to a lower energy state, rather than leaving such transitions to occur at random. The process can only start if there are more atoms in the excited state than in the lower one, an abnormal situation ('population inversion') that has to be artificially created. When it is achieved, spontaneously emitted photons in one part of the medium trigger a 'cascade' of other photons, producing light of a single very pure colour, or *monochromatic wavelength*. By arranging for it to be reflected back and forth through the medium many times between a pair of carefully aligned mirrors, the beam can be brought to a high degree of intensity (power) and directionality.

Laser light has another important property in comparison with light from ordinary sources, namely *coherence*. This means that the phase of its waves is well-defined, both in time and across the cross-section of its beam, whereas the phase in ordinary light is random both in time and space. Coherence properties of laser light are important for several possible SDI-type applications too complex for consideration in this book.

Laser beams would suffer almost no attenuation when operating in the near vacuum of outer space. Space is not totally empty, however, and laser-beam propagation could be adversely affected by rarified gases (Hecht, 1985: 17). More importantly, laser light, though highly directional, is not *perfectly* parallel ('collimated') and spreads out in a manner described by the laws of diffraction. The divergence of the beam is greater the longer the wavelength of the radiation, but can be reduced by using a larger focusing mirror (figure 7.2). Other things being equal, an infra-red laser would need a larger mirror than an ultraviolet one to form the same spot size over the same distance, because the latter has a shorter wavelength (box 7.3).

Chemical lasers are the most mature directed-energy weapon technology at present (Hecht, 1984). The most powerful continuous lasers so far devised use a reaction between hydrogen and fluorine to produce a beam of infra-red radiation, with wavelengths of 2.7 or 3.8 microns, a little longer than visible light. If a sufficiently powerful laser were built, then (in theory) it could be aimed from an orbiting battle station at an ICBM booster. If focused steadily on the same part of the missile it could burn a hole and cause the missile to fail.

The effective range of the laser, against a given 'hardness' of target, would be determined by its power output and the size of its focusing mirror. Some reports have envisaged a combination of a 25 megawatt (MW) laser focused by a 10 metre mirror. However, such technology is far beyond presently (1986) conceived capabilities. The Fletcher Study called for demonstration of a 2 MW space-operable chemical laser in 1987, with almost immediate scaling up to 10 MW (*Aviation Week & Space Technology*, 17 October 1983). In a test on 6 September 1985 the Navy's 2 MW MIRACL (Mid-Infra-Red Advanced Chemical Laser) produced a 3.8 micron wavelength beam from deuterium fluoride, and damaged a pressurized Titan ICBM second stage (*Aviation Week & Space Technology*, 23 September 1985).[1] Increasing the power output further to 25 MW would be neither simple nor cheap, but may not be an insurmountable problem. On the question of scaling single lasers to much higher powers, however, the SDIO's Innovative Science and Technology Office has issued the following note of caution:

> Because of incoherent scaling relationships there is a practical upper limit to the power production capabilities of any space-based laser. This limit, in general, falls short of what is required for envisaged strategic defense scenarios. (US SDIO, 1985: 9)

Thus some means of coherently combining the output of several lasers would be required.[2]

Producing optically perfect mirrors 10 metres across and suitable for launching into orbit is also a formidable technical challenge. This is twice the diameter of the mirror used by the United States' largest telescope (at Mt Palomar) and four times that of the largest mirror ever built for use in space – NASA's 2.4 metre Hubble Space Telescope, which cost over $1 billion, and which has not yet been placed in orbit, because of the loss of the *Challenger* Shuttle in January 1986. Moreover, telescope mirrors are designed to collect rather faint radiation signals. A laser mirror

Box 7.3   Effects of Diffraction on Laser Power Density at Target

Even if we assume an ideal laser source, whose radiation has a near-planar wavefront, the highly directional beam will still spread out (the spot will get larger) according to the distance travelled. This *diffraction* effect results from the basic wave nature of light. The large distances involved in space-based BMD or other weapons, up to thousands of kilometres, make it especially significant. For laser DEW the technical objective must be to maximize the laser power delivered in a given time per unit area of the target, which makes it desirable to minimize the diameter of the laser beam spot, by countering, if possible, the beam-spreading effects of diffraction. In principle, this could be done by using a large curved mirror to focus the beam to a spot of the required diameter. In practice, this may be impossible for some laser wavelengths and exceedingly difficult for others.

Taking into account the effects of diffraction and the properties of ideal laser beams, the minimum spot diameter, d, achievable at a given distance, R, depends on the diameter, D, of the final beam director mirror, and the wavelength, w, of the radiation, according to the *approximate* equation:

$$d = 2wR/D$$

For a wavelength of 1 micron (a millionth of a metre) as proposed for one DEW candidate – the free electron laser (p. 95) – a range of 1000 kilometres, and a mirror diameter of one metre, the minimum diameter of the beam spot at the target would be approximately 2 metres. Given the size of current missile boosters, spot diameters of about one metre are perhaps nearer the requirement. There is however an inevitable trade-off between spot size, pointing accuracy, and available laser power.

Table 7.2 Electro-magnetic Radiation

Wavelength (metres)	Frequency (hertz)	Name	Functions
10^5–10^7	≤ 3 kHz	Extremely LF	Submarine Communications
10^4–10^5	≤ 30 kHz	Very LF	Submarine Communications
10^3–10^4	≤ 300 kHz	Low Frequency	Submarine Communications
10^2–10^3	≤ 3 MHz	Medium Frequency	AM Radio
10^1–10^2	≤ 30 MHz	High Frequency	Short-wave Radio
10^0–10^1	≤ 300 MHz	Very HF	FM Radio, TV, Radar, Satellite Communications
10^{-1}–10^0	≤ 3 GHz	Ultra HF	Radar, TV, Microwave, Satellite Communications, RF Weapons
10^{-2}–10^{-1}	≤ 30 GHz	Super HF	Radar, Microwave, Satellite Communications, RF Weapons
10^{-3}–10^{-2}	≤ 300 GHz	Extremely HF	Radar, Satellite Comm.
10^{-4}–10^{-3}		Far infra-red	
10^{-5}–10^{-4}		Far infra-red	CO_2 laser
10^{-6}–10^{-5}		Short- to Long-wave infra-red	Other chemical lasers
10^{-7}–10^{-6}		Visible and ultraviolet	Excimer & Free-electron lasers
10^{-9}–10^{-7}		Ultraviolet	
10^{-11}–10^{-9}		X-rays	X-ray laser
10^{-13}–10^{-11}		Gamma & Cosmic rays	

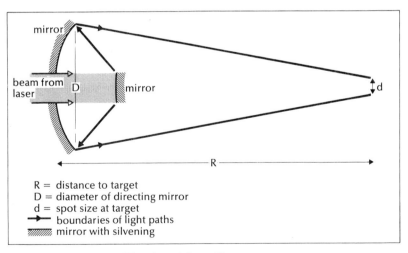

R = distance to target
D = diameter of directing mirror
d = spot size at target
→ boundaries of light paths
▨ mirror with silvening

Figure 7.2 Laser Diffraction : Mirror Size

would need to be capable of handling the very large power densities required for useful weapons. Imperfections in the mirror would not only reduce its beam quality, but might also lead to overheating and damage to the mirror itself. Obtaining perfect, diffraction-limited laser beams would require mirrors machined to within a fraction of a wavelength of the ideal. This is likely to be technically possible, but it will probably be very expensive for large mirrors, robust enough to be launched into orbit. And although they can be smaller, it becomes increasingly difficult to achieve with mirrors for shorter wavelength lasers.

A new technique has been developed at Strathclyde University and elsewhere which may pave the way for cheap, high-quality mirrors. This consists in coating flexible surfaces with perfectly reflecting materials, then shaping them into the required optical parabola by means of suction from a partial vacuum. However, it is not clear how such mirrors could be established in the true vacuum of space, or whether their materials could cope with the high power densities needed for weapons.

The ability to redirect the laser beam rapidly from one target to another will also be important for anti-missile defences. The Hubble space telescope mirror takes three seconds to move only a small amount (0.02 arc-seconds) and then to stabilize. Larger mirrors, even those specially designed for rapid movement (which the space telescope is not), may be severely limited by their settling times when switching from target to target.

Even if optically perfect, 10 metre diameter mirrors with 25 MW chemical lasers were feasible, their technical potential as anti-missile defences would still have to be assessed. Simple calculations show that such a laser beam would need to be directed at the same part of a missile booster for about 10 seconds in order to destroy it at a range of 3000 kilometres (box 7.4). Such a laser would have little potential over much longer ranges because of the reduction in beam intensity due to diffraction. Basing chemical laser weapons in geostationary orbit (35,700 kilometres above the equator) would require improbably large mirrors and laser powers, and is not therefore under consideration. These problems led, at the end of 1985, to a decision in the SDIO to downgrade space-based chemical laser programmes. Whilst still being carried through, if only because of 'contractor momentum', they are not expected to produce weapons for attacking missiles in space. Instead, it is hoped they may provide some kind of 'interactive sensors' (p. 104), operating at much lower power levels (Pike, 1986b).

Box 7.4   Missile Kill Parameters of Chemical Lasers: A
Hypothetical Case

According to the equation in box 7.3 a diffraction-limited
hydrogen-fluoride chemical laser ($w = 2.7$ microns) focused
by a 10 metre mirror would produce a spot of about 1.6
metres diameter at 3000 kilometres range. The spot would
have an area of about 1.2 square metres. (Any more precise
calculation of these values could only give a misleading
impression that technical information had already been
acquired, which in fact is not so.) Using the Fletcher Study
requirement for ICBM destruction, of 200 MJ (megajoules)
per square metre (20 kilojoules per square centimetre), a
beam of 25 MW (25 MJ per second) would need to stay on
the target for almost 10 seconds (the 'dwell-time'). Since the
range might be as much as 4000 kilometres, the power
density problem (box 7.3) for such a laser is considerable.

Lasers working at shorter wavelengths than the chemical kind
have been proposed instead, such as the *excimer* and *free-electron
lasers* first demonstrated in the mid-1970s. Excimer lasers work by
exciting a mixture of two types of atom with a beam of electrons.
The most important ones produce ultraviolet radiation, with a
wavelength shorter than visible light. For example krypton
fluoride and xenon fluoride lasers emit radiation of 0.25 and 0.35
microns respectively. However, current excimer lasers are much
less powerful than chemical lasers, have a pulsed rather than
continuous output, and low efficiencies – 1 per cent or less.

Free-electron lasers work by accelerating a beam of electrons
into a specially designed magnetic field. This causes the electrons
to change direction and emit light. Careful design produces a laser
effect which can be made to operate at wavelengths from
microwaves right down to ultraviolet. Though free-electron lasers
have also yet to demonstrate high power outputs, they seem to
offer the possibility of reasonably high efficiencies, around 20 per
cent or more.

Such short-wave lasers could destroy targets by continuous
heating like the chemical laser, or could instead be designed to
deliver short pulses of energy. Shorter wavelength laser beams
spread out less from diffraction and thus can be effective over
much longer ranges without needing excessively large mirrors.

Lasers operating in the ultraviolet waveband, if powerful enough, could attack ICBM boosters from geostationary orbit. It may even be possible for ultraviolet laser beams to be fired from Earth bases out to such orbits and then reflected back to the targets. No final judgement can be entered at present on the two major technical challenges in this area, which are the development of high-power short-wave lasers, and the production of large enough mirrors to very small optical tolerances, able to direct such beams without themselves suffering destruction. By early 1986 the most promising laser weapon technology in the SDI was reported to be a potential ground-based free-electron laser under development at the Lawrence Livermore National Laboratory in California (Pike, 1986).

A third category of laser has also been considered for use as an anti-missile weapon. This is the *X-ray laser* also developed at Lawrence Livermore. It would consist of thin metal fibres which were energized by a nuclear explosion. The high temperatures produce intense radiation in the X-ray wavelength, which can cause such fibres to lase, delivering a short but intense pulse of directed X-ray energy before the whole device is destroyed. Such beams can be directed, by careful design of the fibres, but cannot be effectively focused by mirrors like other lasers. Not only is radiation of this wavelength – about 14 Ångström (0.0014 micron) – very poorly back-reflected anyway, but also the intense energy pulse released by the nuclear detonation might destroy the mirror before it could have reflected an effective beam (Walbridge, 1984: 180). Despite using an enormous energy source and shorter wavelength radiation, which would reduce diffraction by several orders of magnitude (by comparison with infra-red chemical lasers), the device would need to be reasonably efficient. That is because its relatively unfocused beams would lack brightness. One expert calculation suggests that to be effective against missiles at the comparatively 'short' range of 2000 kilometres, an X-ray weapon might need to be powered by a nuclear explosion with a yield of a few megatons (box 7.5). However, other experts point out that the channelling of X-rays is a familiar technology used in triggering H-bombs. This might help with the brightness problem, by enabling more of the energy from the explosion to enter the lasing process to begin with. 'Theoretical' devices pumped by smaller, kiloton-range explosions are already being conjectured (Allen & Dombey, 1985).

Each X-ray laser device could consist of a nuclear warhead

Box 7.5   X-ray Lasers: Beam Divergence and Yield Requirements

The theoretical minimum beam divergence is determined by the wavelength (e.g. 14 Ångström) and the length of the laser rod. Taking the rod length as 2 metres, Walbridge calculated that an X-ray laser pulse would have spread to over 120 metres in diameter at 2000 kilometres. For a 50-rod X-ray laser attacking Soviet missiles hardened to 20 kilojoules per square centimetre, it might need to be powered by a nuclear explosion of over 3.7 megatons. Enough X-ray lasers to destroy all 1400 Soviet ICBMs in boost phase might therefore arguably require over 100 megatons of nuclear explosive.

Source: Walbridge, 1984.

---

surrounded by a number of laser rods – something of the order of fifty has been suggested (*Aviation Week & Space Technology*, 23 February 1981). Thus each weapon might be able, in theory, to intercept fifty enemy missiles. Destruction would occur not by continuous heating, as with the chemical laser, but by delivering a brief but powerful pulse of X-rays, producing 'a high impulse or shock . . . to break or blow a hole in [a booster] and cause structural collapse' (*Aviation Week & Space Technology*, 23 January 1984). Though an X-ray laser itself could be quite compact, the necessary aiming systems (one for each rod) would add considerably to its size and vulnerability. Each such weapon could only be used once, of course, since it would be destroyed when fired.

In theory the X-ray laser appears to offer a compact, self-contained anti-missile weapon. But apart from problems with generating and directing the beams, it faces other limitations. X-rays cannot usually penetrate far through air and so the X-ray laser might be unusable against targets below an altitude of about 100 kilometres.[3] Also, the notion of using 'third generation' nuclear weapons as part of a defence against nuclear-armed missiles is considered politically dubious by many SDI supporters.

The technical and political difficulties facing Livermore's 'Excalibur' X-ray laser programme were shown on 1 November 1985, when the *New York Times* reported that previously optimistic press stories, about results obtained earlier in the year from a $30 million underground nuclear test explosion, had been invalidated by a serious experimental error. The *Times* recalled that the test, held on the second 'Star Wars' anniversary, 23 March 1985, had led to reports of 'an important advance in increasing the brightness and thus the power of the X-ray device', an increase which was then described in the July issue of *Scientific American*, on the evidence of unnamed Livermore scientists, as having been a matter of 'six orders of magnitude' (a factor of 1 million). The mistake which had apparently produced the overestimate was decribed in *Science* as follows:

> [L]ab researchers discovered that key monitoring equipment had been improperly calibrated, rendering this judgement uncertain. In addition, a new defect in beam collimation cropped up, apparently caused by an acoustic disturbance of the lasing medium. A vigorous search for alternative lasing rod materials is under way, and plans have been set to reduce the laser's considerable mechanical complexity, as well as to boost its relatively low efficiency and power, according to several scientists familiar with the program. (*Science*, 8 November 1985)

The article also pointed out that, of the programme's 'handful of underground tests' since 1980, 'at least three are known to have been either unsuccessful or indeterminate because monitoring equipment failed.'[4]

Whilst SDIO and other Pentagon officials continue to view the X-ray laser as an important, even central element in the programme, the problems referred to have been admitted by Dr George Miller, Livermore's acting associate director for defence systems (*Science*, 29 November 1985). They suggest that when the Assistant Secretary of Energy for Defense Programs referred to the X-ray laser the previous February, *before* the misleading test, as 'the [SDI] project that is furthest along' he was either misinformed or else revealing more than he intended about the rest of the programme (*Aviation Week & Space Technology*, 18 February 1985).

By 1986, it was reported that the SDIO was interested in X-ray laser weapons mainly as asats. For BMD, it was thought they might only be useful as a technology for interactive discrimination of warheads from decoys in the mid-course phase (p. 104).

Particle Beams

Particle beams are the other best-known directed-energy technology proposed in the SDI and, as with lasers, research pre-dates the Reagan initiative. The principle has already been demonstrated by the large particle accelerators used for fundamental physics research, and involves accelerating charged particles such as protons and electrons to very high speeds (close to the speed of light). However, the properties of charged particle beams make them unsuitable for space-based defence, since their charges make the particles mutually repulsive, so that their beams would disperse rapidly in the vacuum of outer space.[5] They would also bend and spread in Earth's magnetic field (Bekefi et al., 1980). In principle, the particles could be neutralized after the acceleration process (for which their charges are necessary). But the physics of the neutralization technique may seriously diminish beam quality (Hecht, 1984: 153–4) leading, in practice, to beam divergences greater than for a chemical laser and smaller than for an X-ray laser (Carter, 1984a: 29). Neutral particle beams would be ineffective below about 150 kilometres, as collisions with air molecules would break them up into charged particles, with the consequences already noted.

If they could be built, neutral particle beam weapons would damage target missiles quite differently from either the surface heating caused by continuous optical lasers or the impulsive jolt delivered by an X-ray laser device. The beam of high-energy particles would penetrate into the target, depositing energy as they collided with particles in the target. This would heat a cone of material within it. If powerful enough a particle beam could melt the metallic structure of the missile. At lower power levels, a particle beam might cause the missile propellant or the conventional explosive primer in the nuclear warhead to explode. Still lower powers might damage or disrupt electronics without causing any visible sign of damage.

Despite this potential for damaging missiles even at relatively low power levels, particle beams are still too immature a technology for their feasibility as BMD to be easily judged. On the same day as Reagan's 'Star Wars' speech, Major General Lamberson testified that 'particle beam technology is currently the least mature of the directed energy technology efforts' (S. Armed Services, 1983: 2653), and little has really changed since then. Neutral particle beam battle stations, complete with power supplies, also look likely to be very heavy, which may limit their practicality as orbital systems for boost-phase BMD.

After two years of study in the SDI, particle beams also seemed more likely to provide a technology for discriminating decoys from warheads than to lead to any actual weapons (Pike, 1986b).

Hit-to-Kill Weapons

Hit-to-kill weapons are those intended to destroy enemy missiles by collision. They include rocket-powered missiles and projectiles from electro-magnetic rail-guns. There is no doubt that a projectile travelling at around 10 km/sec would cause destruction on impact with a target itself moving at high speed. But such accelerations would inevitably involve a trade-off of some sort, with corresponding limitations for space-based systems. Rocket-powered interceptors use chemical fuel which must be disproportionately increased with velocity, because not only the rocket, but also much of the extra fuel, must be accelerated at the higher rate. To reach very high speeds, a multi-stage missile would be needed.

Electro-magnetic rail-guns (also known as 'hypervelocity' guns) work by accelerating a lump of metal through a magnetic field created by producing high voltage differentials between a series of conducting rails. Such guns would have large electrical power requirements to accelerate, for instance, 5 kg projectiles to 10 km/sec (box 7.6). Larger projectiles or greater velocities would involve heavier and more expensive power supplies.

Box 7.6   Electro-magnetic Rail-gun Power Requirements

The energy carried by a moving object, its kinetic energy, can be calculated by multiplying half the mass by the velocity squared. A homing device of 5 kg would need 250 MJ of energy imparted to it to be accelerated to 10 km/sec. This amount of energy is comparable to that needed by the imaginary 25 MW chemical laser discussed above (box 7.4). However, rail-guns require electrical energy, and a capacity for generating or storing this would add to the inefficiency of the system and push up the weight to be orbited. Firing the projectiles at lower velocities would considerably reduce the energy requirements (a 5 kg projectile at 5 km/sec carries 62.5 MJ), but the speed advantage over conventional chemically driven missiles would be lost.

Non-nuclear hit-to-kill weapons are being studied in the SDI for possible use in all flight-phases of enemy missiles. For boost-phase or mid-course interception, a long way from large US-based radars, the accuracy required for collision may be obtained by infra-red homing devices carried by each interceptor. Terminal or very late mid-course interceptors may simply be guided by ground-based radars, as in earlier BMD systems, but these are not currently thought capable of providing sufficient accuracy for hit-to-kill weapons.

Nuclear-Armed Interceptors

Earlier ABM systems, such as the US Safeguard and the Soviet Galosh, were based on interceptor missiles armed with nuclear warheads. The missiles were designed to intercept enemy war-heads either outside the atmosphere in the late mid-course (like the Soviet Galosh and the US Spartan), or in the terminal phase (like the US Sprint). The use of nuclear warheads gave a larger potential 'kill-radius' to the interceptors, so that lower accuracy could be tolerated than with hit-to-kill missiles. The earlier exo-atmospheric ABMs, designed for interception in the late mid-course phase, carried warheads of a few megatons. Despite the lack of blast effects outside the atmosphere, an effective kill-radius of a few kilometres was achieved by means of the very intense pulse of X-rays in the absence of air molecule absorption. Endo-atmospheric ABMs, operating in the terminal phase, would carry lower yield warheads with a much smaller kill-radius. They would destroy or disable enemy warheads by blast or by neutron pulse, or both.

Nuclear-armed ABM interceptors are still regarded as an option for missile defences, but have had a low profile since President Reagan's 'Star Wars' speech. As with the X-ray laser there are political objections to incorporating nuclear weapons into a defence which has the stated aim of making nuclear weapons 'impotent and obsolete'. In addition, ground-based nuclear-armed ABMs have always suffered limited public support because they would involve the detonation of nuclear weapons over the United States. As a former Secretary of the US Air Force, Thomas Reed, expressed it:

> If we want to spend money on things that go bang, then maybe we should spend it on things that go bang over the Soviet Union and not us. (In Smith, R. J., 1982)

And of course, if non-nuclear hit-to-kill weapons were made sufficiently accurate for terminal and late mid-course defence, there would be no point in going any further with 'traditional' nuclear-armed ABMs.

8
The Effectiveness of BMD Systems

Missiles will bring anti-missiles, and anti-missiles will bring anti-anti-missiles. But inevitably, this whole electronic house of cards will reach a point where it can be constructed no higher.
General Omar N. Bradley, 1957

It is one considerable technical challenge to get space weapons to 'work', in the limited sense of making them able to destroy targets at certain ranges and within a certain time constraint. It is quite another, and still more formidable, task to ensure that they and their human controllers would have a good chance of success in actual combat, and not merely under test conditions.

Command-and-Control

Weapons capable of destroying ICBM boosters or warheads are useless without an effective system to coordinate their efforts. Developing such a system is generally considered a more difficult problem than building the weapons themselves.

SATKA – Surveillance, Acquisition, Tracking and Kill Assessment

SATKA research accounted for 32 per cent of the SDI's FY 1985 budget (table AP2.1). It requires the development of sensors and techniques to provide immediate, reliable and detailed information about an enemy missile attack. The intense infra-red radiation generated by the booster rockets can easily be detected at great distances. This type of surveillance is already provided for the United States by three DSP satellites in geostationary orbit (p. 44).[1] But at present the infra-red telescopes carried by these satellites cannot provide sufficiently detailed information for the

103

accurate *tracking* and *target discrimination* required in an ABM system.

Unless boost-phase interception were very efficient, the targets would become far more numerous and smaller thereafter, and would emit far less infra-red radiation. Detection and tracking would become more difficult, and might need active as well as passive means. Passive infra-red sensors would need to be supercooled to detect the faint long-wave emissions from re-entry vehicles. Active systems are those which send out a signal and then measure the reflection to get fairly accurate information about the target's location and velocity. The best known of these is radar (box 3.1), which uses microwaves, but it is also possible to use lasers.

Because of the potentially large numbers of decoys that can be mixed with the warheads, effective discrimination techniques will be crucial to the efficiency of the mid-course defence. One approach would be to observe the deployment of every single warhead and decoy from the buses, because decoys may be easier to spot as they are released. They could then be ignored, whilst the warheads were tracked till interception. But such 'birth-to-death' tracking might have to overcome shielding countermeasures (p. 129). Without it, the prospects for mid-course discrimination would be poor.

Balloon-type decoys, with the same radar profile as warheads, may theoretically be distinguishable by their different infra-red emissions. However, special aerosols could hinder infra-red discrimination just as chaff can hinder radar discrimination. Coping with 'anti-simulation' would also be difficult if not impossible (box 8.1). In theory a low-power laser beam could 'tap' all the objects flying through space. Anything lighter than a warhead would vibrate differently, but this technology would require very complex battle-management software.

Once the components of an enemy missile attack reach the edge of the atmosphere, discrimination should become easier as the lighter decoys slow down more rapidly than the warheads. Defence in this terminal phase is not likely to be limited by the capabilities of the tracking system, but rather by the short time available for interception and the limited effective range of potential weapons.

At every stage of the BMD battle, reliable 'kill assessment' systems would be essential, to confirm kills and to enable efficient re-targeting of surviving warheads. Certain forms of missile 'kill',

Box 8.1   Infra-Red Discrimination of Warheads and Decoys?

All objects warmer than absolute zero ($-273°$ C) emit some level of radiation, with the wavelength depending on the temperature. Objects travelling through space receive radiation from the Sun, if it is not hidden behind Earth, as well as reflections from Earth itself. This radiation warms the objects and is re-emitted as long-wave infra-red. They can therefore be detected against the cold background of space. Because the infra-red signals involved are so weak, however, the defence's sensors would need to be supercooled, otherwise their own background radiation would swamp that from objects in the threat cloud.

Balloon decoys, which probably lose heat more rapidly than warheads, might have a different heat signature. In practice, however, the actual infra-red signatures of the objects in the threat cloud would depend on a number of factors, such as the time of day, the season, the composition of Earth's surface below and the angle the object is viewed from. In addition to the radiated infra-red there would also be some directly reflected from the surface of the object. A discrimination technique using infra-red signatures and capable of taking all these factors into account presents insuperable difficulties, especially if the attacker mixes up decoy and warhead characteristics ('anti-simulation') in the hope that warheads would be 'identified' as decoys.

The SDIO has accordingly closed down this version of its 'space tracking' programmes (p. 291). Instead of a few infra-red sensors at geostationary orbit, it has decided to try for interactive laser-based discrimination, which could require as many as 100 satellites in low Earth orbit.

Far greater emphasis is also being placed on possible airborne sensors, such as the Airborne Optical System, which is intended eventually to combine an infra-red sensor with a laser ranging device. This shift inevitably draws the whole 'architecture' away from space-based boost-phase interception and towards mid-course and terminal systems.

however, might be difficult to detect with the type of sensors under development in the SDI. For example, a relatively low-intensity particle beam weapon could possibly disable a missile's guidance computer without producing any 'visible' effect. Further attacks on such targets could then be a serious waste of effort for the BMD system.

In summary, SATKA requirements impose formidable technological demands. *Detection* and accurate *tracking* of the components of an ICBM attack during its various stages are at least conceivable with the sensor systems currently being developed, even if they are not expected to be achieved for some considerable time. However, reliable *discrimination* of decoys, especially when used in conjunction with chaff and aerosols, would inevitably be a still more difficult task against a determined opponent. Since the decoy problem would be most acute during mid-course interception, high efficiency from the first, boost-phase layer of the defence is recognized as crucial to its overall success.

Aiming (Weapon Pointing and Tracking)

On the basis of the information supplied by the SATKA sensors, the weapons would need to be directed towards their targets. Homing hit-to-kill interceptors could be of the 'fire-and-forget' type, but most directed-energy weapons would require very precise pointing and tracking systems. These would need to be especially accurate for lasers or particle beams with a significant 'dwell-time' – the time required to produce the necessary destructive effect. For example, a chemical laser needing ten seconds to burn through the surface of an ICBM booster (p. 95) would have to be held steady on the same part of the missile whilst it travelled some 65 kilometres. And even though travelling at the speed of light, laser beams would need to be aimed ahead of the target if used over distances of thousands of kilometres.

One scheme for precision pointing would use low-power visible-wavelength lasers. Using the information from the surveillance sensors, the aiming system could scan the target area and 'lock on' to any target reflecting the laser illumination. In theory, this could provide more precise tracking and pointing than radar, because of the greater resolving power of shorter wavelength radiation. The very large structures needed for accurate long-range radars also make them unsuitable candidates for space-based sensor roles of this type.

Besides following its rapidly moving target the DEW beam must also be held steady. Any unsteadiness of aim ('jitter') would disperse the destructive energy and reduce or negate the effectiveness of the weapon. At such long ranges, even vibrations from the operating systems on board the battle station might be significant. To meet these severe requirements the Fletcher Study recommended that the SDI demonstrate a tracking accuracy of better than one tenth of a microradian angle by 1988 (*Aviation Week & Space Technology*, 17 October 1983). That is, the 'line of fire' of some hypothetical beam-weapon must be shown to be capable of being held to within an angle of 0.000 005 73° (presumably in both the lateral and the vertical dimensions) from the actual, constantly shifting line between the weapon and its moving target, throughout the theoretically required dwell-time. Even for a stationary target, that degree of accuracy would be roughly equivalent to an ability to hit just one particular brick in the Kremlin Wall, using a laser fired from a point in space 200 kilometres vertically above St Paul's Cathedral. In reality, the 'brick' would be moving at up to 7 km/sec.

Battle Management – Computers and Software Reliability

The heart of any BMD deployment would be the control system linking all its weapons and sensors. This requires sophisticated, high-speed battle-management computers. Because of the short times available for boost-phase interception there would be little scope for human intervention in the decision to activate the defensive systems, and so they may have to be fully automated. Once in operation such a system 'in most cases replaces human decision-making' as it conducts the complex task of coordinating the defence, with computers which would need to be 'very fast, performing on the order of one billion operations a second' (Fletcher, 1984: 24). Although demanding, this level of computing speed could conceivably be met by the time the weapons are available. Optical 'supercomputers' are one long-term approach currently being explored at the Livermore Laboratory, at Britain's Heriot-Watt University, and elsewhere. And already the continuing advances in integrated circuits, together with the development of array processors and parallel processing systems, suggest that physical processing capacity will not be a major constraint.

However, even assuming that technological progress in weapons, SATKA and computers were adequate for BMD, there would remain a far less easily surmountable obstacle. All these systems constitute the hardware devised to perform certain BMD roles. But software is also required before the computers can carry out their battle-management function. Software comprises the programmes – millions of lines of coded instructions – to be run on the computers. This enables the information received by the sensors to be analysed, and aiming and firing instructions to be passed to the weapons. The Fletcher Study concluded:

> Developing hardware will not be as difficult as developing appropriate software. Very large (order of ten million lines) software that operates reliably, safely, and predictably will have to be developed. (In Lin, 1985: 9)

In 1985 the Phase 1 'architecture studies' (p. 306) confirmed the Fletcher Study estimates of software requirements of 10 to 30 million lines of computer code (Pike, 1986b). Other experts came out with even higher figures.[2] Even ten million lines is very long by current standards. This is about the same overall length as the Space Shuttle programme, but differs in its composition. Only about 500,000 lines of the Shuttle software is devoted to operational performance, with the rest providing support. One estimate for an SDI-type BMD is that 3 million of the 10 million lines would be required for operational software (Lin, 1985: 81, figure 2). This operational software has to work in 'real time' – doing its job at the same time as it gets its information. This means that BMD software will probably be more complex than that used in the Space Shuttle. Indeed the Fletcher Study warned that:

> Specifying, generating, testing, and maintaining the software for a battle management system will be a task that far exceeds in complexity and difficulty any that has yet been accomplished in the production of civil or military software systems. (In Drell et al., 1984: 60, note 100)

Yet even the Shuttle software – designed to assist the human crew in the conceptually straightforward tasks of lifting the craft into orbit, unloading some cargo, and returning to Earth – has not performed flawlessly. In the first space mission of the Shuttle, the launch was delayed for 48 hours by a software error.

The example illustrates some of the problems involved with long and complex software. The error concerned in the Shuttle example would only reveal itself once in every 67 times, and was actually introduced into the software when another problem was sorted out two years earlier (Lin, 1985: 42–3). This is typical of anything other than the very simplest software. Errors can lie 'dormant' for a long time because they may only reveal themselves when the software is run in one particular way under one particular set of conditions. When software is of the order of millions of lines long there is no known way of ensuring it will be totally reliable, especially when it is expected to perform a job (BMD) never attempted before, and in a nuclear-affected environment never previously experienced.

This problem was highlighted in June 1985 when Professor David Parnas resigned from the SDIO's Panel on Computing in Support of Battle Management (the 'Eastport Study Group'). He noted that:

> Because of the extreme demands on the system and our inability to test it, we will never be able to believe with any confidence that we have succeeded. (Letter of resignation to James H. Offut, SDIO Assistant Director for Battle Management and C^3)

Even defining the tasks that BMD software must perform during battle management is a challenge beyond present capabilities.[3] According to Robert Cooper, Director of DARPA:

> Currently we have no way of understanding or dealing with the problems of battle management in a ballistic missile attack ranging upward of many thousands of launches in a short period of time. (In Drell et al., 1984: 60, note 99)

Moreover, though individual components of a BMD may be testable in one-on-one situations, the whole system can never be comprehensively tested in a realistic nuclear war environment. The effects of nuclear explosions high in the atmosphere are not fully understood, as insufficient data were obtained before such testing had to be discontinued under the 1963 Limited Test Ban Treaty. If EMP (box 2.1) and related physical effects of such explosions cannot be adequately described, then they obviously cannot be satisfactorily incorporated into the battle-management software.

By early 1986, the SDIO was explaining that it would seek to cope with the size of the overall software required by subdividing it into shorter, relatively autonomous programmes in separate computers serving different elements of the defensive 'architecture' (Pike, 1986b). There might, indeed, be 'hundreds of individual programs' (Martin, J., 1985), many of which shared large amounts of common code.

It is also suggested that use of a more efficient and error-proof coding language, such as 'Ada', which has been selected for the US Space Station software, would do much to reduce both outright errors and inconsistencies between individual program- mers – another common source of software foul-ups. Computer- aided software engineering might also lead to greater reliability, and new algorithmic programming techniques might result in shorter and thus more checkable programmes. System architectures with multiple sensors and other redundant elements would be resistant to errors generated in only one subsystem. And finally, testing simulation techniques for software have been significantly improved in recent years and the trend is likely to continue (Martin, J., 1985).

But even if absolutely unlimited funds were available (which there will not be) for the development and extensive simulation testing of such a vast battle-management system, and then for its deployment, complete with independent backups for the backups of backups, there would still be two fundamentally intractable problems for its software. One is that a future missile attack would be a novel, unique event, such that it is not possible to conceive of all the tasks which software might have to perform in a complete BMD system. As with the space probes sent to encounter Comet Halley in March 1986, but to an even larger extent, there will be too many unknowns for the sequence and pattern of events to be laid down precisely in advance.[4] Even if that were not so, the second problem is that total, first-time-out reliability is simply unobtainable. No current or foreseeable simulation testing or software-writing techniques could overcome these difficulties, which effectively rule out the development of a BMD system that could be *known*, in advance, to be 'thoroughly reliable and total', in Mr Weinberger's phrase.

Boost-Phase Interception

As already noted, boost-phase interception offers several advan- tages to the defence, and is regarded as crucial by proponents of the SDI. ICBM boosters are conspicuous and valuable targets and efficient interception at this stage would reduce the considerable difficulties faced by later defensive layers. A number of weapon technologies may prove able to destroy an ICBM booster as it rises from its silo. However, even if it were confidently assumed that these technologies will advance significantly, two problems remain. How are the weapons to be based and how are they to be powered? The solutions to both these questions are limited by the laws of physics. In particular the geographical location of the Soviet and US ICBM fields, over the horizon from each other's territory, requires that the boosters must be attacked from space. Three options are available. First, the anti-missile weapons could be permanently based in *low orbits*, continuously moving across the face of the Earth. Second, they might be permanently based in *geostationary orbit*, 35,700 kilometres above a point on the equator. Or else, more probably, lasers could remain on the ground with large mirrors up in geostationary orbit to relay the beams back down to the targets. Or third, they could be *'popped up'* into space on warning of attack.

The SDI is still ostensibly in a basic research phase (lasting from 1984 to 1989) intended to 'provide the evidentiary basis for an informed decision on whether and how to proceed into system development' (in UCS, 1984: 25). The following schemes have been put forward by various proponents as the best hopes for *boost-phase interception*.

Lasers in Low-Earth Orbit[5]

One of the fiercest technological debates between proponents and critics of the SDI has centred on the numbers of low-orbit lasers that would be needed for boost-phase interception. Even using the same assumptions, of a 25 or 20 MW chemical laser with a 10 metre mirror, widely differing conclusions have been reached.

The main problem for low-orbit battle stations is that they are in constant motion relative to the surface of the planet beneath them. But basing chemical lasers in geostationary orbit is not plausible because their infra-red beams would spread out too much when travelling the 39,000 kilometres down (at an angle) to attack the rising ICBMs. In low orbit, a chemical laser would only pass over enemy ICBM fields at intervals (figure 8.1). Most of the time it would be out of range and out of sight – absent over the horizon. Thus for every satellite in range of the enemy ICBM fields at any one time, several more must be spread out around the Earth. Those not in range are known as 'absentees'. The ratio between

the total number deployed and the average number that would be
in range at any time is called the 'absentee ratio'.

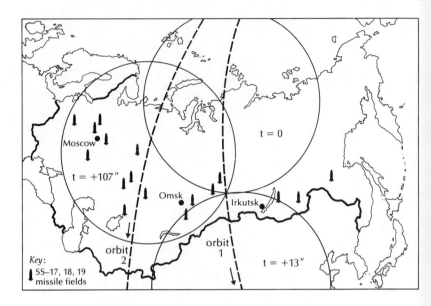

Figure 8.1   Low-Orbit Battle-Station Kill Areas
Source: Union of Concerned Scientists, 1984. Reproduced with permission.

The total number of laser battle stations required in orbit
depends on the assumptions which are made about the scenario,
and this in part explains the wide range of estimates. Another
reason is that some early estimates were based on very rough
calculations.[6] The lowest credible estimate produced was that of
60 satellites at 300 kilometres altitude, from Christopher Cunningham of Lawrence Livermore National Laboratory. According to
his computer model this number of satellites 'could enforce very
low leakage' against the current number (1400) of Soviet ICBMs
(Cunningham,1984: 6). However, his model allowed negligible
switching time for the lasers to redirect their beam from one target
to another, with the improbable result that just four laser battle
stations would have been needed to account for most of the 1400
ICBMs, each destroying about 300.

Because of the extremely precise pointing requirement for
effective use of laser DEW, however, the time needed for laser

mirrors to slew and settle between targets may be a serious limiting factor on such a weapon system's performance (Garwin, 1985). Taking re-targeting times into consideration, the number of satellites required for a similarly effective boost-phase defence would probably be at least 100, perhaps several hundred. That is without allowing for the various plausible countermeasures which the other side might undertake (p. 127).

A hundred laser battle stations would be extremely expensive to build and put in orbit. According to one SDI supporter, Robert Jastrow, 'Every laser-equipped satellite will cost about as much as a Trident submarine – several billion dollars' (in Bethe & Garwin, 1985: 343). So several hundred billions of dollars might have to be spent, even before lifting the weapons into orbit, which would be an additional major cost. One estimate puts the weight of each such satellite at 100 tons, giving a total of 10,000 tons to be lifted into orbit (*Aviation Week & Space Technology*, 17 October 1983). Using the Space Shuttle the predicted cost of lifting one ton into suitable near-polar orbits used to be about $6 million, or about $60 billion to orbit the whole constellation.[7] In fact the Shuttle has a maximum load of about 30 tons. Hence one of the major support programmes within the SDI is for an improved space transportation system (Project 0013 in PE 63224C – appendix 2).[8] This would need to be considerably cheaper per ton for space-based lasers to become an economically feasible prospect.

In 1981 a mistakenly declassified piece of congressional testimony from the Defense Department put the cost of an imperfect or 'damage denial' space-based laser defence at $500 billion in 1981 prices (Drell et al., 1984: 48, note 80). Although technical advances may drive down some of the costs this still seems like a minimum estimate, which might be far exceeded in practice. Since the Soviet Union could probably double the size of its ICBM arsenal, so threatening to overwhelm the defence, for less than this, such a defence does not seem likely to be 'cost effective at the margin', in Ambassador Nitze's phrase (1985). Edward Teller, a leading advocate of the X-ray laser option for the SDI, has noted that:

> To put objects into space is expensive . . . infra-red lasers, the presently known chemical lasers, do not seem to me to fulfil the basic requirement of good defense, and that is that the defense must be considerably less expensive, must require considerably less effort than the offsetting effort in offense. (In Drell et al., 1984: 50, note 84)

Geostationary Relay Stations

Because of the prohibitive costs of orbiting a constellation of chemical laser battle stations, an alternative approach might be to use short-wave lasers in geostationary orbit, thereby reducing the replication of space platforms and fuel. The smaller diffraction in short-wave – excimer or free-electron – laser beams suggests it may be feasible to base part of any system using them in GEO. With the battle stations always in the same position relative to Earth, there would be no unnecessary absentees. However, there are significant drawbacks to this approach, quite apart from the immature state of the laser technology.

Firstly excimer lasers are certainly and free-electron lasers probably far too heavy to be placed in any orbit, let alone GEO. Even if it could be done – and one free-electron laser is being put forward as a possibility – the cost advantage of no longer having wasteful absentees might be swallowed up in the enormous extra cost of lifting lasers to this high orbit. It is usually proposed, instead, to base such lasers on the ground, with only their relay mirrors and auxiliary pointing equipment in GEO. This would greatly facilitate their construction, maintenance and power supplies. The idea would be to put the laser stations on mountain tops to reduce atmospheric attenuation of their beams as far as possible. But several laser stations would have to be built, in case one or more were put out of action by bad weather.

Transmitting the laser beam up through the atmosphere without loss of power and quality would be a problem. Theoretical solutions exist, but they are still just that, and there are serious doubts about whether they would work with the high power-levels needed for missile defence.[9] If this can be done there then remains the task of reflecting the beam back from GEO to destroy the enemy ICBMs, some 39,000 kilometres away.

The same problems of mirror size and diffraction that rule out chemical infra-red DEW in geostationary orbit also pose insuperable obstacles to precise tracking of the infra-red signals from target boosters at that altitude. One solution might be to service the GEO relay mirrors needed for short-wave ground-based DEW with exact target tracking by short-wave lasers, after rough positional fixes had been provided by early-warning sensors. The reflections from such low-power lasers could possibly provide sufficient resolution to aim the main beam. But holding the beam steady on a moving ICBM would also be a formidable task at this distance.

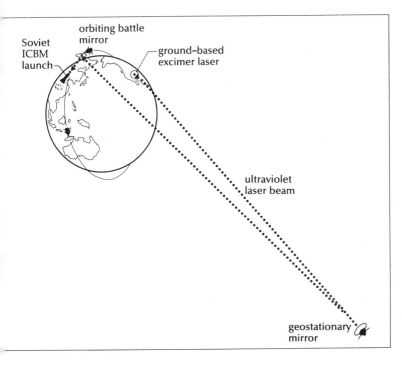

Figure 8.2   Geostationary Relay of Short-Wave Laser Weapon
Source: French, 1985. Reproduced with permission.

Another option generally favoured by proponents of this concept, like the former Presidential Science Adviser George Keyworth, is to have low-orbit 'mission mirrors' to track the ICBMs and redirect the beams sent down to them from GEO (figure 8.2). These would suffer from the absentee problem, but each could be cheaper and lighter than a chemical laser battle station.

The feasibility of such a system is hard to judge because the concept is not yet clearly defined and a number of technological problems are still some way from resolution. Multi-megawatt excimer or free-electron lasers, large space-based mirrors capable of handling high-power short-wave beams, and the necessary atmospheric propagation techniques have all yet to be developed. However, the theoretical power requirements for a system using such ground-based lasers suggest that the cost would be high.[10] Again the cost of developing, building and deploying such a system, should it prove technically possible, seems likely to exceed the cost of conceivable countermeasures.

Pop-Up X-Ray Lasers

'Pop-up' anti-missile systems avoid the absentee problem without the need to develop weapons capable of GEO operation. On warning of enemy missile attack the defensive weapons would be launched (popped up) to intercept the ICBMs during their boost phase. For this to be a practical proposition the weapons must be reasonably light and robust, so that they can be rapidly accelerated by rocket boosters. So far the only weapon which looks at all possible for this role is the nuclear-pumped X-ray laser.

The pop-up method must operate under severe time constraints. After being launched on warning of enemy missile attack, the weapon would need to reach an altitude from which to engage attacking missiles which had a considerable head-start. This problem is particularly serious for X-ray lasers which probably have to climb high enough to fire over the top of the atmosphere. (But see chapter 7, note 3.) Allowing for 100 kilometres' clearance, the X-ray laser would need to climb to about 2500 kilometres to hit a current generation Soviet ICBM, launched at a point 8000 kilometres away, and burning out at an altitude of 400 kilometres. In other words, it would be impractical to base such a system in the United States. It might be possible, instead, to base the launcher rockets for X-ray laser weapons on submarines patrolling closer to the enemy missile fields, or on the territory of a US ally in Europe or elsewhere. With the distance between launch-points reduced to 3000 kilometres, for instance, the launcher rockets would need to climb to about 175 kilometres in the three or so minutes available. With automated launch-on-warning this may just be conceivable, but Soviet missiles with a shorter boost phase would make it far more difficult. Not only would the X-ray laser have to ascend to a greater height to hit the ICBM before it burned out, but it would have to do so in even less time. At the same launch-point distance, the X-ray laser would have to be lifted to about 350 kilometres in less than three minutes, to be in position to hit an MX-type booster burning out at 200 kilometres altitude (figure 8.3).

Very powerful rockets could perhaps cope with this if the command-and-control system were fast-acting enough, but any further reductions in the duration of the enemy ICBM boost phase

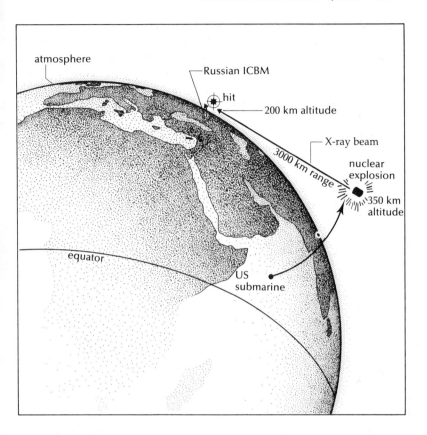

Figure 8.3   Pop-Up X-Ray Laser in Action
Source: French, 1985. Reproduced with permission.

would probably rule out pop-up X-ray lasers for boost-phase
interception. Studies presented to the Fletcher Panel indicated
that ICBM boost phases can be reduced to less than a minute with
only 10 to 15 per cent reductions in their payload (Drell et al.,
1984: 55). Such fast-burning ICBMs would burn out below 100
kilometres, and so probably be protected from X-ray interception
throughout the boost phase.

   In conclusion, pop-up X-ray laser weapons could be effective
against present-day Soviet ICBMs and most current US ICBMs, at
least in theory. The X-ray lasers would probably need quite large
nuclear warheads to power them because of their beam divergence, unless the brightness problem can be overcome, but have the

advantage that precision aiming would not be required (box 7.5). Such weapons would need to be based not much further than 3000 km away from the target ICBM fields, and would need rapid, probably automated, command-and-control systems. Using them could mean detonating many megatons of nuclear explosive in outer space, creating a radiation-affected environment that might pose problems for other parts of the defensive system. Against future ICBM arsenals, specially designed with short boost phases and perhaps launched on depressed trajectories, pop-up X-ray lasers seem unlikely to have much operational value. Hence their recent 'demotion' within SDI architecture studies (p. 98), even if their budgetary support, for other roles, remains high.

Particle Beam Weapons

If space-based neutral particle beam weapons were ever technically feasible, they could be placed in low orbits for boost-phase interception. Much the same sort of considerations apply as for chemical lasers. A constellation of battle stations, with many absentees, would be required to ensure adequate coverage. Particle beam weapons are likely to be quite large – consisting of an accelerator, a power supply, and steering and pointing equipment – and would be even more expensive to build and orbit. Moreover, they could not penetrate the atmosphere below about 150 kilometres, and so could only operate during the later part of the enemy ICBMs' boost phase. Only if they could disable each missile very quickly indeed (with short 'dwell-times' and rapid re-targeting), would particle beam weapons be feasible. Like the X-ray laser their utility could fall rapidly against high-thrust ICBMs burning out below 150 kilometres. With these considerations in mind, the SDI's particle beam programme has been largely reoriented towards developing sensors.

Hit-to-Kill Weapons on Low-Orbit 'Space Trucks'

One proposal for boost-phase interception, championed in the United States by the High Frontier pressure group, does not rely on developing such exotic technology. The basic idea is that enemy ICBMs could be intercepted during their boost phase by non-nuclear homing missiles fired from low-orbit satellites. Like many BMD approaches the concept is quite old, dating from the 1950s when it was known as BAMBI (BAllistic Missile Boost Intercept).[11]

The 1982 High Frontier proposal envisaged a constellation of 432 'space trucks' (satellite platforms) at an altitude of 600 kilometres, each carrying about 40 to 45 homing missiles (Graham, 1982). These were to be capable of a velocity of 1 km/sec relative to the platform, which meant they would have a range of only 300 kilometres at their own altitude during a 300 second boost phase. The scheme looked inadequate even against the current Soviet ICBM arsenal. Since SS-18-type ICBMs burn out at 400 kilometres, 200 kilometres below the proposed orbit, the actual horizontal range of the interceptors would have been only about 225 kilometres. With such a short range the absentee ratio required to ensure enough interceptors over the enemy missile fields at any one time would be very large.[12] The proposal was carefully studied by the Pentagon, but attracted no official support. Robert Cooper, director of DARPA, estimated 'project expenditures on the order of $200 to $300 billion in acquisition costs alone for the proposed system' (S. Armed Services, 1982a: 4635).

A more effective arrangement might be to use a lower orbit with higher-velocity interceptors. Because of atmospheric drag 300 kilometres would be the minimum feasible altitude for permanent basing. In theory a constellation of only 48 satellites might provide adequate coverage. This would ensure eight satellites over the enemy ICBM fields at any one time, each of which would need to be capable of handling 175 ICBMs. Interceptors of 10 km/sec velocity would have a range of 3000 kilometres during a 300 second boost phase. This greatly reduces the absentee ratio, but at a price. Each interceptor missile would have to be much heavier to carry the fuel needed to attain the higher velocity, probably between 200 kg and 400 kg.[13] Each platform would therefore be carrying 35,000 kg to 70,000 kg of interceptors. Adding this to the platform structure, power units (to maintain orbit), and control systems, gives an estimated total mass of 50 to 100 tons per satellite. Putting an entire constellation of such units into near-polar orbit could cost $15 to $30 billion at past Shuttle prices (Carter, 1984a: 33).

Since such a system is unlikely to be cheaper to build than the less effective High Frontier proposal, lifting it into orbit would be a relatively small part of total costs. Also, even with highly optimistic assumptions, its theoretical capabilities would be extremely sensitive to changes in the composition of the enemy ICBM force. Against MX-type ICBMs, with a boost phase lasting

180 seconds, the effective interception range would drop to less than 2000 kilometres, and the absentee ratio would rise steeply. Both development and deployment costs would of course increase. If ICBMs were developed with a very short boost phase (less than 60 seconds), which is conceivable, then satellite-borne missiles would be no use at all for boost-phase interception. Such high-thrust ICBMs could burn out below 100 kilometres, within the atmosphere, so that the interceptor missiles would require improbably high velocities (over 20 km/sec) just to hit them if firing straight down.

Although this technology is still the prime SDI candidate for a boost-phase and mid-course 'kinetic energy weapon' (KEW), the 'Porcupine' design which emerged in late 1985 would require 'several thousand carrier satellites, each with between 1 to 10 kill-vehicles' (Pike, 1986b). This suggests it was drawn up entirely free of cost considerations.

An alternative route to meeting the acceleration requirement for space-based KEW in the early missile trajectory phases has been looked for from electro-magnetic rail-guns. One report spoke of 'interest' in velocities above 10 km/sec with projectiles of up to 1 kg, firing rates of one shot per second and energy conversion efficiency of 50 per cent (*Aviation Week & Space Technology*, 27 January 1986). So far small projectiles of a few grams have been accelerated to velocities of 10 km/sec. But the best firing rate is still only two shots a day for a laboratory device, and the projectiles have shown a tendency to shed pieces on leaving the launcher rails. Although able to penetrate a steel plate at two metres distance, such systems are still far from operational capabilities involving ranges of several thousand kilometres.

According to SDIO Chief Scientist Gerold Yonas, however, the initial goal is a 2 kg *homing* projectile, or 'smart rock', accelerated to 10 km/sec (*Newsweek*, 17 June 1985).[14] The kinetic energy which must be imparted to get the projectile to this velocity is 100 MJ (box 7.6). And projectiles incorporating such complex machinery would necessarily be more susceptible to damage during firing than simple metal slugs.

Hypervelocity guns seem likely to be more expensive to develop than interceptor missiles. Only if they were able to achieve velocities in excess of 20 km/sec might they prove a useful option, though requiring even greater power supplies. Thus, estimates of 100 ton satellites capable of firing 150 shots (a total energy requirement of 15,000 MJ) appear optimistic, unless nuclear power units are used. The cost of such hypervelocity gun platforms would hardly be less than for the simpler chemical missiles considered above, yet their capabilities would be no greater. Indeed they would be rather less, because whilst a satellite's whole consignment of missiles could be launched virtually simultaneously, a hypervelocity gun could only fire its projectiles one at a time. Even if firing rates became high, the whole battle station, weighing something like 100 tons, would need to slew round for re-targeting.

The 'macro-particle stream' accelerator, to which General Abrahamson referred enthusiastically in 1985 (chapter 9, note 8), is a sort of cross between a rail-gun and a particle beam weapon. The technology is being developed by the USAF Armament Division, at Eglin Air Force Base, and by scientists at the University of Texas. It may avoid some of the problems of solid rail-gun projectiles, but is unlikely to be either quick or cheap to develop and deploy.

In general, hit-to-kill weapons offer only limited capabilities for boost-phase interception. Although the technology either is available or can be reasonably envisaged it would still be very complex and expensive, and probably not of high reliability. Moreover, their low velocity, compared to lasers and particle beams, makes such systems relatively easy to thwart by reducing the length of ICBM boost phases. ICBMs which burn out below about 100 kilometres allow little time for interception, and may be hidden by the blinding effect of friction on the infra-red sensors in the homing devices ('red-out'). Thus although hit-to-kill weapons appear to offer the possibility of early deployment of a boost-phase interception BMD, this would be expensive, of dubious effectiveness, and liable to early obsolescence.

Other Defensive Layers

Interception During the Bus (Post-Boost) Phase

When the last rocket booster stage finishes burning and drops off, the ICBM enters the post-boost or bus phase. In current ICBMs this lasts for about five minutes, during which the nuclear warheads, the decoys and other penetration aids like chaff and aerosols are dispensed. The bus constitutes a rather different target than the boost-phase ICBM, being much less conspicuous and somewhat less fragile. As the warheads are released the bus becomes a less and less valuable target.

Post-boost-phase interception could use the same kinds of weapons as during the boost phase, but in almost all cases the performance requirements would be more demanding. Besides the problems of simply tracking the bus (p. 128), maintaining an accurate aim spot on it as it travels at about 7 km/sec, whilst manœuvring to one side or another, is a formidable technical requirement.

The seriousness of this difficulty depends on the type of weapon proposed. Weapons which require a significant dwell-time to cause damage, like infra-red lasers or particle beams, would require very high quality pointing. Weapons capable of causing destruction virtually instantaneously, such as the X-ray and pulsed excimer lasers or hit-to-kill devices, might be less demanding. For most weapons other than hit-to-kill interceptors, however, the bus would be a harder target to damage. Even existing types would be significantly more resistant to damage from lasers than ICBM boosters. Only particle beams might find the bus a relatively softer target, because it contains the guidance and control electronics.

Interception during the post-boost phase is more difficult than boost-phase interception in all respects save one. The extra few minutes currently required to deploy the warheads would give a little more time for the defensive weapons to operate. In the case of weapons severely constrained by atmospheric shielding (such as X-ray lasers and particle beams) the post-boost phase may be the only opportunity for effective interception. However, like the boost phase the post-boost phase could be much shortened, with similar results for the defence. Since the bus is also much easier (and cheaper) to harden against attack than the rocket boosters, the prospects for interception during this phase seem quite low. The current ICBM bus, in action for some five minutes, is a difficult enough target as it is. Future ICBMs may offer even less of a target during their post-boost phase. Perhaps for this reason both proponents and critics of SDI have tended to concentrate on the other layers of the defence – boost-phase, mid-course and terminal.

Mid-course Interception

In current generation ICBMs there is considerable overlap between the post-boost and mid-course phases. Over a period of about five minutes the bus dispenses its payload of warheads and penetration aids onto their free-fall trajectories. Some have thus begun their mid-course phase whilst others are still in the bus phase. By the end of the post-boost phase the bus would have been replaced by a large, confusing threat cloud, consisting of warheads, decoys, chaff and aerosols. Mid-course BMD would either have to find and destroy the real warheads, or else attack indiscriminately everything that might be a warhead.

The second option, destruction of all warhead-like objects, would probably require too many homing collision weapons or too much DEW capacity to be either militarily or economically feasible. Further proliferation of decoys could always push up the cost and reduce the effectiveness of such a defence.

The basic problem for mid-course interception is therefore not so much destroying the warheads, but discovering them amongst the many accompanying decoys. Although the mid-course phase is relatively long, lasting up to twenty minutes, this would be very difficult. A layered defence of the United States might at present have to cope with some hundreds of thousands of objects, free-falling through space towards the defended territory. Several sensor technologies may theoretically be able to gather enough information to determine the nature of an object, but complex software would also be needed to interpret it. 'Birth-to-death' tracking might allow decoys to be identified as they are deployed from the bus, and then ignored whilst the warheads are attacked. The battle-management computers would need to assign a file to every object detected, update it as more information was gathered, and make decisions as to whether it should be attacked, and if so, how and when. Reliable confirmation of its destruction would also be needed, to enable re-targeting if necessary. However, prospective countermeasures against BMD discrimination techniques make the outlook for the mid-course layer seem rather bleak (p. 000).

It should also be noted that the important targets in this phase, the actual warheads, would be much harder than the rocket boosters or the bus. Designed to re-enter the atmosphere at about twenty times the speed of sound, they are coated with carbon or carbon compounds which burn off slowly on re-entry. Interception during the first part of the mid-course phase could be accomplished by weapons operating from space, as for boost-phase interception. During the later part of the mid-course phase, ground-based direct-ascent weapons might be used. In particular, infra-red homing interceptors, launched on ground-based missiles, could destroy warheads in this phase. Their technical feasibility has already been suggested by an experimental one-on-one interception. On 10 June 1984 a dummy warhead from a Minuteman ICBM launched from Vandenberg AFB in California was intercepted by such a device (*Aviation Week & Space Technology*, 18 June 1984). The interceptor itself, also launched on a modified Minuteman, was initially aimed towards the target by ground-based computers which had calculated the warhead's flight-path from radar readings. Some ten minutes after launch from Kwajalein Atoll the homing sensors locked onto the warhead's infra-red signal and a 5 metre diameter metal net was unfurled. Moments later the 'Homing Overlay Experiment' (HOE) interceptor crashed into the warhead, over 160 kilometres above the Pacific. Within the SDI, HOE technology is being applied especially to a smaller experimental weapon, known as ERIS, for Exo-atmospheric Reentry-vehicle Interception System. However, though effective weapons are of course necessary for BMD, demonstrating one under test conditions hardly proves that an overall mission such as mid-course interception can be reliably achieved.

Only if the attack had already been heavily thinned by previous layers would mid-course defence seem likely to be able to cope with the numbers of real and potential targets. Nevertheless, the difficulties being encountered with boost-phase systems did lead to a significant increase, in 1985, of SDI efforts directed towards mid-course discrimination systems, principally with interactive laser and particle beam technologies formerly proposed for weapons.

Terminal Defence

The basic problems of terminal BMD are well understood, and have not changed much over the twenty-five or so years since the first ABMs were deployed. The main limitation of terminal defences is that they cannot defend large areas, except at quite impractical cost. Once the warheads start to enter the edge of the atmosphere they become easier to distinguish as the decoys and chaff are filtered off by air resistance, but only a minute or so would then remain before detonation. Ground-based missiles are thus limited by the distance they can travel in the short time left after they are fired by the battle-management computers. If a decision is made to attack a particular target soon after it enters the atmosphere then perhaps 30–40 seconds may be available for the missile to reach the target. However, the earlier a commitment is made to interception the more likely it is that a sophisticated decoy may be mistaken for a warhead. Current ABM interceptors are capable of accelerations up to 100 times that of gravity, to a final velocity of 3 km/sec. If launched when the incoming warheads were 40 kilometres away, they could intercept them at a distance of 20 kilometres (Weiner, 1984: 71). But clearly even if much higher acceleration ABMs were deployed they would only be capable of defending the immediate locality around their base.

Terminal defence of the entire United States (or just the major cities) would therefore require enormous expenditure to build adequate numbers of ABM sites. For example, the HEDI (High Endo-atmospheric Defense Interceptor) project, for a terminal interceptor missile possibly armed with a neutron atomic warhead, is being considered in terms of several *hundred* 'missile farm' launch sites (Pike, 1986b). Even so it would have low prospects for preventing intolerable damage to urban areas, which would result from quite low leakage rates (figure 8.4). Even extensive deployments of such terminal systems could not guarantee total defence, because they are relatively easy to overwhelm with numbers. Also, cities are very fragile and could be massively damaged by high-yield warheads exploded at high altitudes, early in their terminal phase. Whether the ABM radars can function as required in a nuclear war environment is also a problem (box 8.2). The most that could be hoped for is that a terminal defence might force an attacker to pay a high 'entry price' to destroy a protected target. Currently available systems for defending hardened targets with nuclear warheads are thought able to charge an entry price of somewhere between two and eight attacking warheads per target (Carter, 1984b: 9).

Although of little value for protecting cities, such an approach may be relevant for defence of missile silos. An ABM system sited at a missile field could be programmed for 'preferential defence'. By defending strongly only certain silos, whose identity is unknown to the attacker, the entry price for all silos might be raised to a level the attacker could not match, or could not afford to match.

Because they are usually the most fragile part of a terminal defence the missile-tracking radars would be a sensible target for the enemy to destroy first, as the missile silos would then be left defenceless. So long as a short tracking range is adequate, the defence might try to counter this by using many small, relatively cheap and expendable ('Kleenex') radars, perhaps based in hardened shelters.

Box 8.2   Effects of High-Altitude Nuclear Detonations on Radar Performance

Air-burst nuclear explosions can black out radar signals in two ways. At lower altitudes the high temperature of the fireball causes ionization, creating a volume almost opaque to radar with a diameter of about 10 kilometres for a one megaton detonation. At higher altitudes, above about 60 kilometres, the beta radiation produced is the main ionizing agent. The 'beta blackout' can affect a volume more than 100 kilometres in diameter for several minutes. High-altitude nuclear explosions also produce a large electro-magnetic pulse (EMP) effect over a wide area, which might damage or disable the electronics of an ABM system (box 2.1).

Such radars are probably capable of guiding a high proportion of nuclear-armed ABMs close enough to the warheads for the detonation to cause disabling damage, but may not be adequate for non-nuclear interceptors. These must either hit the warhead or get so close that a fragmentation explosive would cause sufficient damage. Whilst the feasibility of such homing interceptors has begun to be demonstrated for late mid-course defence, this has not yet been done for warheads in the terminal phase (Weiner, 1984: 73).

Another approach is to launch a cloud of perhaps 10,000 small, cheap, unguided missiles into the predicted path (or 'threat tunnel') of the incoming warhead only some 500 to 1000 metres from the defended silo. Still another suggestion, even less attractive to the local population, is that of exploding buried nuclear weapons a kilometre or so in front of the defended silos. The column of earth and dust so lifted would for a short time protect the silos from re-entering warheads.

No other technologies offer any better prospects for making terminal defence of populations feasible. Directed-energy weapons, such as lasers and particle beams, are not practical alternatives in this situation whilst they remain limited by poor propagation through the atmosphere. Used alone, terminal defences could not effectively defend the Soviet Union or the United States against a large-scale ballistic missile attack. Only if the earlier parts of a layered defence were very efficient could terminal defence play a useful role in protecting the population and other unhardened targets. On the other hand, terminal defences could probably increase substantially the number of offensive warheads required to destroy hardened targets. But if preferential hard-point silo defence is the aim then investment in the exotic space-based systems described above would be unnecessary. Little could be gained by attempting interception early in the ICBM's flight when its aim point was still unclear. The pre-SDI 'preferential defence' approach, of designing the system only for late mid-course and terminal interception of warheads on course for selected missile silos only, would be more effective.

Countermeasures

There are several possible ways in which an opponent might reduce the efficiency of an ABM system, known as countermeasures. The offence could simply increase its arsenal of ICBMs in an attempt to overwhelm the capabilities of the defence, particularly if this were cheaper than any offsetting increase in defence. But countermeasures aimed at disrupting the performance of the defence by actively interfering with some part of it – for example, by attacking satellite battle stations – are also a threat. Another, potentially quite effective approach could be to make the offensive missiles harder targets to destroy.

Passive Countermeasures

A wide variety of passive countermeasures could be developed in the attempt to thwart a defensive system, though most involve some sacrifice in payload. One consequence of the ABM developments of the 1960s and 1970s is that modern ICBMs incorporate the use of decoys and protective hardening. Decoys can be used during all parts of a missile's flight, but offer the best value during the mid-course where even the lightest decoys travel at the same speed as the warheads. Decoys for the boost phase must mimic the large infra-red output of the booster and would therefore be expensive. However, old ICBMs could be used, or new ones could be built without the expensive guidance system and warheads. Such 'decoy boosters' would be extremely difficult to distinguish during the boost phase and so would have to be attacked. Ground-based lasers might also be used to mimic the infra-red signal of real ICBMs.

In most phases the BMD weapons' aiming techniques would be susceptible to deceptive countermeasures. If the large plume of infra-red radiation produced by an ICBM were being used to aim a laser at the thin-skinned booster, then irregular burning of the rocket flame could cause the tracking computer to wander in its aim. Such irregular burning could be caused by injecting various additives into the flame in an unpredictable pattern (UCS, 1984: 126). A laser weapon designed to destroy its target by continuous heating would be far less effective if its aim could not be kept steady.

Shortening the boost phase of ICBMs so that they burn out within the atmosphere, less than a minute after launch, has already been mentioned. Current generation ICBMs were designed without consideration of boost-phase interception, and have rather long boost phases. The latest US ICBM, the MX, burns out after about three minutes at an altitude of 200 kilometres, whereas the latest Soviet ICBM, the SS-18, burns out at 400 kilometres after five minutes (Carter, 1984a: 8). The next generation of Soviet ICBMs could presumably match the performance of the MX. However, according to studies done for the Fletcher Panel an MX-type ICBM can be built which burns out below 90 kilometres after only 50 seconds (Carter, 1984a: 8). This would involve a reduction of about 25 per cent in payload as compared to the MX, but would make boost-phase interception extremely difficult.

It has been argued on the other hand that the relative denseness of the atmosphere at burn-out altitudes as low as 90 kilometres would require unacceptable losses in warhead accuracy for such an ICBM, and that its decoys could be revealed by their distinctive responses to the residual atmospheric friction. Thus even if the bus were harder to detect than the booster it would be a more or less equivalent target for a similar period and along the same path. However, technologies are already being studied for reducing the bus phase by rapidly deploying the warheads and penetration aids after burn-out, within the upper atmosphere. The payload could be sub-divided into a number of 'microbuses', each carrying penetration aids and one or two warheads for another 10 to 15 seconds, to a height of about 110 kilometres (Carter, 1984a: 8). This is just one of the clever ideas being studied or developed in the Pentagon's twenty-year-old Advanced Strategic Missile System (ASMS) programme, which is directed to develop countermeasures to possible Soviet ABM systems. During the many years it would take the opponent to develop and deploy anti-missile defences, such fast-burn ICBMs could be introduced as part of the regular modernization process, if that route to countering an emerging ABM 'threat' was preferred.

Such developments would make the tasks of boost-phase and post-boost-phase BMD extremely difficult. Warheads might be deployed well before X-ray lasers, particle beams or hit-to-kill weapons could be brought into the battle. And even ABM systems capable of penetrating into the upper atmosphere would have far less time to attack the ICBMs before the warheads were released.

A determined opponent can do much to hinder the discrimination of warheads from decoys which is essential to the mid-course layer. Against birth-to-death tracking, large screens could be the first objects dispensed from buses or microbuses, to conceal the subsequent deployment of other objects. Hit-to-kill homing devices may be confused during mid-course by decoys tethered a few metres away from the real warhead (Bethe & Garwin, 1985: 360). The correct target could only be destroyed if the homing device's infra-red sensor was sensitive enough to differentiate between the radiated heat of the two objects, and if its guidance computer was sophisticated enough to plot the necessary manœuvres in the short time available.

The ASMS is also working on penetration aids other than decoys, for use during the mid-course where all objects in the threat cloud would free-fall at the same speed regardless of differences in mass. Anti-simulation techniques could confuse the appearance of warheads and decoys with a bewildering range of characteristics, hampering the decision process in the defender's battle-management computers. One such ASMS technology, already tested on an ICBM, enables the warhead to be coated with an insulating layer of carbonized foam, greatly reducing the radiated heat (Smith, R. J., 1983: 134). The effect could be to defeat infra-red sensors designed to discriminate decoys from warheads by their relative coolness. Special aerosols have also been developed to thwart mid-course discrimination by infra-red sensors. By reflecting radiation rising from earth, such aerosol clouds produce a messy background of infra-red glare. They are to infra-red sensors what chaff is to radar. The latter, familiar penetration aid consists of large quantities of strands of lightweight metal wire designed to reflect radar signals, thereby certainly confusing radar readings and possibly hiding the warheads altogether. Currently deployed on some US Minuteman ICBMs, it would be released from dispensers flying through space alongside the warheads.

Decoys intended for the terminal phase must be more expensive than mid-course ones, if they are to 'look' like a warhead to the defending side's sensors. Unless they have the same ballistic co-efficient (weight-to-drag ratio) they would slow down quicker than the real warheads due to air friction (Bunn, 1984). If the decoys are to be lighter than the warheads then they must also have lower drag, but if they have a different shape the terminal defence radar might still be able to distinguish them. (Clearly there is little military value in using decoys of the same size and weight as the warheads. It would be more effective simply to use more warheads.)

Several ingenious techniques can be employed to make lighter decoys act like warheads during re-entry. Light decoys with the same radar profile as warheads can be powered by a small amount of solid propellent so that they behave like warheads, at least in the upper atmosphere. Or they can be designed in a 'V' shape, with most of the weight at the join, so that they have a lower drag than real warheads whilst producing a similar radar image (Bunn, 1984: 25–6). Another option is for decoys to pump out an ionizing substance, such as salt, in order to create an ionized wake like real warheads.

As well as developing terminal-phase decoys the ASMS also finds ways to make warheads themselves more deceptive. Manœuvring warheads which change course during the terminal phase obviously create problems for defensive systems. If the warhead changes direction after the interceptor missile has been fired then it will probably penetrate that defence. Such a warhead, the Mk-500, has been developed for US Trident SLBMs, ostensibly as a hedge against further Soviet ABM developments (Bunn, 1984: 33–44; also p. 164 below).

Although Soviet penetration aid development is thought to lag well behind the work of the ASMS, there can be little doubt that the prospect of a large-scale American defensive effort will spur them on. Given that even the most expensive penetration aids would probably be cheap compared to the cost of reliable and effective defences, a highly protective BMD system would be difficult to achieve. In incorporating penetration aids the offence must, however, make some reduction in the amount of explosive power that could be delivered, for any fixed number and 'throw-weight' of ICBMs.

Another type of passive countermeasure is to harden the target against destructive effects. This can be effective throughout a missile's trajectory, but offers little protection against hit-to-kill weapons, which would probably destroy anything they struck. Part of the technical attraction of boost-phase interception is that ICBM boosters are so fragile at present (p. 86). Against DEW, current generation liquid-fuelled Soviet ICBM boosters may be vulnerable to an energy intensity as low as 1 kilojoule per square centimetre (kJ/cm^2), if delivered in a few seconds (Carter, 1984a: 18). However, a small layer of heat-shielding material, such as carbon, coupled with the use of a solid fuel propellent to diminish structural stresses, could easily push the required energy level up to 20 kJ/cm^2. This was specified by the Fletcher Study as the energy needed for boost-phase destruction of ICBMs, sufficient to evaporate a layer of carbon about 3 millimetres thick (Bethe et al., 1984: 43). The extra weight of such heat-shielding would involve only a small reduction in payload, perhaps one warhead from the ten normally carried on an SS-18 (Drell et al., 1984: 46, note 72). Further hardening is obviously feasible if an additional reduction in payload capacity is acceptable.

ICBM boosters could also be made more resistant to lasers by putting a shine on their surfaces to reflect most of the laser energy. There is the problem that the shine might be removed by air friction during take-off. One possible way round this, being developed by the LTV Corporation in research for the US Navy, is to protect the highly polished aluminium booster skin initially with a layer of 'ablative' material (Karas, 1983: 182). When this protective substance, probably made of carbon or a carbon compound, is burnt off by the laser, the shiny aluminium could then, in theory, reflect about 97 per cent of the energy from an infra-red laser. Another hardening technique against a continuous laser is simply to spin the ICBM, roughly tripling its hardness because the laser energy would then spread out around the missile.

Whether such hardening techniques would be of significant value is not yet clear, but research in this area is currently being pursued in the United States by both the Defense Nuclear Agency (DNA) and the Defense Advanced Research Projects Agency (DARPA), (Smith, R. J., 1983: 135). None of them, however, would be effective against pulsed lasers such as the X-ray laser or some short-wave designs.

Hardening against such weapons, designed to destroy missiles by impulsive damage, would be difficult to achieve and require structural redesign of the ICBM. Particle beams might also have another 'kill mechanism', by penetrating deep into the target to disrupt its electronic systems. Shielding against them would require dense material, such as lead, which is necessarily very heavy. This would not be practical to protect the whole missile, but might be used to shield its electronics against low-intensity particle beams, if they were what was envisaged. The electronics could also be directly hardened, perhaps by a factor of 1000, by using gallium arsenide components. Given the immature state of particle beam weapons technology, however, countermeasures against them are not a high priority. Edward van Reuth, former head of the materials science branch of DARPA, has stated that: 'Particle beams are considered way out. We're not all that worried about them yet' (in Smith, R. J., 1983: 135).

The bus and warheads are already hardened to some degree on most modern ICBMs. They were re-designed for protection against radiation and EMP effects in response to the development of nuclear-armed ABMs in the 1960s, and so already tend to be harder than their boosters. Further hardening of both bus and warheads is quite feasible.

To sum up, a wide range of passive countermeasures is available. Just how effective they would be is not clear in all cases, but there is little doubt that some would add considerably to the already difficult tasks facing any defence. Shortening the duration of the ICBM boost phase seems likely to be especially effective. Virtually all passive countermeasures would involve some reduction in payload, plus time and expense to develop and deploy. This only implies a reduction in offensive arsenals, however, if it is assumed that the numbers of ICBMs and other delivery systems could not be increased (p. 167). Since the assumption begs the main question, this predicted effect of BMD on the opponent's offensive force hardly constitutes a significant step towards 'rendering the Soviet weapons useless' as some of its supporters have claimed (Jastrow, 1984: 24).

Active Countermeasures

Active countermeasures are those designed to reduce the effectiveness of a BMD system by directly attacking or interfering with some of its components. At lower energy levels, radiation can be emitted to jam or 'spoof' (deceive) the vital BMD sensors.

Jamming is less efficient, because it leaves sensors still able to function if it is removed. Spoofing is a better option, because the attacker knows more about the signal-image of his own warheads, which has to be imitated, than he does about the characteristics of the energy sensors that jamming would have to overwhelm.

One spoofing decoy is already being developed by MIT and General Electric, though it is unclear how successful it is. Though smaller and lighter than a real warhead, it has the same weight-to-drag ratio and therefore decelerates at the same rate. Although only about 'the size of a half-gallon milk carton', it contains advanced electronics which can calculate how a real warhead would appear to any defensive radar beam. Within a couple of microseconds a deceptive signal, tailored to look like a warhead to radar, is emitted (Smith, R. J., 1983: 133–35).

In the complex kind of layered defence envisaged in the SDI programme, damage to even a few parts of the system might have drastic consequences for the performance of the whole. Indeed, the vulnerability of parts of the system might prove to be its greatest flaw, as has been recognized by the first 'Nitze criterion', that BMD must be 'survivable' (p. 264). Whilst the defence must be able to perform in a highly coordinated and effective manner, counter-attacks against the defensive system need not have a high individual success rate to achieve their overall objective. As Richard Garwin has noted, 'it is far easier to break a fine watch than to make one' (*Newsweek*, 17 June 1985). In short, the defensive system should not only be able to perform the difficult task of intercepting ICBMs, but also to do this in the face of attacks against itself. The Fletcher Study found that: 'Survivability of the system components is a critical issue whose resolution requires a combination of technologies and tactics that remain to be worked out' (in UCS, 1984: 150). In other words they did not know how survivability could be ensured, which is hardly surprising given the range of possible active countermeasures.

For example, any effective strategic BMD seems almost certain to involve the use of satellites. Apart from the pop-up X-ray laser, virtually all schemes for boost-phase interception require a constellation of low-orbit battle stations. These would be continuously moving across the face of the Earth, with some coming into range of the enemy ICBM fields as others moved out. Thus the whole constellation would not need to be attacked by the side launching ICBMs, just those moving into place over its missile fields at the time. Punching a 'hole' in the BMD satellite constellation would allow the ICBMs to be launched unimpeded by this layer of the defence.

Anti-satellite developments have already been described (chapter 4). Though current capabilities are limited, they could be enhanced by further development of systems like the US F-15 MHV. (The commonality of asat and ABM technology is exemplified by the similarity between this weapon and the mid-course HOE interceptor (p. 124).) Large-scale deployment of such relatively cheap and versatile asats would present a threat to LEO satellite survivability. As to the future, Donald Kerr, Director of the Los Alamos National Laboratory, has noted that:

> Many of the laser weapon technologies being developed for ASAT weapons are identical to BMD technologies, although since satellites are generally 'softer' than ballistic missile boosters, post-boost vehicles or warheads, ASAT technology requirements are usually thought to be less demanding. (Kerr, 1984: 114)

One candidate for such an asat is the X-ray laser, should it ever prove to be a feasible technology, since atmospheric attenuation would not be a factor. Louis Marquet, the SDIO's chief of directed-energy weapons, accepts that a 'Soviet pop-up X-ray laser would be very lethal if it were used against our satellites' (*Newsweek*, 17 June 1985).

In general, satellites are larger and softer targets than warheads, and their orbits are predictable, so that virtually all weapons technologies considered for the SDI would be equally applicable to this less demanding role. Once again, whereas the defence must be ready to work 'perfectly' at any time, the offence could pick its time and method of anti-satellite attack.

Satellites can be hardened and defended to some extent. Hardening, however, may be limited by the job which the satellite is designed to perform. If large mirrors must be continuously ready, pointing down towards the ICBM fields of a potential attacker, they would be an especially vulnerable element. Hardening would also make satellites heavier, adding to the cost of placing and maintaining them in orbit.

Self-defending satellites (DSATs) may become possible, since anything capable of destroying ICBMs should also be able to destroy current asats. This would be particularly true of satellite defence against anti-satellite systems based on some form of missile, like the US MHV. However, satellites which combined self-defence with BMD functions would be more complex and would need additional energy to power their weapons. Even an 'unsuccessful' anti-satellite attack might drain sufficient power from them for the BMD system to which they belonged to be fatally weakened. And DSAT capabilities would naturally be eroded, to the extent that much faster-acting asats were developed.

The advances in weapons technologies required for space-based BMD also make the survivability of such systems harder to ensure. The fundamental reality, that satellites are easier to destroy than missiles, is thus a major obstacle to their development.

Satellites may be the most obviously vulnerable part of a space-based defence, but others would probably be weaker. Unless each component of the defence were autonomous, which is extremely unlikely, there would have to be communication links, which it would be very difficult to make invulnerable against jamming, spoofing, or direct attack. Any satellite ground-stations in the defensive system, for example, would be prime targets for sabotage. The US 'DSP – East' early-warning satellite, which currently watches over Soviet ICBM fields, relays its information to the United States through a single ground-station at Nurrungar in South Australia.[1] And the main US satellite ground-control centre, at Sunnyvale in California, stands 'within bazooka range of a highway', as one expert puts it (Ford, 1985: 64).

Although difficult to assess in quantitative terms, the opportunities for active countermeasures against a BMD system raise serious questions. Satellites are high-value targets which are difficult to harden and expensive to replicate. At best the attempt to ensure survivability would push up the cost of a space-based defence, at worst active countermeasures could drastically reduce a BMD system's effectiveness.

Changes in Nuclear Force Structure

One obvious type of countermeasure is to thwart the defence by increasing the capability of offensive nuclear forces, either by increasing their size, or by changing their composition and the contingency plans for using them. If an opponent were determined that, no matter what the cost, it must be able to 'penetrate' a defence, then there is little doubt that increases in numbers of missiles, coupled with other countermeasures, would assure this. As Richard DeLauer, then Under-Secretary of Defense for Research and Engineering, once testified before the House Armed Services Committee, 'any defensive system can be overcome with proliferation and decoys, decoys, decoys, decoys' (in Bethe et al., 1984: 45). Such proliferation could involve either a huge build-up in ICBMs, or the diversification of strategic delivery systems for nuclear weapons. In the first case, the crucial question would not be whether defence was effective so much as whether it was cost effective. If increases in the offensive forces were to cost less than the additional defensive systems needed to intercept them, then, once again, the defence could not be 'cost effective at the margin'. Although judgements about the relationship between future offensive and defensive systems must be tentative, the inherent complexity of defence and the novelty of many of the technologies involved point to very high costs.

The wide range of possible delivery systems only goes to emphasize the difficulty of defence against nuclear attack. At present the Soviet Union is heavily reliant upon land-based ICBMs, and concepts for US defence have naturally concentrated on this threat (in the public debate, at least). Intercepting SLBMs, which may be able to hit targets within five minutes of launch, is a similar, but far more difficult problem.

Defence against nuclear weapons delivered by means other than ballistic missiles is not an explicit objective of the SDI programme (p. 332). Cruise missiles, in particular, which can be launched from many kinds of platform (including seemingly non-military ships) would require another kind of defensive system. Although some laser wavelengths can penetrate the atmosphere fairly effectively, delivering destructive power levels to low-flying targets from orbital battle stations seems likely to be quite impractical. The Reagan administration has announced plans for increasing air defences, but the scale of deployments that might be needed to protect the whole of the United States would be very costly.

Another rather obvious approach which an opponent might adopt in preparing to overcome a BMD system would be to change the way it plans to use its nuclear forces. For example, if faced with an apparently quite effective defence, the opponent might choose to target more missiles on enemy cities to ensure their destruction. Just exactly how targeting plans would be changed must remain hypothetical, but it is unlikely that they would stay the same (p. 172).

Prospects for Defensive Systems

Although the SDIO is formally responsible only for its current

five-year research programme (appendix 2), with no commitment
to production or deployment thereafter, this is rarely reflected in
the public pronouncements of its supporters. Along with others,
the President and the Secretary of Defense have been uncom-
promising in stressing the necessity for such a defensive system.
The President declares that:

> [I]n the long term, we have confidence that the SDI *will be* a crucial
> means by which both the United States and the Soviet Union can
> safely agree to very deep reductions, and eventually, even the
> elimination of ballistic missiles and the nuclear weapons they carry.
> (US WH, 1985a: (i), emphasis added)

Weinberger is not only 'absolutely convinced it can be done', but
also that 'it must be done' (*The Scotsman*, 11 February 1985). His
optimism is echoed by Lieutenant General Abrahamson, Director
of the SDIO: 'Anything the nation decides it must do . . . it can
do' (*The Observer*, 26 May 1985). But can it?

To answer this the requirements of the proposed defence must
first be specified (p. 161). If the objective is a perfect defence, as
has frequently been declared by both the President and Mr
Weinberger, the answer is almost certainly: No! Building a perfect
or even nearly perfect ballistic missile defence against a determined, technically advanced and heavily armed opponent like the
Soviet Union is virtually impossible. Even General Abrahamson
has conceded that 'there's no such thing as a perfect defence' (*The
Times*, 16 May 1985). And SDIO Chief Scientist Gerold Yonas has
said that the Fletcher Study 'agreed at the start that there was no
perfect defense against a determined adversary, and it is not likely
there ever will be' (Yonas, 1985: 28).

Technological breakthroughs could never be enough to change
this. Perfect or near-perfect population defence will almost
certainly remain unachievable even if all the looked-for technical
advances were made. Analogies with great technical achievements
of the past are misleading, as those consisted in triumphs of human
ingenuity in manipulating the forces of nature. Mr Weinberger, for
example, has cited the judgement of Einstein in 1932 that there
was not 'the slightest indication' that nuclear energy could be
produced. According to Weinberger:

> Another mistake of critics of Strategic Defense is to contend that
> effective defense is technically unobtainable. History is filled with

flat predictions about the impossibility of technical achievements that we have long since taken for granted. (*Arms Control Reporter*, January 1985)

Developing and deploying an effective defence against nuclear-armed ballistic missiles, however, is a different and far more difficult task than, for instance, putting men on the Moon. The scale of the technical challenge is of course enormous. According to Richard DeLauer, it requires breakthroughs in eight key technologies, each 'equivalent to or greater than the Manhattan Project', which developed the first atom bomb (in UCS, 1984: 26). But more fundamentally, it is not a challenge posed by nature, but by human opponents who can be expected to be equally ingenious in devising new offsetting technology for the offence. If ballistic missile attack were a static threat then it is just about conceivable that an effective defence against it could be built, though it might still turn out to be impossibly expensive. The whole history of warfare, however, shows that the relationship between offence and defence is dynamic:

> We may say in conclusion that in considering active defences, a realistic analysis does not first assume an offence and then design a defence to counter it. In actuality the order is reversed. A defence is built, and the offence seeks to exploit its weak spots. And the history of the [human] race thus far suggests that there is always a hole, an Achilles heel. (Brodie, 1959: 202)

In short, any talk of perfect or near-perfect defences against a full-scale missile attack is clearly misleading. There seems very little prospect that the United States could prevent the Soviet Union from being able to deliver a significant part of its nuclear arsenal to the American homeland. An attack could be targeted so that if 500 half-megaton warheads – less than 10 per cent of the explosive power carried by Soviet ICBMs alone – got past the defences, then all US cities and large towns would be destroyed, killing half the urban population and injuring most of the rest (Carter, 1984a: 66; figure 8.4). As few as ten such warheads, about one thousandth of the explosive power carried by Soviet strategic missiles (ICBMs plus SLBMs), would kill several million people and injure over 10 million more, if exploded over the ten largest American cities.
Few would consider such losses to have been a satisfactory

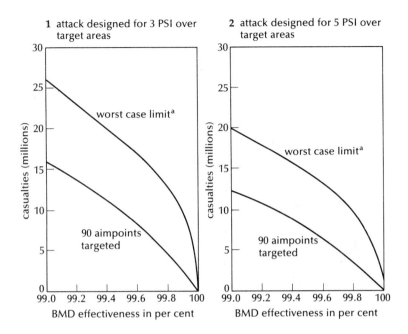

Note:
[a]Soviets know exactly how capable US defence is and target accordingly.
They hit every city they target.

Figure 8.4   Casualties with Various BMD Efficiencies
Source: Office of Technology Assessment, 1985a.

defence, even though the actual warhead 'leakage' might be low in percentage terms. Absolute perfection is not something that could be expected of a very complex, technologically 'state-of-the-art' system, which could never be fully tested, operating in a novel environment against a determined enemy. Yet anything less than perfection exposes the fragile fabric of society to devastation by the extremely efficient explosive power of nuclear weapons.

Whilst 'perfection' is not to be expected, there is little doubt that single ICBM warheads can be intercepted. Indeed technology for the interception of single warheads has already been 'demonstrated' for both the mid-course and terminal phases. Achieving boost-phase interception, before the warheads have been released, is far more difficult and would probably require space-basing, making it much more expensive. Available technologies are not up to the job of boost-phase interception against the current Soviet ICBM arsenal. By the time 'futuristic' technologies have been developed – and two years of the SDI does not seem to have brought them much closer – their usefulness may have been limited by reductions in the duration of ICBM boost phases and other countermeasures.

Taken singly or in a layered system all these interception technologies offer the prospect of some partial defence. This would reduce the opponent's certainty of destroying any particular target, but could lead him to assign extra warheads to important targets. More positively, from the defender's point of view it means that certain targets might be protected to some degree, depending on the hardness of the target and on how many warheads the offence is willing to 'pay' to destroy it. If limited hard-point defence of missile silos or other hardened targets is required, then mid-course and terminal defences are technically feasible, in this sense. From the mid-course onwards it might be possible to determine the aim-point of the warheads, unless manœuvring warheads were used, so that only those on course for the defended sites would need to be intercepted. Such preferential defence is not possible with boost-phase interception, which is therefore of doubtful utility in such a system.

In summary it is worth quoting the conclusion reached by Dr Ashton Carter, in a study for the OTA based on full access to classified data:

> The prospect that emerging 'Star Wars' technologies, when further developed, will provide a perfect or near-perfect defense system, literally removing from the hands of the Soviet Union the ability to do socially mortal damage to the United States with nuclear weapons, is so remote that it should not serve as the basis of public expectation or national policy about ballistic missile defense. (Carter, 1984a: 81)

Two years later, and after a storm of protest from every part of the BMD lobby in Washington, the OTA had not budged from this position:

> Assured survival would probably be impossible to achieve if the Soviets were determined to deny it to us. By improving or adding to their offense, they could increase the number of weapons penetrat-

ing to the United States. Another basic problem would be the difficulty of knowing with high confidence how well our defense would actually perform against their offense, since it could never be tested and we could never know in great detail the working of their offensive weapons. Many who advocate assured survival envisage it being achieved by agreements that limit offensive levels far below defensive capabilities. (OTA, 1985a: 113)

For all its importance, however, the immediate technical argument about BMD is clearly subordinate to wider economic and military considerations, and to both foreign and domestic political objectives.

9
Current Soviet Space Weapon Programmes

I don't think . . . they have any kind of leak-proof umbrella over Moscow . . . it is a limited system, it isn't the kind of thing that the United States is talking about. It would only be partially effective. We do know they have invested heavily in directed energy systems, but it is very hard to see concrete results . . . Are they dramatically better off? No. In the key technologies needed for a broader defense – such as data processing and computer software – we are far, far ahead.

Lieutenant General J. Abrahamson, October 1984

Although the American SDI programme is a major and significant new development in the militarization of space, it should not be taken out of context. Prior to President Reagan's 'Star Wars' speech both the United States and the Soviet Union had active research and development programmes geared towards space weaponry. Perhaps justifiably, some SDI supporters are dismayed that the American development, more visible than the usual Soviet military profile, has taken the brunt of the criticism. Partly as a result, the argument is now being vigorously advanced that the SDI is necessary as a response to Soviet space weapons program- mes. The contention is that the Russians have unscrupulously and perhaps illegally exploited the years since the signing of the ABM Treaty. And that whereas the Americans have respected not only the letter, but also the spirit of the Treaty, their rivals have 'spent as much on strategic defense as on strategic offense over the past 12 years' (Yonas, 1985: 27).[1]

This confusing and misleading statement is not untypical of the Reagan administration's arguments in support of its claims about a growing Soviet military space advantage, reminiscent of the supposed bomber and missile 'gaps' of earlier times, which were both later discovered to have been in the Americans' and not in the Russians' favour (Kaplan, 1983). Detailing the Soviet strategic defence capabilities, Dr Yonas lists 7000 air-surveillance radars, some 10,000 surface-to-air missiles, and 12,000 interceptor aircraft. Impressive indeed, but not perhaps surprising considering that the Soviet Union faces a very credible 'threat' from US aircraft and cruise missiles.[2]

As for the air-defence 'gap' itself, the United States has not spent equivalent amounts on such systems in the past, for the simple reason that it faces virtually no airborne threat, from the Soviet Union or anywhere else. However, the Reagan administration believes the Soviet long-range bomber threat is now increasing, and it is beginning to direct major efforts towards modernizing and expanding US air-defence systems, partly in cooperation with Canada.[3]

An assessment of current Soviet anti-missile defence activities must cover three major issues. First, those Soviet surface-to-air missiles (SAMs) which might conceivably take on an ABM role. Next, the one Soviet system designed specifically for defence against ballistic missiles, the Galosh ABM around Moscow, which is currently undergoing modernization. And third, such other, more futuristic space weapon programmes as may also be in progress. Finally, when trying to reach some conclusion about the Soviet Union's own view of its military space programmes, it will be necessary to refer once again to existing and possible future Soviet anti-satellite weapons. Although virtually all information on Soviet weapons development comes through the filter of the American intelligence services, a reasonably fair, though tentative, picture of recent activities can be pieced together.

In doing so, it will be important to see whether any light can be thrown on the official Soviet position, which was repeatedly asserted in the latter part of 1985, that the Soviet Union is not developing, and perhaps does not possess, 'attack space weapons'.

Soviet SAMs as Potential ABMs?

US officials have long been concerned about the possible use of Soviet SAMs against ballistic missiles. Clearly missiles designed for intercepting modern high-altitude military aircraft or (as yet non-existent) supersonic cruise missiles might be of some use for the more demanding job of destroying ballistic missile re-entry vehicles in their terminal phase. The most modern Soviet SAM, known as the SA-X-12, has now become the focus for American concern. This truck-mounted system is still undergoing tests, but deployment is expected in the next year or two. Capable of supersonic speeds and high-speed manœuvring, the SA-X-12 can reportedly engage targets at altitudes between as low as 90 metres and as high as 30 kilometres or more (*Aviation Week & Space Technology*, 16 January 1984). The SA-X-12's radar, carried on another truck, is said to have a range of around 240 kilometres. This would mean it could not 'acquire' an incoming ICBM warhead until about the last 60 seconds of its trajectory.

If armed with nuclear warheads the SA-X-12 might have some utility for terminal defence, intercepting re-entry vehicles after the atmosphere had filtered off the decoys. However, it has nothing like the more than 100 g acceleration of the United States' old endo-atmospheric Sprint ABM, the lower layer of the short-lived Safeguard system, which could overtake a modern machine-gun

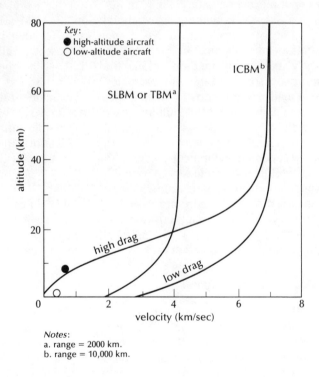

Figure 9.1   Missile Velocity–versus–Altitude Curves
Source: Carter & Schwartz (eds), 1984. Reproduced with permission.

bullet in less than three seconds. Nor does the SA-X-12 have Sprint's manœuvrability. Its performance, whilst formidable against the aircraft it has been designed to intercept, would be far less impressive against incoming warheads travelling at ten times their speed.

Nevertheless tests of the SA-X-12 have apparently been carried out against slower re-entry vehicles from the SS-12 Scaleboard intermediate-range ballistic missile, though it is not clear how successfully the system performed. The SA-X-12 might have little effectiveness against the low-drag, high-velocity warheads of modern US ICBMs, but its utility could be greater against short-range ballistic missiles, including the SLBMs which might be used for a 'precursor' attack on command-and-control systems.

US officials have also drawn attention to the Soviet SA-10 air-defence missile system in recent years. Though at first claims were made about its possible capability against ballistic missiles, it is now thought to be too slow for that, and is currently seen as a possible defence against relatively slow-moving cruise missiles, perhaps in conjunction with an air-borne 'look-down' radar system. In general, widespread deployment of the SA-X-12 and other Soviet SAMs might just take a little of the edge off the accurate 'counter-force capability' of some parts of the US nuclear arsenal, by reducing the probability that specific US warheads would hit specific defended targets. This could not, however, amount to an effective 'force defence' (chapter 10), let alone to defences capable of protecting the Soviet Union as a whole.

If Soviet SAMs have any ABM capabilities, however minimal, that only highlights the need to strengthen the ABM Treaty by removing 'grey areas' in it that can currently be exploited by either side (chapter 11).

The Galosh Modernization

The Galosh complex around Moscow is, as the Americans are fond of pointing out, the world's only operational ABM system. Since 1980 work has been under way to up grade and expand the system, which had been generally assessed as technically primitive and of little operational value.

As readers have doubtless understood by now, the front-end of any BMD is its radars. In the new Moscow system the overall battle-management role is to be filled by a very large phased-array radar nearing completion at Pushkino, north of the city, the only one of its kind in the world. It stands 36 metres high, with a 20 metre diameter receiving antenna array, together with a smaller, separate array for signal emission, on each of its four 150 metre long sides, giving all-round (360°) coverage. After receiving missile-tracking data from the early-warning systems, the Push-kino LPAR could track large numbers of incoming warheads and other objects from some distance, and guide longer-range, exo-atmospheric interceptor missiles, like the new SH-04, to meet their targets. It will greatly enhance the current battle-management and guidance system, centred on the older Dog House and Cat House LPARs at Naro-Fominsk and Chekhov, south-west and south of the capital.

The second type of ABM radars, those which handle the precise tracking and guidance needed for separate, successive intercep-tions, were represented in the earlier Moscow system by 24 mechanically rotated Try Add radars. These are now being replaced by 'Flat Twin' 'hybrid' radars, in which multiple phased-array aerials are mounted in mechanically rotated frames. Although an improvement on Try Add, the new tracking radars will probably still be unable to handle more than one or two incoming warheads or rising interceptors at a time. Another interception radar, known as 'Pawn Shop', consists of three mechanically rotated aerials housed in a van-sized container, each capable of providing guidance information to a single defensive missile. This is probably intended for use with the endo-atmospheric SH-08. Both the Flat Twin and the Pawn Shop are, to some extent, transportable and modular. With pre-prepared sites they could be relocated in a few weeks.

The 32 remaining above-ground Galosh ABM-1B interceptors are being replaced by the two new interceptors just referred to, both based in silos. The SH-04 is probably a slightly smaller version of the old Galosh design, carrying a nuclear warhead to about 300–400 kilometres. The SH-08 'Gazelle', by contrast, is a two-stage high-acceleration endo-atmospheric interceptor, with a range of less than 100 kilometres. Like similar American ABMs it would probably carry a low-yield nuclear warhead, but it too is reportedly not as fast as the US Sprint, which preceded it by over ten years. Together the two new Soviet ABMs will comprise a layered defence similar to the 1975 US Safeguard deployment (p. 35). It is expected that 100 interceptor missiles may be deployed altogether, as allowed by the ABM Treaty.

Whilst it may be an improvement on the ABM-1B, the prospective 'ABM-X-3' Galosh system hardly looks up to the job of defending Moscow against a large-scale attack. Nothing has occurred lately in Soviet anti-missile capabilities to invalidate the judgement of one Pentagon ABM analyst, speaking in 1983: 'If both sides use missile defense and penetration aids – if we went all out with the technology in hand, we could eat them up' (in Smith, R. J., 1983). As for widespread deployment of current Soviet ABM systems, it could only be very slow, very expensive, and make very little difference to the present strategic situation. At most it could cause some additional uncertainty for US nuclear attack planners, about whether a few of their offensive counter-force 'requirements' might be frustrated. But that uncertainty exists already, for reasons having nothing to do with any active Soviet defences.[4] Thus, evidence for the view that the Soviet Union is preparing such an overall ABM 'break-out' appears to be lacking. The absence of any sufficient or plausible motive also strongly suggests that no such intention exists.

The New Soviet Early-Warning Radars

Since the mid-1970s the Soviet Union has been building a nationwide network of large phased-array early-warning radars, called in the West after the location of the first to have been identified, near the city of Pechora at the northern end of the Urals (figure 9.2). In the early 1980s four more construction sites were discovered by means of US reconnaissance satellites. Each is a 'bistatic' installation, with two separate large antenna arrays, one each for emission and reception of signals, set at an angle of 20° to the vertical, and giving a coverage fan of 120°. The Soviet Union has clearly been building a network 'providing almost complete coverage of potential missile attack trajectories' (Longstreth et al., 1985: 40).

In the summer of 1983 an American Big Bird satellite produced the first pictures of a new Pechora-type radar construction site near the Siberian village of Abalakova in the Krasnoyarsk region. At the end of July its findings were checked out with a special launch of one of the CIA's KH-8 low-altitude photographic satellites, used only to provide information about intelligence targets with the highest national priority. After some months assessing the resulting data, US administration members began to make public accusations that the Soviet Union had broken the Treaty. That charge is discussed, with others, in appendix 3. At

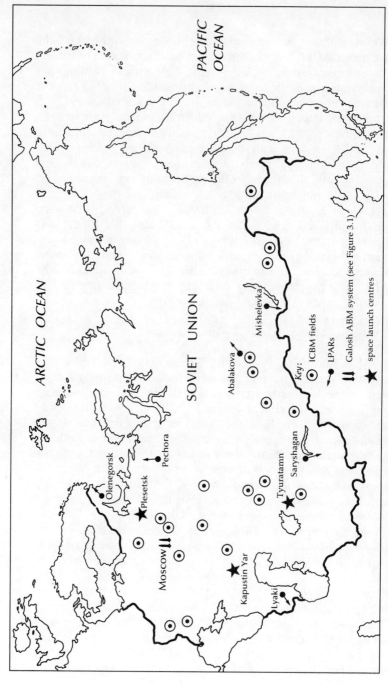

Figure 9.2 Map of Soviet Union with Sites of Interest

this point it is only appropriate to consider the possible military functions of the Abalakova radar.

Unlike the other 'Pechora' sites, this one is between 2000 and 4000 kilometres away from the edge of the Soviet Union towards which it faces, the coastline of the great Kolyma Peninsula of north-eastern Siberia. It is by no means obvious what the role of such a radar in such a place could be. It is sited about 750 kilometres from the nearest international border, and lies to the north-east of, and facing away, north-east, and usually 'down range' from, some of the Soviet Union's most important space-launching facilities. Some independent analysts in the West remain unconvinced by the Soviet contention that it is for space-tracking, little though they may be disposed to accept the Pentagon's account of Soviet military activities at other times. It seems that aspects of the Abalakova radar's design – its low angle to the horizon and relatively imprecise long wavelengths – suggest that it is not well-suited for such a role. Longstreth et al., however, suggest that it 'may be suitable to support space tracking activities for a space shuttle . . . [or for] an advanced anti-satellite weapon' (1985: 54). But these authors also point out that Soviet authorities have said nothing to bear out their second conjecture.

Few experts believe it could be much use for ABM battle management and interceptor guidance, since it is pointing in the wrong direction for dealing with most conceivable attacking warheads, which would arrive from the north or slightly west of north, rather than from the north-east.[5]

If the Abalakova radar is for early-warning the site may have been chosen because of practical difficulties with construction in the far north-east of the Soviet Union (see appendix 3). There is, however, one serious problem for this interpretation, namely that of assessing just what type of missile attack the Abalakova radar might be able to warn about, that was deemed important enough to justify an action that was certain to be criticized as an infringement of the ABM Treaty. Large phased-array radars at inland sites can be suited for detection and tracking of SLBMs launched from close in to the defender's coastline, which would almost certainly be an important element in any strategic nuclear attack. But the Abalakova radar is so remote from the relevant coastlines that even if it has sufficient range to cover the East Siberian and Okhotsk Seas, the curvature of the Earth would mean it had little chance of picking up US SLBMs, fired on a depressed trajectory, until they were almost halfway through their brief flight. Accounts of Abalakova published by the Pentagon and by independent US defence analysts imply that it has no such range, and are therefore hard to reconcile completely with the notion of its being for early-warning.[6] It therefore seems inappropriate, at this stage, to suggest any final conclusion as to which of its potential military uses is the least unlikely.

Directed-Energy Weapon Programmes

Soviet programmes to develop lasers and particle beams as ABM or anti-satellite weapons would be limited by the same scientific and technical problems as confront the SDI. The progress of any Soviet effort is difficult to assess with certainty, but none of the available evidence suggests that Soviet scientists have come nearer to solving any of the major technical problems involved than their Western colleagues. The fact, if it is a fact, that there are three directed-energy weapons complexes at Sary Shagan, each reportedly testing a different laser device, proves very little. Similar programmes have been under way in the United States for many years. US laser weapon enthusiasts may be sufficiently illogical to complain that American research programmes never produce any actual weapons, whilst at the same time claiming every possible Soviet programme as certain evidence for perfected Soviet ones. But no unbiased person is likely to follow their example.

Recent US intelligence estimates[7] of Soviet programmes for potential directed-energy weapon technologies can be summarized as follows:

The AST-1200 1.2 metre diameter segmented *telescope* mirror is claimed, unconvincingly, to be the prototype for a 25 metre diameter *laser weapon* component.

The magneto-hydrodynamic (MHD) generator at Sary Shagan is driven by controlled chemical explosions to deliver bursts of electrical energy of over 15 megawatts along superconducting cables to the pulsed iodine lasing assembly. No similar technology on this scale exists outside the Soviet Union.

Soviet development of ion induction linear accelerators has concentrated on such goals as high current levels, reliability, efficiency, reduced size and simplicity of operation. The contrast is with admittedly more powerful Western accelerators, which have concentrated rather less on optimizing beam generation technology, rather more on fundamental physics experiments with high-energy particle collisions.

Previous US intelligence claims about a betatron prototype atmospheric particle accelerator at Sary Shagan (*Aviation Week & Space Technology*, 28 July 1980) have now been quietly dropped, in favour of the view that the device in question is the pulsed iodine laser noted above.
As for 'macro-particle stream' kinetic energy technology for electro-magnetic acceleration of particles of heavy metal (e.g. tungsten or molybdenum) to speeds of 25 km/sec in air, 60 km/sec in vacuum, no details of Soviet progress with this technology, first identified in the 1960s, have been volunteered for many years[8] (US DoD, 1986: 47).

Soviet technology for propagating beams of high-power radio microwaves is said to be 'on a par, if not superior' to its US equivalent.

In view of all the hawkish fuss about a 'Soviet lead' in space weapons, this catalogue of Soviet beam-weapon capabilities is not particularly impressive. Past Western alarms about the early deployment of a Soviet space-based laser defence have been consistently premature:

> I submit that the Soviets, on the basis of what I have examined, have every expectation that well before 1980, if they don't blow themselves up – and they may – will perceive that they have technically and scientifically solved the problem of the ballistic missile threat. (General Keegan, in *Aviation Week & Space Technology*, 28 March 1977)

Upon the non-fulfilment of this particular millenarian alarm, the General is reported (*Washington Post*, 3 April 1983) to have reset it to 1983. Three years later he is presumably still as clear about it as ever. Less versatile US analysts are not so sure. Their worst fears can only be stated as a tangle of reservations and qualifications: 'US sources believe that in the late 1980s there could be [Soviet] prototypes of space-based lasers capable of use as anti-satellite weapons' (Pretty, 1985). The CIA's March 1985 report, however, set that date for 'the 1990s if their technology developments prove successful.'

Pretty and other Western experts seem slightly more confident when claiming that the Soviet Union has developed ground-based lasers already powerful enough 'to interfere with US satellites', if not actually to damage them. And the latest paper from the Committee of Soviet Scientists for Peace, Against the Nuclear Threat (Sagdeyev & Kokoshin, 1985) does give some prominence to the possibility of high-powered ground-based lasers as an 'active countermeasure' to the space-based elements of any large-scale US BMD system. The authors of this paper recognize that no hard and fast distinction between 'offensive' and 'defensive' weapons can really be made in respect of anti-satellite systems. At the same time as denying that there is any Soviet effort going into developing space-based anti-missile defences at present, the Soviet scientists draw attention to the destabilizing interaction whereby the SDI, as a US attempt to deprive the Soviet Union of its retaliatory deterrent capability, could provide the motivation for Soviet work on 'anti-BMD' weapons which would necessarily have multiple anti-satellite capabilities:

[I]n case of SDI implementation there will emerge the necessity and therefore the opportunity . . . of developing a whole class of new types of weapons, specially designed to destroy and neutralize elements of a large-scale anti-missile system, above all its space-based echelons. (Sagdeyev & Kokoshin, 1985: 52)

Their prediction was reinforced by General Secretary Gorbachev, in a post-Summit speech to the Supreme Soviet:

[W]ill the US feel more secure if our weapons in space will be added to the echelons of space weapons planned by Washington?
Indeed, the US cannot really hope to achieve a monopoly in outer space. (*Morning Star*, 16 December 1985)

What can be said is that Soviet work in the basic technology of directed-energy devices seems to be broadly on a par with Western achievements. Soviet scientists have, indeed, made significant contributions in the field of particle beam technology. Their innovations – the radio-frequency quadrupole and the negative ion source – were reported in the open Soviet literature, and have subsequently been incorporated into US weapons programmes. The first detailed description of nuclear-pumped X-ray laser technology in the open literature was published by Soviet scientists. But in the Soviet Union its applied development may not have kept pace with American work.

In general, the large Soviet effort in basic physics and in weapons research does not appear very well suited to the effective incorporation of innovatory concepts. Systems may be rapidly deployed, but they often prove to have been put into production prematurely. US systems, on the other hand, tend to undergo much more development before they are even tested, never mind deployed. For example, the Soviet Sputniks were an early and impressive achievement, but the components of their ill-fated contemporaries, the American Vanguard experimental series, were really far more advanced. And since those early days Soviet space technology has been consistently out-done in many respects by its American rival.

The 1985 Office of Technology Assessment Report to Congress noted that the Soviet Union was 'vigorously developing advanced technologies potentially applicable to BMD'. However, the report went on to add that 'in terms of basic technological capabilities the United States clearly remains ahead of the Soviet Union in key areas' (OTA, 1985a: 11–12).

In 1986 the Pentagon's annual comparison of US and Soviet military technology (table 9.1) continued to show the United States ahead in most of the technologies – such as computers and software, sensors, signal processing and automated control – which are essential for BMD. The exceptions were lasers, optics and power sources. The report, by Undersecretary of Defense for Research and Engineering Donald Hicks, concluded that 'it will remain difficult for the U.S.S.R. to close many already existing technology gaps, and new ones are likely to emerge' (*Washington Post*, 2 April 1986).

Is there a Soviet Space Threat?

In the late 1960s and early 1970s the Soviet Union met with serious setbacks in its efforts to develop a very large launcher, comparable to the US Saturn V. By the early 1980s, however, a similar launcher began to be reported, with speculation that it might shortly proceed to initial trials. Western estimates that it would have a lift-off thrust of 4.5–6.3 million kg were later revised to 3.6–4.0 million kg, which would mean it could place very large payloads of between 130 and 150 thousand kg into orbit at 180 kilometres, by comparison with the 127 thousand kg achieved

Table 9.1   Comparison of Soviet and US Military Technologies

United States Lead

Increasing	*No change*	*Decreasing*
Computers & software	Electronic signal processing Electro-optical sensors Guidance & navigation Life sciences & biotechnology Microelectronics Production & manufacturing Robotics & machine intelligence Stealth designs Telecommunications	Ground & aerospace propulsion High-strength materials Radar sensing devices Submarine detection

US-Soviet Parity

Aerodynamics & fluid dynamics
Conventional warhead design
Lasers
Nuclear warhead design
Optics
Power sources & energy storage

Soviet Lead

No areas found

Source: FY 1987 Department of Defense Program for Research and Development.

eighteen years previously by the American launcher (Christy, 1984).[9] When completed, it would therefore be an important element in any Soviet capability for building either a space-based defensive system or large manned space stations which might have military functions. However, it has yet to be tested successfully. A recent report in *Aviation Week & Space Technology* (27 May 1985) assessed the entire Soviet heavy booster and shuttle programmes as 'in serious development trouble'.

The fact is that the current state of high technology in the Soviet Union is not able to compete with US efforts. The traditional Soviet approach to military problems, of compensating for possible deficiencies in quality by the use of large numbers, cannot be so successfully applied to the demanding roles of space weapons. For the near future at least, though not because of lack of effort on their part, the Soviet Union will not be able to develop a credible ballistic missile defence. As for the less demanding job of attacking satellites, Pentagon officials have started referring to Soviet fighter tests which might be the first faint sign of a programme for something like the F-15 asat, or which might instead be merely an effort to gain a better understanding of the latter's capabilities. There is a small chance of the Soviet Union fielding an operational asat based on infra-red homing technology, or on an effective laser, before the end of the century. But the likelihood is that they will be confined to their handful of existing, cumbersome and limited asats for some years to come.

'Attack Space Weapons'?

The Soviet Union's claim that it is not now developing 'attack space weapons' is closely linked to the position taken in its 1983 draft treaty deposited at the United Nations (p. 226), which proposes to ban 'the use of force' in or from outer space, and to ban all weapon systems which might make such actions possible. The expression 'attack space weapons', or 'space-strike arms' (Sagdeyev & Kokoshin, 1985), seems to encompass all weapons capable of attacking objects in space, either from within space or else from Earth, together with all space-to-earth weapons. And the usual Soviet position is that whilst their F-LV asat means they still possess a few copies of one such weapon, they would be prepared to dismantle them under article 2.4 of their draft treaty. Furthermore, when pressed to explain just what is meant by Soviet appeals for action to prevent the 'militarization' of space, Soviet experts tend to suggest that a banning of all space weapons, in this comprehensive sense, is all that is being called for, rather than an end to the non-weapon military uses of space, reviewed in chapter 4.

Unfortunately, the official Soviet position is not always clearly maintained. Some pronouncements go so far as to claim the Soviet Union already has no 'attack space weapons' (p. 7). Others give the impression that weapons that might be placed or employed *in space* are what really matter, and that ground-to-space weapons like the current asats of both sides are less important. In view of US concerns about alleged Soviet BMD programmes emphasizing *ground-based* anti-missile and possibly anti-satellite systems, such language as the following, from Soviet Chief of Staff Marshal Akhromeyev, can cause serious misunderstanding:

The limitation, still more the reduction, of nuclear arms is inconceivable in conditions of the militarization of space. *The creation and deployment in space of strike arms* will inevitably lead to an increase in the quantity of, and to the qualitative improvement of, strategic nuclear arms . . . (*Pravda*, 4 June 1985, extracts in OTA, 1985a: 313–4 – emphasis added)

On 13 January 1985, the then Soviet Foreign Minister Andrei Gromyko had given a more careful explanation of Soviet policy to Moscow television, when interviewed about the agreement to open new bilateral arms talks with the United States. Instead of limiting the unwanted militarization of space to Akhromeyev's later 'creation and deployment *in space* of strike arms', Gromyko spelled out the aims of Soviet policy as more comprehensive:

[A]rms intended for use against targets in space should be banned categorically and also . . . arms intended for use from space against . . . targets on the ground, in the sea and in the air should be banned categorically. (Excerpts in OTA, 1985a: 313)

However, Gromyko's reference to subjective intentions could give rise to problems.[10] Even if the US administration were to accept that the Soviet Union is not now developing any ground-based 'attack space weapons', which of course they do not, they would doubtless still raise the issue of the modernized Galosh system at Moscow, on which considerable effort has been expended over the past decade. The primary anti-missile purpose of this, together with its permissibility within the ABM Treaty, appear to exclude it from those 'new anti-satellite systems and . . . systems already possessed' which General Secretary Gorbachev recently confirmed his government was willing to scrap (1985). Yet, armed with nuclear warheads as it seems to be, Galosh retains an anti-satellite capability against LEO targets, just as the US Nike-X system had twenty years ago (p. 22). If a trustworthy clean sweep is to be made of space-strike arms, it may be necessary to amend the ABM Treaty to ban even the single, limited deployment of ground-based systems that it now permits, and with it the Soviet Union's new, layered terminal-phase BMD at Moscow.[11]

Whilst the Soviet Union clearly is willing to sacrifice its very modest and rapidly disappearing 'advantage' in existing asats, it may perhaps be less ready to do the same with its new Galosh system. Gorbachev's letter to the Union of Concerned Scientists (1985), for example, seems carefully worded so as not to leave any such impression. The opportunity which this presents, for challenging the bona fides of the comprehensive Soviet opposition to space-strike arms, has probably not been lost upon the hawks within the US defence community.

Part 3
Ends

10
Who Needs Space Weapons?

I think we have a feeling that we are in many ways naturally superior. And that superior-inferior relationship is something the Soviets constantly seek to overcome.
I'm not sure that our country is willing to accept genuine equality. What we define as equality almost always is superiority.

Ambassador Paul Nitze, 1985

The survey of space weapon programmes and technologies in previous chapters suggests that asats will be restricted for some time by relatively low operational altitudes, the vulnerability of some ground-based and all space-based components, and inadequate overall reliability. Advances in directed-energy weapon technologies, however, might in the medium term bring powerful and multiple anti-satellite capabilities, reaching for thousands of kilometres within near-Earth space, well before they could provide effective BMD systems.

Missile defence programmes offer two conceivable developments. First, within a few years there could be more or less feasible options for partial defence of military targets. And second, it would be imprudent to dismiss absolutely the possibility that *if* massive effort and expenditure continued to be given to the SDI to the end of this century and beyond, *if* ways were found to defeat the many offensive countermeasures that presently appear both cheap and feasible, and *if* strategic ABM systems could be adequately tested in peacetime, then eventually some form of layered defence might be devised to provide at least some degree of protection for civilian populations and their essential life-supporting infrastructures. The probability may be minimal that all three of these conditions could be met, and any defensive protection realized would certainly be imperfect.[1] But for purposes of discussion in what follows, the minimum level of BMD to count as 'population defence' is taken to be one that ensures that significantly more people on the defending side would live than would die *during* a nuclear war – a level of defensive capability that advocates of BMD suppose would render major nuclear attack unlikely in the first place.[2]

This hypothetical approach is intended to focus attention on the 'strategic' question, namely: If space weapons were in due course shown to be *feasible*, would they even so be *desirable*? The feasibility issue is made up of a range of technical questions, discussed in previous chapters. These may, or may not, be resolved within the SDI and similar programmes. But the desirability of space weapons is a separate, non-technical issue that can and should be discussed without waiting for all the technical questions to be answered. Indeed, if space weapons are thought wholly undesirable, there is no need even to try to answer the technical questions. It is therefore important to discuss the two issues of feasibility and desirability separately.[3]

All parties to the debate on space weapons agree that the central strategic question about them is whether they make nuclear war less or more likely. This was recognized by the United States and the Soviet Union when establishing the framework for their bilateral talks in January 1985, aimed at resolving the 'complex of questions concerning space and nuclear arms . . . in their inter-relationship.' It is a problem usually addressed, in the West, through words like 'deter' and 'deterrence'.

One problem with the language of deterrence, however, is common to all discussions of this kind. Those who advocate space weapons claim they would make nuclear war less likely. But they are often vague as to whether the improved 'deterrence' would be of the two-way (mutual, West-East : East-West) or the one-way (unilateral, West-East) variety, and the confusion is so common as to suggest that it may be intentional. Many citizens in many countries have become resigned or even positively converted, over the past twenty-five years, to the mutual, 'balance of terror' version of nuclear deterrence. But they may be far less acquiescent, under the nuclear circumstances, towards pretensions of 'winning' the arms race by vastly enlarging one side's military capabilities and undermining or even supposedly negating those of the other. Whichever side of the argument one may lean to, clarity and honesty require that these two, very different senses of 'deterrence' be kept distinct at all times. Unfortunately, however, the Western space weapons lobby is in large part an internal ideological crusade, intent on replacing such values as 'co-existence' with a vision of one-sided deterrence of the Soviet Union by means of a decisive US 'superiority'. And such crusades are less about truth and consistency, more about victory at any intellectual or other price.

Defending Military Targets?

The claim that ABMs can assist deterrence and stability has always been strongest in respect of their deployment as a partial and selective protection for elements of an offensive nuclear capability, such as hardened command bunkers, communications 'nodes', and the silos of land-based missiles. Such ABMs, it is argued, would not negate the 'ultimate deterrent' of any nuclear power, namely its ability to destroy an opponents society. They would merely increase the *survivability* of each side's forces, reducing the likelihood of a 'counter-force' strike against them because the opponent would be terrified of trying, only to fail. As such, they would supplement existing sources of 'retaliatory force survivability', such as missile submarines or reinforced concrete silos.[4]

This view of the stabilizing character of 'force-defence' ABM systems[5] rests on questionable assumptions. One is the 'Pearl Harbour' expectation, that nuclear war might well begin through a sudden and deliberate all-out strategic nuclear attack by one side against the nuclear forces of the other, usually referred to as a 'disarming first strike'. Another is the view that, whether one-way or two-way, the deterrence it takes to prevent this depends absolutely on retaining a capability for massive 'counter-city' retaliation.

Surprise Attack

As long as nuclear weapons exist, so must the physical possibility of one side deliberately initiating an all-out nuclear attack on the other. But the preoccupation with this unlikely scenario amongst some American strategic theorists has long since departed from the shores of common sense.[6] Much expert opinion (Frei, 1983; Bracken, 1983) now takes the alternative view that nuclear war is more likely to result, not from positive but from negative military motivations; not out of deliberate and cold-blooded choice, but from pressures built up during an acute international crisis. At such a time, various military moves might begin to be made on both sides, the rationality and objectivity of decision-makers might rapidly 'degrade' (as for example in Britain during the Suez crisis of 1956), and a *pre-emptive* first strike might then be launched by whichever government first grew convinced that nuclear conflict could no longer be avoided:

> Most U.S. military leaders have, however, discounted the likelihood of a 'bolt from the blue' – a coolly calculated effort to achieve military superiority by a bold disarming first strike. The more plausible situation would be a mounting crisis, in which both sides would wrestle with the fear that the opponent might strike first on rational or irrational calculation of advantage. And in the face of the apparent inevitability of nuclear war, a preemptive strike could appear better than awaiting the adversary's initial blow. There are grounds in Soviet military writing for taking this scenario as one which their strategic policy has contemplated. (Drell et al., 1984: 68)

The Counter-Force Requirement

The other proposition mentioned above was that the essence of all deterrence (one-way or two-way) lies in the preservation of 'counter-city' forces. This no longer seems to be accepted in such authoritative statements of nuclear doctrine as the report of the Commission on Strategic Forces, chaired by General Scowcroft, which appeared just after President Reagan's 'Star Wars' speech:

> In order to deter such Soviet threats we must be able to put at risk those types of Soviet targets – including hardened ones such as military command bunkers and facilities, missile silos, nuclear weapons and other storage, and the rest – which the Soviet leaders have given every indication by their actions they value most, and which constitute their tools of control and power. (Scowcroft, 1983: 6).

This assessment of American offensive nuclear force requirements was *not* superseded by the SDI, and is not thought likely to be so in the next few decades:

> What it [SDI] represents is an extended transition from today's total reliance on offensive deterrents to an increased dependence on defensive means to deter attack. What we're talking about is a

changing mixture of the two over a period, no doubt, of many years. That means we're still going to have to rely on strategic nuclear weapons as deterrents during that period, and *SDI in no way lessens the importance of continued modernization and deployment of strategic systems*, such as Trident submarines.

This dual emphasis on both SDI and *continued deployment of modernized offensive weapons* makes perfect sense when we remember that we are talking about this extended transition . . . from offense to mixed offense and defense . . . [during which] our security will continue to rest to a significant degree on *the potency of our retaliatory deterrents*. (Keyworth, 1985b – emphases added)

The 'potency' of US offensive nuclear weapons which it is intended to preserve by means of their modernization is nothing but their estimated counter-force capability, which comes down to their ability to evade and penetrate ABM defences on the other side, and yet to deliver nuclear warheads with sufficient accuracy to ensure destruction of even 'super-hardened' military targets.

Actual or alleged Soviet ABM programmes (chapter 9) are viewed by the Pentagon as a type of force defence for 'key targets' (US DoD 1985c: 12). The US military response to this 'challenge' has been to declare their intention to preserve a capability to penetrate such defences, together with their confidence of being able to do so. Since the 1960s, the Pentagon's Advanced Strategic Missile Systems (ASMS) programme (formerly 'ABRES', for Advanced Ballistic Re-entry Systems) has been about little else (p. 128). The present US Secretary of Defense expressed his satisfaction with it a few years ago, when explaining why the MX missile was a vitally necessary addition to the US nuclear arsenal:

One cannot exclude the possibility of a breakthrough in lasers which would make a spaceborne laser ABM look more attractive. However, there are a wide variety of counter-measures we could take ranging from hardening of the missiles to space borne ASATs which would reduce the effectiveness of a [Soviet] laser ABM . . .

If the Soviet Union were to deploy an ABM system, we would still have confidence in the ability of MX to destroy hard targets through the use of chaff, decoys and other penetration aids that we have developed through the Advanced Ballistic Re-entry Systems Program (ABRES) combined with tactics such as saturation. Our SLBM force could be equipped with MK-500 maneuvering re-entry vehicles and other penetration aids as well and the defended area could also be saturated with MK-4 re-entry vehicles. (S. Appropriations, 1981: 122, 131)

Today, of course, such techniques of penetration are being studied intensively by the 'red teams' that simulate possible offensive tactics within the SDI, by way of anticipating the toughest varieties of missile attack that it might have to counter. And doubtless, in view of the SDI, the Soviet equivalent of ASMS is not idle.

Authoritative pronouncements from the Soviet Union confirm that there, too, the maintenance of a counter-force deterrent *capability* by the Strategic Rocket Forces is regarded as essential. Notice, for example, the reference to 'qualitative improvement' by Marshal Akhromeyev:

> The creation and deployment in space of strike arms will inevitably lead to an increase in the quantity of, and to the qualitative improvement of, strategic nuclear arms . . . How is the other side, the Soviet Union, supposed to behave under these conditions? It is left with no choice; it will be forced to ensure the restoration of the strategic balance and to build up its own strategic offensive forces . . . (*Pravda,* 4 June 1985 – extracts in OTA, 1985a: 313–14)

In a speech to the Supreme Soviet at the end of November 1985, Mr Gorbachev made the point even more explicit:

> To restore the balance the Soviet Union will have to enhance the efficiency, accuracy and strike power of its arms, in order to neutralize, if we must, the electronic space machine of Star Wars. (*The Guardian,* 28 November 1985)

Soviet nuclear targeting also seems to have given priority to counter-force options from early on. And available information about Soviet nuclear weapons suggests they have been increasingly provided with a technical capacity for counter-force roles, notably through MIRVing and improvements in warhead accuracy in the late 1970s. That was certainly the impression given in 1980, for instance, by a Soviet colonel's publication in *Foreign Military Review* of a map of 'military facilities' in Britain (in Campbell, 1984: 333). Soviet policy-makers may have little faith in 'windows of vulnerability' opening up in face of the accuracy of their ICBMs (Holloway, 1983: 49–50), but they have done little to remove

Western fears in that respect. And the long-standing Soviet arms control policy, to the effect that only a comprehensive 'parity' between the forces of both sides will suffice to ensure their own security and global peace, underlines their insistence on a broad technical and numerical equivalence, at least, if not in every respect a doctrinal one.

Whether or not the Soviet requirement for 'nuclear deterrence' has a counter-force emphasis which exactly parallels its American equivalent, it is certainly perceived in that light by the US administration and by those who support its current boosting of missile defence:

> In a global conflict, Soviet strategic policy would be to destroy Western nuclear forces before launch or in flight to their targets . . . From these policy directives come several overarching strategic wartime missions:
>> eliminate Western nuclear capabilities and related supporting facilities; . . .
> These missions would involve:
>> disruption and destruction of the West's essential command, control and communications capabilities;
>> destruction or neutralization of the West's nuclear forces on the ground or at sea before they could be launched; . . . (US DoD, 1985a: 26)

When evaluating Western arguments for space weapons, therefore, it must be accepted as a premise that, whilst both sides undoubtedly do in some sense confront each other with an 'ultimate' threat of virtual annihilation by counter-city nuclear attack, both also set great store by the 'deterrent' role of their counter-force nuclear capabilities. Defence by threat of cataclysmic 'mutual assured destruction', amounting to suicide by whichever side initiated it, has long seemed unconvincing to military planners on both sides. Neither seems likely, therefore, to welcome anything which would diminish its supposedly more 'credible' counter-force capabilities, thereby throwing it back on 'massive retaliation' as the only available nuclear posture.

Competing Force Defences

Established patterns within the nuclear arms race strongly suggest that the immediate strategic consequence of ABM deployments in defence of hardened military targets would be that each side would continue doing its utmost to minimize the effectiveness of the opposing defences. As the budget for the SDIO rises, so does that for ASMS. And the SDIO comes to Britain to buy the Chevaline penetration aids as well (table 12.1 §12), so that US warheads can be on the 'safe' side.

Faced with a prospect of strategic BMD deployments by the opponent, the response on the offensive side would be threefold: 'anti-BMD' countermeasures, the technologies for which were reviewed in chapter 8, would be developed, to preserve the penetrativity of the attack; nuclear strikes would be planned to be as difficult and disruptive as possible for the defences, and would include elements of 'defence suppression'; and the size of the offensive arsenals would be increased, to cope with a possibly adverse 'exchange ratio' (p. 62). Using the latest authoritative US and Soviet information, Dr Paul Rogers has calculated that the Soviet Union could almost certainly expand its strategic nuclear forces from the present level of about 9000 warheads (including those on bombers) to 31,000 by the end of 1995 (Rogers, 1986).

Precisely because of their, by definition, limited capability, force-defence ABM deployments would strongly encourage attempts by the opposition to develop ways to beat them, and are therefore certain to increase arms race instability. Amongst their dangerous effects, however, the intensification of the 'qualitative' dynamics of the competition is perhaps even more disturbing than the quantitative, 'far more of the same' aspect analysed by Dr Rogers. In a two-horse technological race between the United States and the Soviet Union, pitting missile defences against offensive 'penetrativity', there could be only three militarily different possible outcomes: either (a) both sides might retain levels of effective counter-force capability comparable to those they now possess; or (b) both sides might see their counter-force deterrent capabilities diminished; or (c) one side might keep its offensive capability whereas the other, broadly speaking, would lose it.

Since this chapter is directed to the question of whether ABMs and other space weapons would be worth having if in some sense they 'worked', the first of these outcomes, (a), is strictly irrelevant for the present purpose. But it is worth pointing out, in passing, that even if the technological competition between offence and defence continued to be won by the former, so that ABMs made little or no significant difference in practical terms, they could and probably would undermine crisis stability by radically confusing the perceptions of both sides. Their very deployment by one side would require the other to suppose that they might 'work' in some sense after all, however sure its military scientists had previously been that this was impossible. The US and British responses to the Soviet Galosh system have been an actual case in point over many years. Furthermore, the fear that the opponent might seek to improve on the performance of his marginally capable ABMs, by combining them with a pre-emptive strike to 'thin' the attack they would have to cope with, could itself provide the fateful trigger to war.

The possibilities with 'effective' force-defence ABM deployments, then, are either, (b), that both sides get them, or (c), that only one does so. Neither outcome seems likely to assist in the prevention of nuclear war. It is of course impossible to predict the full effect of ever-increasing technical complexity, within both the actual and the perceived strategic confrontation, on the likely behaviour of the two sides during any severe international crisis. But it seems most unlikely that the process would work in only one direction, *away* from the resort to war.

Possibility (b), that with or without any agreed disarmament a technology race for increased ICBM 'survivability' might be won by everyone, is part of the declaratory rationale of 'increased security all round' advanced in favour of the SDI. What it sometimes fails to notice is that, in order to 'deter' satisfactorily, ICBMs have to be sure of surviving not merely in their defended silos, but also throughout their threatened journey over to visit the other side with destruction. If the purpose of force defence is to ensure that ICBMs are not vulnerable targets, that can after all only be so as to preserve their capacity to threaten the opponent. So if the other side is envisaged as deploying force defences also, the main point of the first side's move into force defence, hoping to ensure the survivability of its own ICBMs, would appear to be lost.

There is an easy way past this objection for the pro-SDI argument, however. Whilst by definition force-defence ABMs would remove the 'survivability' of some proportion of the ICBMs attacking them, not being adequate for population defence they would not have this effect against ICBMs aimed at cities. Therefore, it can be argued, the mutual deployment of force defences would not amount to each side's depriving the other of its 'deterrent'.

The point is well made, on its own terms, provided those arguing for the SDI are genuinely prepared not only to frustrate the 'offensive counter-force requirements' of the other side but also to abandon their own, a decision which seems unlikely whilst the other side still has a large arsenal of missiles capable of destroying the first side's society. It would be, in effect, a decision to license the restoration of MAD, and with a vengeance. That would be because, for the sake of protecting its own 'retaliatory capability' against the other side, each would be licensing, so to speak, the enhanced survivability of a comparable 'assured destruction' aimed against itself.

For in case (b) a far higher proportion of the already enormously redundant arsenals of both would be expected to survive. In the absence of effective population defences on either side, nuclear war would mean each was able to visit the other side with even greater retaliatory devastation than it had ever previously been possible for them to threaten. Perhaps there would be force reductions, with or without negotiations. But if not, two-sided BMD at this level seems likely to do far more to enhance the threat posed by nuclear missiles than to eliminate it. To which might be added that if force-defence BMD could only play any useful role in the context of negotiated disarmament, it would be safer and simpler to let the negotiations do the eliminating instead.

(And of course, the pro-BMD lobby can *only* claim that missiles could retain their 'deterrent' properties even when both sides have force defences, provided the terms of the discussion are restricted to force defence. Once population defence is brought into the picture, the SDI's supporters must take a different tack, since at that point they are no longer, apparently, in favour of keeping ICBMs in business in any way, but only of rendering them 'impotent and obsolete'.)

Compound Instabilities

The main question raised by possibility (b), of mutual force defences, is whether they really would make nuclear war any less likely by making the 'effectiveness' of any offensive counter-force first strike seem highly uncertain. This can be made to seem an easy question to answer, if one accepts the assumption that only one type of nuclear war outbreak needs preventing, that which might start 'out of the blue', as a result of deliberate premeditation. Lessen the military and economic 'gains' for the 'aggressor', it is argued by SDI supporters, and you diminish the risk of war. However, it was suggested above that nuclear war is far more likely to arise out of interactions between the military vulnerabilities, political inadequacies, and irrational cultural dynamics of *both* sides. And in such situations there simply is no single, calculating 'aggressor' available, to be rationally deterred. Both sides, to all intents and purposes, would have 'started it'. Both would also have been driven to it. And part of what might have driven them could well have been anxieties about each other's force-defence ABMs.

There is a certain military equivalence, for 'damage limitation' purposes, between offensive pre-emption, on the one hand, and a defensive 'denial' of targets to the attack, on the other. In World War 2 the Royal Air Force could shoot down Hitler's flying-bombs over Kent, or they could try to bomb their factories and launch sites on the Continent. Today, the Pentagon thinks that 'Soviet strategic policy would be to destroy Western nuclear forces before launch or in flight to their targets' (p. 166). But, as already noted, force-defence deployments are necessarily of limited effectiveness. Their true capabilities would be unknown, and would tend to be estimated, by the usual military double-think, as lower for strategic planning purposes than it was when seeking development funds in earlier years, and also as lower for the estimators' own BMD than for that of their opponents. There might therefore be a motive, if ever war seemed imminent, to improve the BMD system's limited performance through a pre-emptive first strike against enemy missiles. With the attack thus 'thinned' before it could even get off the ground into 'boost phase', the ABMs might have a far better chance of coping with whatever missions they had been assigned.[7]

For the exactly converse reason, a side facing such defences, whether or not it had any of its own, would be anxious to make doubly sure of the survival of its own forces, not only against the other's defences, but also against 'defensive' pre-emption from the other side. (Even if it had force defences, it might fear the offence had found some particularly penetrative countermeasure against them.) Precisely because the 'exchange ratio' had worsened against it, it would want to be sure that all its missiles were launched, rather than letting many of them be destroyed in their silos, and it could only achieve this by making sure of being the first to strike, once all hope of avoiding war seemed lost.

But above all, what matters for preventing nuclear pre-emption in some future acute East-West crisis is not whether either side might accept all or part of the 'logic' of such thinking for itself, but whether each could remain reasonably confident that the *other* side had not done so. Nothing could better illustrate the fact that the widely used distinction between 'arms race instability' and 'crisis instability' is not absolute, but only a matter of degree. There are many reasons for attempting to control the arms race, and later to end it. It is criminally wasteful; it damages the lives of the millions who inhabit it; and in the name of avoiding war it makes war more and more destructive, should it ever come. But the main problem with an accelerating and unpredictable arms race instability, based on very 'high' and unfamiliar technologies, is that it increases crisis instability too.

The third possibility, (c), is that one side begins to lose the technological and strategic competition, or sees itself as doing so. That would effectively be a simplified case of the interactions already discussed. Faced with a 'final choice between a *kamikaze* apotheosis on a national scale or a "conditional surrender" in the form of a realignment with the policies and social structure [of the 'winner']' (Golovine, 1962: 118), it is barely conceivable that the loser could manage to accept what would seem, and might perhaps in truth be, its permanent subordination and defeat, without doing anything desperate and suicidal. It seems most unlikely, on the balance of historical experience, that such a clear-headed accept-ance of what would be considered an unprecedented national disaster could occur before a great deal of actual nuclear war had taken place, a war which it would be difficult or impossible for anyone to stop (Bracken, 1983). Both the civilian population of the 'loser' in the offence-defence arms race, and, more impor-tantly, its military commanders, would have been indoctrinated for decades with an absolute faith that the possession and threatened use of nuclear weapons made such a defeat a historical impossibility. More than that, the nuclear forces aimed at preventing the arrival of such a day would have been prepared and designed and trained for *use*, should 'deterrence fail', as the euphemism goes.

The point is that deterrence could well be perceived as 'failing', not merely if the other side resorted to large-scale military action as such, but also if they were thought to be achieving an irreversible 'strategic superiority'. For military action need not consist in actual combat and destruction. The arms race itself is perhaps a kind of war, for which, of course, the expression 'Cold War' was invented. And it is now 2500 years since Sun Tzu laid down the unchallenged strategic maxim that 'supreme excellence consists in breaking the enemy's resistance without fighting' (1953: 48).

In the light of this survey of its only possible outcomes, BMD for defending nuclear retaliatory forces no longer seems an unambiguous contribution to international peace and stability. Even if one accepts, for the sake of argument, that (for as long as rationality prevailed) force defences on one or both sides might help to deter a calculated 'disarming first strike', there remains a more relevant and more difficult question. Does it make any sense to acquire an enormously expensive and technically dubious way of preventing a type of war outbreak that is already very improbable, if that involves *increasing* the probability of what is already the more likely way for nuclear war to start?

Force Defence and Counter-City Targeting

For one side to deprive the other of its counter-force capability seems unlikely to have the effect of leaving the other reluctant to pose any nuclear threat at all. On balance, given the superpowers' conviction that *some* form of nuclear deterrence is essential to their continued security, such a 'success' on the part of one might only induce the other to restore all-out genocidal attack to the status of its primary option, however much of a 'kamikaze apotheosis' this might appear.[8] That is, indeed, the implication of the pro-SDI argument, already noted, that force defences do not undermine 'deterrence' because they do not prevent the other side from targeting the cities and other undefended 'values' of the side whose missiles would now be protected. The question therefore arises as to whether this probable result of deploying force defences is a sensible one for the side doing so to bring about. As Britain's Foreign Secretary has wondered:

> [W]ould the establishment of limited defences increase the threat to civilian populations by stimulating a return to the targeting policies of the 1950s? (Howe, 1985)

Many major cities are already targeted on both sides, whether or not because of their military significance, by today's supposedly more selective nuclear planners. But it is not necessary to accept misleading claims about the comparative 'benefits' to civilian populations of being subjected to a counter-force rather than to a counter-city nuclear war, to see the point of Sir Geoffrey Howe's concern.

An effective force-defence BMD may in actuality, or in the view of the side deploying it, shift the opponent towards placing greater emphasis on the ultimate sanction of his remaining nuclear capability, that of annihilating the other society despite its limited BMD systems. The defence-deploying side could not, indeed, regard its force defences as worth having unless they made *some* difference to the opponent's targeting policy. Perhaps the opponent might find other options than a return to counter-city targeting, though they are hard to imagine. But even if he did, he would still have the problem of persuading the other side (the one deploying force defences), that he had not adopted such a posture.

He would need to do so, however, if both sides were not to be drawn into ever greater danger of nuclear war. It cannot be held an advantage but only a drawback of force defence that, even if deployed only on *one* side, it could possibly move *both* sides to emphasize 'massive retaliation' in their preparations for nuclear war. A side without force defences would be expected to re-target in that way because it had no other choice. But it is important to realize that any 're-targeting effect' of force defences is unlikely to be a global shift in the potential attacker's operational planning, simply substituting urban population targets for the now defended nuclear forces of his opponent. The response would probably be more complex, involving plans for heavier attacks on some of the defended military targets, together with an assessment of their likely 'counter-city' side-effects ('collateral damage'), plus a tendency to assign a proportion, only, of any strike to undefended, densely populated areas that would not necessarily be devastated by the militarily-motivated portion of the target list. There might also be a reflexive re-targeting effect, for the side actually deploying BMD. But this would certainly be even less predictable, because less rational. Obviously the first priority for that side would be to maintain its counter-force targeting against enemy forces which now threatened it with even greater and more immediate devastation. Because the re-targeting effect would not be simple and clear-cut, however, it would still have to reckon with the possibility of some counter-force attacks, and would probably be reluctant to place absolute confidence in its own BMD systems.

It would therefore have to consider that, in the event of war, it might possibly 'lose' if it refrained from counter-city attacks, because its capability for resorting to such measures later on might, after all, be removed. (Though it could be argued that force defences should never have been deployed in the first place, unless they meant that this scenario could be decisively ruled out, such rationality is not how such decisions are usually taken, still less how they are adhered to after the event.)

If the result of erecting limited defences will be that the side (or sides) that does so may face an even more devastating attack, does that represent a voluntary acceptance, on its part, of a larger burden of 'deterrence' from the other? Can deterrence be increased in that way? The reduction in the total explosive power of US nuclear weapons over the past twenty years suggests that no previous administration has accepted any such crude equation between the quantity of deliverable 'megatonnage' and the amount of 'deterrence' obtained therefrom. The present administration, also, cites that long-term trend towards 'smaller' warheads as a virtue, but not in any way as a diminution of US security.

It therefore seems unlikely that any supporters of force-defence BMD would claim it was a direct way to enhance or seem to enhance 'deterrence', simply by raising the 'costs' on one or both sides through the re-targeting effect. To adopt such an argument would also be to forget the effect which such a shift or shifts in targeting policy would probably have on international relations, both before and during any severe crisis. 'Absolute' weapons and threats of total annihilation both produce and are produced by absolute hostility and absolute mistrust. One reason given for NATO's shift away from 'massive retaliation' and towards 'flexible response' in the 1960s was that the former was judged to pose a severe threat to crisis stability. If the worst ever did come to the worst it was felt vital that there should be lesser, more controllable options in nuclear conflict, preferably on both sides. It was also thought that if both sides had reason to hope that, even if war broke out, they would not be utterly destroyed in the first hour or two, then they might be slightly less 'driven' towards mutual pre-emption in the first place. If however it were to be replied, on behalf of force-defence ABMs, that 'deterrence' does now need improving or restoring by a reversion to massive counter-city threats, it is not exactly clear why force defences are required for such a strategic posture, which was adopted perfectly well without any, thirty years ago. (The usual argument, that they are needed to thwart the threat of a counter-force first strike by the other side, requires an assumption that such an attack could be far more 'successful' than there is any reason whatsoever to expect – see chapter 9, note 4.)

By making unattractive the so-called 'lesser' option of limited or counter-force nuclear attacks in a severe international crisis, force defences would certainly appear to raise the stakes. In the absence of any crisis, having such heavy penalties attached to 'going nuclear' may seem sensible enough. But the effect on crisis stability of sharply increased nuclear anxieties would not be the simple one of imposing greater caution. It might be better and more appropriately described in terms of mathematical 'catastrophe theory', which deals with situations in which relationships between variables can be suddenly inverted. People may usually grow more cautious as risks are increased, but this is by no means a simple causal relationship. Nor is the related one, between levels of caution and the severity of losses actually experienced. Raising the nuclear devastation stakes might make both sides slightly more determined to avoid war, in times when the risk of war is present but not high.[9] Within the crisis itself, however, the increased severity of the immediate price to be paid could lead, paradoxically, to war breaking out from motives of pre-emption, when it might otherwise have been avoided.

By their own pretensions, the acid test for nuclear forces and anti-nuclear defences must always be, not their probable role in the normal 'peacetime' jostlings betwen the superpowers, but how they would lead governments to perform during a really dangerous international situation, with war already a daily or an hourly possibility. So far as can be judged, BMD for the protection of nuclear retaliatory forces fails that crisis stability test. Since it obviously fails the arms race stability test also, it should not be adopted.

Defence of Populations?

Any discussion of the strategic value, or lack of it, of BMD systems for population defence must be even more hypothetical than for the force-defence option. It seems sensible to consider the possible effects only of an imaginary population defence that afforded a considerable degree of anti-missile protection to 'soft' civilian targets, such as cities.[2] 'Perfect' defences are simply not worth discussing, and seem unlikely to become so (p. 140).

This evaluation has been shared by the SDIO (p. 137) and by leading figures elsewhere in the Reagan administration, past and present. Their case for BMD is regularly made within its constraints:

> [A]chieving perfect defenses [is] an unnecessarily stringent require-ment. The President's ultimate goal of impotence and obsolescence for nuclear weapons isn't based on the need for such perfection . . .
>
> [B]oost-phase strategic defenses . . . create . . . a *symbolic* dome over the East that prevents ballistic missiles from getting out . . . [it may be] a *leaky* dome . . . [but] it's more than effective enough as a deterrent against a first strike . . . (Keyworth, 1985b – emphases added)

The question to be discussed at this point, therefore, is whether an anti-missile defence capability which provided an imperfect but significant degree of population defence would be worth having, even if it were to leap suddenly into existence, with no dangerous and unstable 'transition period', and even if it were equally effective against all types of nuclear weapons and not just long-range missiles.

Dispensing with Retaliatory Deterrence?

The SDI's case for population defence has been summarized as follows:

> Finally, in conjunction with air defenses, very effective defenses against ballistic missiles could help reduce or eliminate the apparent military value of nuclear attack to an aggressor. By preventing an aggressor from destroying a significant portion of our country, an aggressor would have gained nothing by attacking it in the first place [sic]. In this way, very effective defenses could reduce substantially the possibility of nuclear conflict. (US WH, 1985a: 3)

Would such defences really be 'a move away from a future that relies so heavily on the prospect of rapid and massive nuclear retaliation' (US WH, 1985a:(i)), as the President expressed it? The very language used suggests the contrary. For it is assumed that the main point to any nuclear attack on the United States would be to destroy 'a significant portion of our country', and that an attack which could not achieve this would have 'gained nothing'. In short, the re-targeting effect of US force defences, assumed to have been deployed some time before the 'very effective' population defences proper, is being duly taken into consideration.

But large scale 'counter-country' attacks are not only expected to continue as a *Soviet* option, hopefully frustrated. The State Department has explained that:

> For the foreseeable future, offensive nuclear forces and the prospect of nuclear retaliation will remain the key element in deterrence. (US State, 1985: § 11)

Even after 'a mix of offensive and defensive systems' had been deployed, and even if that had occurred on both sides (to judge from the report's earlier remarks about 'continued Soviet defensive improvements') the administration expects that, in addition to a defensive shield for denying the enemy any military 'gains', the United States would retain an offensive capability to impose punishing 'costs' upon any 'aggressor':

> We would deter a potential aggressor by making it clear that we could deny the gains he might otherwise hope to achieve rather than *merely* threatening him with costs large enough to outweigh those gains. (US State, 1985: § 10 – emphasis added)

For all the talk of dispensing with deterrence by assured destruction, the looked-for offence-defence 'mix' is expected to be one in which, on the one hand, US ABMs could stop most Soviet missiles reaching their targets, whilst on the other, the United States would retain sufficient and sufficiently 'capable' missiles to preserve its offensive deterrence indefinitely. For the latter to be the case, of course, Soviet defences would have at each stage to be somewhat less capable than US ones. Despite the State Department's superficially even-handed treatment of the 'more stable basis for deterrence' which is being looked for from the SDI, the truth is that alleged 'coercion' could not be any better resisted in an era of *mutual* BMD than it can be now, so long as the Soviet Union remained able to inflict much the same amount of damage on the United States as vice versa, even if at reduced levels.

The 1985 OTA report is only one of many studies to point out that the Reagan administration has no intention of carrying through a strategic revolution towards 'defence dominance' until the very last remote, utopian moment. The US offensive capability would not be gradually abandoned, as a defensive 'alternative' was acquired:

> The capability to protect our cities would mean a major shift in our strategy away from retaliation and toward assured survival. However, we could not abandon retaliation until defenses gave us confidence that they could assure a high degree of protection. Protecting cities requires an extremely capable defense. Opinion differs as to how many nuclear explosions in populated areas in time of war would lead to unacceptable or intolerable damage. However, that number would be at most tens of weapons out of an attack measured in thousands. *A defense that let through no more than 1 percent of the attack – and perhaps far less than that – would be required before the basis of our strategy could shift away from retaliation.*[10] (OTA 1985a: 103 – emphasis in original)

In short, *it is not so much the assuredness as the mutuality of the destruction that the SDI lobby finds so objectionable about MAD.* There is, of course, nothing particularly 'new' about the role assigned to strategic defences in this perspective.

Deterring the Soviet Union More Effectively?

The long-standing aspiration towards BMD of conservative strategic thinkers in the United States was noted above (p. 68), but their reasoning has yet to be examined. The views of D. G. Brennan, who championed the strategic virtues of population defence both during the 'ABM Debate' of the late 1960s and in the early stages of the renewed 'BMD debate' ten years later, are representative of the 'school'. In Brennan's view, Robert McNamara, US Defense Secretary for the Kennedy and most of the Johnson administrations, was wrong to argue that in order to deter a nuclear attack by the Soviet Union the United States had to be permanently capable of some fixed high level of 'assured destruction' in retaliation, after suffering the aggressor's first strike. Instead, claimed Brennan, one-way 'deterrence' would always be secured by *any* level of US strategic nuclear capability, relative to its current Soviet counterpart, that:

> would imply that, if the Soviets started a strategic war with us, they would be guaranteed to come out worse, a powerful deterrent to starting such a war. (Brennan, 1969c: 105)

Like Gray and Payne more recently (1980; 1981) Brennan was also convinced that relative strategic superiority required defensive as well as offensive strategic systems, and that the United States had a far better chance of achieving it at *lower* levels of offensive nuclear forces on either side:

> [This posture] indicates that both the U.S. and the Soviet Union might reasonably engage in some measures to limit the possible damage of a war without necessarily impairing U.S. security in the process. Such measures might include . . . direct reduction of strategic offensive forces . . . [Even if such reductions were for the time being unlikely] . . . a symmetric increase in active defenses deployed by each of the superpowers would have approximately the same kind of potential impact on possible war outcomes that percentage cuts in offensive forces would have. (Brennan, 1969c: 105–6)

Brennan was no more counting on anything like a perfect anti-missile shield than does the SDIO today. He assumed that extensive levels of 'assured vulnerability' (as he preferred to call it) would continue to feature on both sides of the strategic confrontation:

> [There] is the theory that anything that makes war more tolerable makes it more likely. But "more tolerable" in this instance means 20 or 30 million American people killed instead of 120 million . . . It is very unlikely that you could find any American decision-maker whose behavior in a crisis, so far as his propensity for starting a nuclear war is concerned, is going to be significantly altered because he is told that his action will cost the lives of 'only 30 million', instead of, say, 100 million. And my reading of the Soviet bureaucracy, which is very much a committee-type government, is that it is going to be in the same position and is going to react in the same way. (1969a: 33–4)

A US retaliatory capability measured in tens of millions of 'assured' Soviet casualties was not just unnecessary, in Brennan's view, but also inadequate by itself to deter the Soviet Union. Only the certain prospect that the Soviet Union would 'come out worse' than the United States, by suffering higher and militarily more damaging losses, could do that. His argument turned on the issue of whether or not partial population defences, on both sides, would transform the psychological content of one-way, US-Soviet 'deterrence'.

Brennan expected the advent of partial defences on both sides to remove from the minds of Soviet leaders their former, supposedly insufficient inhibitions against starting a nuclear war, sustained by fear of *massive US retaliation*, and to substitute for them some new and allegedly more effective ones, focused only on the *militarily disadvantageous results of a nuclear exchange* with the United States. Provided that thought could be maintained in the Kremlin, all would be well, and 'deep cuts', which would help the defences to look and to be more effective, could also proceed.

The assumption obviously was that the defences would be fairly effective, since otherwise the vulnerability of the Soviet Union would be unchanged – and apparently, changing it was desirable. But Brennan was also ready to allow that reductions in expected casualties of the order of 75 per cent (120 to 30 million in his example) would *not* make nuclear war seem 'more tolerable' to Soviet decision-makers, any more than it would to American ones, whether or not for the same reasons. This was an inconsistency. Either the new defences would make it possible for both sides, and especially the Soviet Union, to see the risks of war in a different, less all-or-nothing light, or else they would not. Brennan seemed to be saying both things at once.

But if offensive nuclear deterrence was admitted to 'work' against the Soviet Union, in some sense, what need of BMD to replace it? How was Brennan able to concede that Russians were just as inhibited by their assured vulnerability as were Americans, but nevertheless to call for the additional contribution, to deterrence, of effective population defences? What indeed could *defences* contribute to deterrence at all?

The answers depend on understanding the problem Brennan and most other US advocates of BMD have been trying to solve. The real concern has never been with some mythical imperviousness of Russians to nuclear terror.[11] The difficulty lay closer to home, in the fact that nuclear deterrence worked both ways, and inhibited American power quite as effectively as it did the Soviet variety, by Brennan's own admission. It was this *relative* psychological situation, not just the hypothetical one in *Russian* heads by itself, that Brennan and others have always wanted to alter by introducing BMD.

Once the American side of the strategic relationship is placed clearly in the foreground, Brennan's position becomes easier to grasp. The rule is: any combination of offensive and defensive forces on the two sides is permissible, provided it is such that, if ever there were a war, the Soviet Union would 'lose'. Under those conditions, US deterrence of the Soviet Union may move up or down a little, but Soviet deterrence of the United States can never move up, only down. Seen as a 'zero sum' situation, with one side's 'losses' counting as the other's 'gains', BMD is thought to provide opportunities for increasing US-Soviet deterrence simply *by lowering Soviet deterrence of the United States.* The method is more convenient and more reliable, by far, than trying to bring off speculative changes in the psychology of the Soviet leadership. Hence Brennan's seeming inconsistency about the latter was not important, because that was only on the decorative and 'declaratory' side of his proposal, not on its 'business' side.

Today, as it always has been (Possony, 1963), the demand for population defence is a demand to relieve the United States of the burdens imposed upon her in international relations, actually or potentially, by the existence of Soviet strategic nuclear forces. Whilst the Soviet nuclear arsenal remains as unstoppable as at present, there is no way for even the hardest-headed hawk to imagine that, in the case of major war, there could be much chance 'to terminate the conflict on terms favorable to the forces of freedom' (US DoD, 1984a: 29). If nuclear war could become slightly more fightable, however, through a reciprocal process of lessening the offensive threat (at any rate against the United States) and introducing anti-missile defences, then America's economic and technological strengths might enable it to establish and maintain a convincing version of strategic superiority once again (Gray & Payne, 1980). As has frequently been pointed out, the Soviet Union is really only a 'superpower' at all, on terms of some kind of parity with the United States, thanks to the levelling effects of 'absolute' nuclear weapons in an epoch of zero BMD. Population defence would be about removing that levelling effect, if possible for good, and with it the resented 'distortions' in international affairs which, according to the American Right, it makes possible.

The unexpected conclusion is that the whole point of population defences is precisely that *they will always be imperfect.* For offence-defence combinations on the two sides therefore admit of 'less and more', in a way that strategic offensive missile arsenals have not done. An epoch of parallel, licensed and competitive BMD can therefore be one-sidedly announced – by those who expect to win the race they thus initiate. And having restored the possibility of decisively overmatching the other side, sincere offers can also be made to put an end to 'overkill'.

Most American advocates of BMD are so *only* because they think it plays to US strengths. It would be a different matter if they thought for one moment the competition might go the other way. (Whether the converse applies completely on the Soviet side is hard to say; but Meyer's interpretation (1985) suggests it probably does, more or less.)

Improving Stability?

Just as with force defence, there are only three possible relationships between the superpowers with respect to a hypothetical future population defence capability. Either neither gets it, and things remain much as they are; or both get it; or only one does. The first case, once again, does not concern us here. To answer the central question about the prevention of nuclear war, with respect to the other two possibilities, it is necessary to pretend, as before, that the technical problems may some day be solved, though in fact, as we have seen, they would be even more daunting than in the case of force defence.

Neither of the two strategically relevant cases appears to promise much for arms-race stability. In the case where both sides get population defences of some kind (call it (b) again) 'mutual assured destruction' might just conceivably be replaced by a somewhat lower level – 'mutual assured disaster' – but the basic situation would probably be unchanged, as Brennan recognized: '20 or 30 million American people killed . . . the lesser prospect is hardly likely to lead to dancing in the streets' (1969a: 33). As with force defence, both sides seem likely to do whatever they can to prevent such a change, or at least the part of it which goes against their own offensive deterrent capability – arms race instability again. As shown above, the US administration's stated policy is that until its defences became very capable indeed it would need to preserve the striking power of its retaliatory offensive missiles. But that could not be done by sitting back and allowing Soviet defensive capabilities continually to increase, relative to US ICBM 'penetrativity'. The very similar Soviet determination to retain its *relative* offensive power was also noticed.

In short, the technological arms race between offensive missiles and population defences, should they start to enter the picture, would not be ended by any level of defence short of something inconceivably 'reliable and total', along the lines of Mr Weinberger's persistent aspiration.

The third possibility, that one side might get some degree of population defence whilst the other still had none, is again the limiting case of the second. US supporters of BMD may in some cases comfort themselves with a version of the familiar 'last step' fallacy, in which the likely Soviet response to some recommended new US posture is deliberately overlooked, thereby placing the whole enterprise in a deceptively promising light. But the real course of events is likely to be different. If the US defence was still less than 'perfect', there would be plenty for the Soviet Union to do by way of old-fashioned counter-city threats, in order to restore the high confidence which Soviet leaders require, for their own security, that the United States will never attack. One-sided US population defences might of course *aim* at depriving the Soviet state of such MAD capabilities, but on the basis of the technical assessments of chapters 7 and 8 they would almost certainly remain quite unable to do so.

As for crisis stability, that too seems likely to suffer, unless the defensive capabilities were to reach unexpectedly high levels after all. And one new consideration that would certainly begin to play a part, indeed has already begun to do so since 1983, is the growing fear, however irrational in the light of the above assessment, that the opponent might be able to bring off a final decision, on the basis of very advanced military technology alone, in the political contest between East and West.

The Reagan Version

The Reagan administration has added nothing to the content of the various long-established contentions and arguments for the merits of an effective population defence. But it has brought to them its own particular emphasis and style. The emphasis has been on the manageability of East-West relations, and to some extent of intra-NATO relations also, during the years it would take to develop the technology and to introduce 'a better, more stable basis for deterrence'. The style has been that whenever doubts and objections grow too many and too serious, resort is had to rhetoric about a mysteriously perfect BMD, that will solve everything. This fairy-tale, happy-ending approach is very much part of the administration's overall defence philosophy, under which all threats can be 'eliminated', if Congress would but follow its prescriptions.

'We do not Seek Superiority'

The President's initial guarded acknowledgement that BMD deployments for population defence might diminish crisis stability, in the March 1983 speech (p. 79), has been widely noted and welcomed, for instance by Sir Geoffrey Howe (1985). It is difficult, however, for the administration to make credible its claim not to be seeking a decisive strategic superiority over the Soviet Union, not merely because of what population defence would have to be about, but also because of its own record in nuclear strategic policy.

It is principally the second of these two factors that has led commentators on the SDI to doubt whether the introduction of BMD into the East-West confrontation could be handled as easily as its supporters suggest, with or without negotiations:

> To the extent we succeed in deploying an effective but imperfect [population] defense, what we are challenging in the first instance is not just the Soviet incentive to a first strike, but the efficacy and reliability of their retaliatory deterrent forces which they count on to deter a U.S. first strike. They will look at the emerging U.S. mixed defense-offense posture to judge whether the United States remains vulnerable to, and deterred from initiating, nuclear war. In so doing, they will apply the test put vividly during SALT 1 and in other Soviet statements: *does the United States still recognize that to start a nuclear exchange would be for it to commit suicide?* (Drell et al., 1984: 66)

The question comes with increasing urgency today not only from the Soviet Union but also from the United States' allies and from world public opinion, and needs a convincing answer. It has needed it, in fact, since well before the SDI's appearance on the scene, for it began to be asked in 1982, when the *New York Times* (30 May) revealed that the Pentagon's Defense Guidance document had specified that in any nuclear war, which could well be 'protracted', 'the United States must prevail and be able to force the Soviet Union to seek earliest termination of hostilities on terms favorable to the United States.' Despite the objections of its critics,[12] the administration has persisted with this overtly 'unilateralist' policy. The Defense Secretary reasserted it, for example, in his 1984 report: 'credible deterrence, either nuclear or conventional, requires that we have the ability, in case deterrence fails, to halt any attack and restore the peace on terms favorable to us and our allies' (US DoD, 1984a: 38). Nor has authoritative and bipartisan domestic support for the Reagan view of 'deterrence' been entirely lacking:

> By attaining strategic nuclear parity with the United States, the Soviet Union has severely undermined the credibility of US strategic nuclear forces as a deterrent to a conventional attack on Europe. (Senator Nunn, S. Armed Services 1982b: 2)

> Deterrence is the set of beliefs in the minds of the Soviet leaders, given their own values and attitudes, about our capabilities and our will. It requires us to determine, as best we can, what would deter them from considering aggression, even in a crisis – not to determine what would deter us. (Scowcroft, 1983: 3)

In such a context, the President's oft repeated line that 'nuclear war cannot be won and must never be fought' is not an answer to humanity's anxious enquiry about the SDI and related US developments, merely the welcome but trite preliminary to one that he still refuses to give.

The credibility problem arises from the fact that, for the United States to be able to change the terms of its strategic relationship with the Soviet Union *no matter how the latter responds*, something as near to strategic superiority as makes no difference must have been attained, or at least brought within sight. And change of that order is the declared purpose of the SDI, which has always been considered at least as important for its strategic as for its technical possibilities:

> [T]he overriding importance of the SDI is that it offers the possibility of reversing the dangerous military trends cited above [Soviet force improvements] by moving to a better, more stable basis for deterrence . . . (US State, 1985: 3)

Though prepared to negotiate about it, however, the Reagan administration is also prepared to do without Soviet consent, if necessary:

> This commitment [to negotiations] should in no way be interpreted as according the Soviets a veto over possible future defensive deployments. (US State, 1985: § 8)

Such statements have made it hard for the Soviet leadership or anyone else to believe that the parity result (b), whether in its force- or its population-defence form, is what the SDI is aiming for. By definition, if both sides ever developed equally effective defences against missiles there would have been some slight move away from the regime of mutual assured vulnerability, whether of missiles or of populations. But this revised version of the strategic 'balance' would surely diminish rather than increase one-way US-Soviet 'deterrence', and would complicate and destabilize the superpowers' relationship by injecting into it a set of military factors that would be, and be perceived as being, both vitally important and utterly incalculable.[13]

There is also the consideration that, if improvements to parity and two-way deterrence *were* what the SDI was really about, the most that might be necessary would be to preserve US retaliatory deterrence, with weapons programmes and arms control negotiations focused solely on the survivability of nuclear forces and their continued ability to penetrate Soviet defences. However, parity and MAD have been consistently resented by the Reagan administration, which appears dedicated to ending what it sees as Soviet lack of respect for US interests and demands around the globe, by means of a 'strength' which could only be based on some form of superiority.

'The Transition Could be Tricky'

The dangers inherent in building defensive systems at the same time as retaining, or increasing, strategic offensive capabilities, were clearly recognized by Ambassador Nitze at the start of 1985:

> [T]he transition period – if defensive technologies prove feasible and if we decide to move in that direction – could be tricky. We would have to avoid a mix of offensive and defensive systems that, in a crisis, would give one side or the other incentives to strike first. (Nitze, 1985: 4)

Mr Nitze stressed the many years it would take to deploy and, somehow, test any major BMD system. The risks of a 'transition' in this primary sense are those which always accompany any major alteration in the composition and disposition of military forces. It may be interrupted or somehow pre-empted by a manœuvre, or even by an attack, by the other side. And obviously, the longer, more comprehensive, more complex and more public such an alteration is, the more likely it is to be countered in some way.

But there has also been considerable discussion, not only by the SDI's critics but also from the State Department and the White House, about the problems of a very different sort of transition period, one that begins even *before* any military changes have been made, and even before 'defensive technologies prove feasible'. This is the subjective and anticipatory change in the focus of strategic and political attention which would certainly precede by many years the completion of full-scale development of the BMD system, let alone its deployment. Quite rightly, the administration has been acting, in some ways, as if that other, more comprehensive transition period had begun on 23 March 1983.

The probable damage to both arms race stability and crisis stability during such a period has already been sufficiently examined. Like much else about BMD, it was recognized long ago. The extent to which deployment of unique or of very superior BMD by one side might be regarded as not just politically but also militarily unacceptable by the other was emphasized, for example, in McNamara's comment to Kosygin in 1967, with reference to the first stages of the Moscow Galosh system,:

> You are trying to deprive us of our nuclear deterrent. And we shall not let that happen. (*New York News*, 25 March 1983)

It was no surprise when such, originally American, strategic views on BMD were repeated in response to President Reagan's speech by the Soviet President, Yuri Andropov:

> [T]he intention to secure itself the possibility of destroying, with the help of the ABM defenses, the corresponding strategic systems of the other side, that is, of rendering it unable to deal a retaliatory strike, is a bid to disarm the Soviet Union in the face of the U.S. nuclear threat. . . (*Pravda*, 27 March 1983 – extracts in OTA, 1985a: 312)

Mr Weinberger has also warned, in December 1983, about at least one version of such instability:

> I can't imagine a more destabilizing factor for the world than if the Soviets should acquire a thoroughly reliable defense against these missiles before we do. (In *F.A.S. Public Interest Report*, March 1985)

Both superpowers thus seem to appreciate that movement by either side towards large-scale BMD would seriously upset the nuclear balance and generate strong pressures for a pre-emptive strike during a crisis.

Within the Reagan administration two nominal solutions have been offered for this difficulty, which it would not be inappropriate to dub the 'State solution' and the 'Pentagon solution' respectively. The State solution is 'to make the transition a cooperative endeavour with the Soviets' (Nitze, 1985: 4), by means of 'arms control agreements . . . to manage and establish guidelines for the deployment of defensive systems' (US WH, 1985a: 5). Besides seeking to avoid the kind of moves into BMD that might drive the US-Soviet relationship towards some final catastrophic breakdown, this approach would also try to meet the additional goal, explained above, of using the arms control process to increase the feasibility of any likely BMD system by restraining the strategic offensive competition:

> [T]he preconditions for moving toward the President's vision of a dominant defense extend beyond presently unforeseeable technical and operational breakthroughs, and require a framework of effective restraints on offensive forces. Also required would be a prior agreement on the ABM systems to be deployed, and an underlying deep mutual respect and trust which some desire but few anticipate soon. (Drell et al., 1984: 73)

The State solution recognizes a strong common interest, therefore, between the SDI and at least a version of arms control, a version which would permit the United States to reorganize the whole military-technological competition along lines that would almost certainly be to its advantage. The obvious problem comes with getting the Soviet Union to go along with the idea, and thus to drop its present imperfect but nonetheless real capability to deter the United States, in favour of what General Graham once so refreshingly pointed out would be 'assured survival for the United States' – period (*Aviation Week & Space Technology*, 8 March 1982).

The Pentagon solution is to claim that the recalcitrant Soviet Union, and for that matter dissenting US allies, will be obliged to come to heel, because the defence is anyway going to be a perfect one – or at least so nearly perfect as to ensure the Soviet Union would have no choice but to accept a position of permanent inferiority. This view involves paying as little attention as possible, for the time being anyway, to the awkward question raised by Nitze and many others about how any defence, perfect or not, could be acquired quickly enough to preclude pre-emptive moves by the other side. Perhaps this is because, in the mind of the Secretary of Defense, the system has already been achieved.

'Thoroughly Reliable and Total – Yes'

Statements to the effect that the SDI is about achieving a 'total' defence, that would make all Soviet missiles, and even eventually all Soviet nuclear weapons of any kind 'impotent and obsolete', have come most frequently from the President, from Defense Secretary Weinberger, and from the erstwhile Presidential Science Adviser, Dr Keyworth. General Abrahamson at the SDIO has tended to be more reserved, declaring only that the SDI can certainly achieve whatever it sets out to achieve, but not confirming that its goal is in fact one of rendering *all* Soviet missiles impotent.

It would be a mistake to isolate the administration's technological optimism about the SDI from all its other optimisms. Despite the enormous problems of the US budget, for instance, no difficulty has ever been admitted to be possible with finding funds for full development and eventual deployment of a large-scale BMD system, usually estimated in terms of trillions rather than mere billions of dollars. As for arms control, despite clear signs to the contrary it is regularly insisted that the SDI will help convince the Soviet Union to agree to deep cuts in offensive nuclear forces. The programme has also been given a spotless bill of legal health (chapter 11). And so on.

It is, however, the total inadmissibility of technical failure, shown for example by the recent $100 million increase in the budget for the X-ray laser programme after its rather unimpressive record came to public light, that gives the enterprise a truly unique status in the modern, and indeed perhaps in the entire history of the United States. The SDI goes far beyond the familiar technologically-minded 'can do' spirit of US society. Never before has an administration managed to convince itself, out of pure ideology, that both the forces of nature and the will of the Soviet and most other advanced governments could be so completely commanded.

This sort of single-minded 'tunnel-vision' approach is not

confined to politicians, but President Reagan's career provides more examples than most. In the case of SDI, several personal motives may be postulated for his decision to 'go for broke'. Before the 1985 Summit, one anonymous Presidential aide was quoted as follows:

> The only thing he can accomplish in the last three years to set himself apart from other presidents is a big arms agreement. This would be the kind of big movie star happy ending Nancy likes. (In Barnes, 1985)

But Nancy Reagan, or more probably the aide, may not have understood the President correctly. Perhaps, like the young 'star warriors' at Livermore, he hopes 'to eliminate offensive nuclear weapons' himself 'without having to wait for arms control' (Broad, 1985: 51). Before leaving office in January 1989, he may hope to have initiated the full-scale development stage of a major BMD programme, involving multiple and very costly weapons projects together with a global strategic shift in the US military posture. After such a stage had been reached it would be extremely difficult for any future administration or Congress to prevent a major BMD deployment of some kind. As the British Foreign Secretary so unenthusiastically expressed it in 1985, 'research may acquire an unstoppable momentum of its own' (Howe, 1985). On the other hand, the President's ambition also benefits from the fact that after only three full years there would be little danger of the programme having been so comprehensively discredited as to destroy his reputation. After his departure from office, the responsibility for failure, however defined, would pass to others.

Whatever may have been the personal factors, other, less conjectural influences have also been at work. There can be no 'half-way house' in population defence, not, at least, on its 'declaratory' side. (The value of the necessary incompleteness of a population defence was suggested above with reference to the objective military balance, a rather different aspect of the matter.) If the attempt is risked, therefore, it must 'succeed'. It is all very well for Livermore's head of defence programmes, Dr G. H. Miller, to complain:

> I'm very alarmed at the degree of hype, promises, and a failure to focus on what this national program really is – a research program with lots of unanswered questions. I'm afraid the public is losing

sight of how difficult this job is. (*International Herald Tribune*, 19 December 1985)

But his remarks seem both naïve and addressed to the wrong people. The 'public' take the SDI in this way for the very good reason that that is how it has repeatedly been presented to them, by political leaders with total access to the most thorough and comprehensive assessments of its technical chances.

What such leaders choose to hear, and from which sources within the US weapons development community, owes much to their quasi-religious commitment to the complete achievement of all US strategic (and economic) objectives. That 'faith' is a key component within the style, if not the content, of the Reagan foreign policy. For the hostages of nuclear terrorism (MAD), just as for all other hostages, a way *must* be found to go right in and get them out to safety.[14] If reality seems likely to obstruct the venture, it is reality that is out of step. And if the technical capability is not yet available, it is certain to become so, if enough dollars get thrown at enough problems. Meanwhile, technical 'demonstrations' are to be produced, or laboratory heads will roll.

In short, optimistic technical predictions about the SDI's feasibility are for the most part *internally and subjectively generated*. Which was why the 'question' to which the 'research' programme was supposedly addressed had already been answered in advance, by people with no technical but only an ideological role to play:

Well, the defensive systems the President is talking about are – are not designed to be partial. What we want to try to get is a system which will develop a defense that is – [Total? – interviewer] – thoroughly reliable and total, yes. And I – [Against all incoming missiles of any kind?] Yes. And I don't see any reason why that can't be done. (Weinberger, 'Meet The Press', *NBC*, 27 March 1983)

Not that the technocrats have been slow to support such claims:

I guess my experience as a technologist and as a manager with a long career in this [space] effort is that we indeed can produce miracles. I think that is what the shuttle program has recently shown me, and I believe that American technological muscle can, over a long period of time, do precisely what has been laid out.[15] (Abrahamson, in S. Foreign Relations, 1984: 30)

The conclusion has to be that statements about the already assured attainability of a more or less total population defence for the United States and her allies would have been highly irresponsible, *unless* they were intended to convey a single message to the Soviet Union: that the United States has begun, and intends to continue, drawing decisively ahead in the arms race, thereby achieving a growing military superiority. In other words, the intended message has been just what the Soviet Union has angrily understood it to be.[16]

There is one further important move within the Pentagon's solution to the political and other problems of the SDI. When statements about the attainability of perfect BMD are failing to convince, it is always possible to shift the ground of the discussion slightly, into an area which is highly relevant to dealing with possible countermeasures against the SDI, and which offers slightly less daunting challenges. The area of 'space control' itself.

Asats and 'Space Control'

The idea that overall military control of our planet might be achieved from space is now so venerable that references to it as 'the new high ground' are ludicrously inappropriate. In February 1957, not long after the appearance of von Braun's prospectus for imposing an imperial peace by means of a single bomb-toting space station, and before a single satellite had been placed in orbit, General Bernard Schriever, commander of the USAF's Ballistic Missile Division, had given public expression to much the same kind of thinking:

> In the long haul our safety as a nation may depend upon our achieving 'space superiority'. Several decades from now the important battles may not be sea battles or air battles, but space battles, and we should be spending a certain fraction of our national resources to ensure that we do not lag behind in obtaining space supremacy [sic]. (In Caidin, 1960: 56–7)

One early analytical treatment of space combat recognized that there were two basic strategic alternatives for any possible weapons designed to operate within space. Their technology and intended roles might tie them very closely and immediately into warfare down on the planetary surface, for which they might become a potential catalyst or trigger. Or else 'incidents', armed clashes, and even 'wars' could occur in space, usually between lifeless machines, without necessarily degenerating into World War (Golovine, 1962: 113–8).

Would asats and other true space weapons help us to avoid the destabilizing aspects of rival anti-missile defences? Or would they merely extend and intensify that instability? Arising from the previous discussion, such questions obviously concern the first alternative posed by Golovine.

It should be noted, first, that interest and support for the idea that 'space control' is the missing key to 'Earth control' is now greater than ever within the United States, particularly amongst advocates of anti-missile defence:

> It is time that we lay the phantom of MAD to rest and that we turn our attention to the realistic task of affording maximal protection to our society in the event of conflict . . . Any nation which deployed two dozen . . . first generation chemical laser battle stations would command the portals of space against the rockets of any other nation. (Wallop, 1979)

> U.S. military superiority in space could, conceivably, achieve a technological 'end-run' around the Soviet strategic competitive challenge unfolding today. Moreover, if both superpowers extended every effort to race as vigorously as possible for command of the 'high ground' of space, the U.S. would probably win – for a while at least. (Gray, 1983: 13)

> Our goal is to build a military flying machine [the Trans-Atmospheric Vehicle or 'TAV'] that will be able to take off from a military airfield, insert itself into the upper reaches of the atmosphere and the lower reaches of space, and go around the planet in ninety minutes. We're not looking for a cargo machine. We're looking for a killer Air Force weapon that can go out and get the enemy . . . Wouldn't it be great if the Soviet Union suddenly found itself faced with the United States Air Force having a machine that could operate on its own, totally free from counteraction, capable of rapidly delivering weapons anywhere on the globe? (S. A.Tremaine, USAF Deputy Director for Development, Planning & Aeronautical Systems, in Canan, 1984)[17]

Such thinking is an established part of US strategic policy, dating back at least to preparatory planning in the Carter years, when major studies initiated by Air Force Secretary Hans Mark and by the Defense Science Board decided amongst other things upon the crucial military requirements for the Space Shuttle, without which, Canan remarks, 'Congress would have killed the program long ago' (1982: 170). Carter's Defense Secretary, Harold Brown, explained that:

> We see some Soviet space capabilities that could directly threaten our terrestrial forces and some of our critical satellites. The Soviets are operating satellite systems that could perform targeting of U.S. naval and land-based forces, and they have an ASAT system . . . These Soviet activities could threaten our access to space . . . We will have to develop an equivalent capability to destroy Soviet satellites if necessary. (In Canan, 1982: 175–6)

Under the Reagan administration there has been more of the same, but with important additions, particularly with respect to possible space-to-earth weapons like the TAV. The Defense Guidance documents expressed a determination, in 1982, to 'pursue technology and systems both to provide responsive support and to project force in and from space as needed', and in 1983, to 'achieve capabilities to ensure free access to and use of space in peace and war; deny the wartime use of space to adversaries . . . and apply military force from space if that becomes necessary' (Stares, 1985: 219).[18] And the official Fact Sheet on National Space Policy (US WH, 1982) laid stress on programmes to increase 'the survivability and endurance of space systems' and to develop 'a United States ASAT capability to deter threats to space systems of the United States and its Allies and . . . to deny any adversary the use of space-based systems that provide support to hostile military forces.'

The USAF's long-term Space Plan, drawn up in 1983, divides all space combat into 'space control' and 'force application', with the latter covering both anti-missile defences and STEW. Its purpose is frankly expressed:

> To prevail in theater conflict, the Air Force must seize the initiative and quickly achieve both air and space superiority . . .
> Space superiority is required to ensure that our space-based assets are available to support theater forces. Superiority in space will require a robust force structure and the capability to destroy hostile space systems. (In Canan, 1984)

The extent of current US policy for 'space control' would not be clear without a mention of some of the major organizational changes introduced by the Reagan administration. Hans Mark, bent on ensuring that the Air Force should no longer have to accept the 'leavings' of the Shuttle programme, became Deputy Director of NASA in 1981 (Canan, 1982: 172). In 1982 and 1983 separate operational Space Commands were established by the USAF and US Navy, with the latter determined as ever not to succumb to Air Force attempts to gain overall control of the US military space effort. However, by 1985 the Air Force appeared to be winning the inter-service contest, with the opening of a supposedly 'unitary' Consolidated Space Defense Operations Centre (SPADOC) close to the underground Colorado headquarters of the North American Aerospace Defense Command (NORAD).

On the Soviet side, the nature of strategic doctrine for space is far less certain. The extension of air defence to include anti-missile defences was recognized in the third and final edition of the classic post-war Soviet work on military strategy – said by some Western analysts still to be required reading for all Soviet officers:

> [I]t is now necessary to assure, essentially, 100 percent destruction of all attacking enemy airplanes and missiles. Even one airplane or missile with a nuclear warhead, which has broken through the air defense system, can cause tremendous destruction and damage. The high effectiveness of modern means of air defense permits . . . the complete destruction of all attacking enemy airplanes and missiles . . . (Sokolovskiy, 1975: 299)

As the American editor of this now twenty-year-old text points out, however, all direct expression of the idea that such a capability might require the deployment of some variety of *Soviet* space weapons was revised out of its final version in the mid-1960s, even before the arms control process leading to the ABM Treaty had been set in motion. References to work on 'antimissile and antispace defence satellites for intercepting ballistic missiles' and to 'the creation of offensive space systems of strategic designation' – on the other side – were retained or rewritten (1975: 87–8). The statement that 'all this work which is being conducted in other countries deserves great attention' was dropped (1975: 454).

On the whole, Western experts have not been inclined to take such early Soviet claims for the total efficacy of their anti-missile defences too literally (Gehlen, 1967: 101; Peebles, 1983: 97). But together with recent authoritative Soviet warnings about the probable action–reaction consequences of the SDI, noted above (p. 152), they do seem to evidence an established military interest in the long-term possibilities of weapons in space, if no legal limitations on them can be agreed. And Western intelligence reports of Soviet programmes for a variety of space vehicles, such as a small TAV-like 'space-plane', large boosters for launching space stations or their modular components, and 'space tugs' to service and re-position long-life satellites, suggest that the Soviet Union should be *able* to deploy considerable military force in space by about the end of the century, if she feels the necessity to do so. Gray summarizes the present situation as follows:

> What we know for certain is that the U.S. and the Soviet Union use space for military purposes that would be critically important in the conduct of war. Unless the U.S. alters that condition, the Soviet incentive to fight in space cannot be reduced. U.S. military space activities, which are increasing in scope and significance, create that incentive. Military conflict in space is not a matter for U.S. policy choice today – the choice has already been made. (1983: 49)

But a necessary, or unnecessary, military competition and confrontation in space is one thing. The idea that there is some way to achieve a decisive advantage in that competition, and thereby turn the arms race 'corner' down on Earth as well, is quite another.

The region of near or planetary space is an extremely public one. Military uses of it cannot easily be concealed – there is no 'dead ground'.[19] Offensive space operations would be slow and awkward by comparison with much terrestrial military action. For at least the remainder of this century, nearly all military space activity, except possibly at low altitudes, could only be carried out by satellites in Earth orbit, most of them unmanned. Some systems, such as the new US DSCS-III military communications satellites, two of which were deployed in 1985 on the first mission of the new *Atlantis* Shuttle, are already being given a capability to defeat attempts at jamming. Others still being developed, such as the US MilStar communications system, are to be hardened against nuclear and directed-energy effects. But for some time many will remain more or less vulnerable to any weapons able to reach them. Protection may therefore be sought in ever higher orbits, out to 200,000 kilometres or more (Gray, 1983: 31), especially if this is combined with on-board capabilities for evasive manœuvres without too great a loss of functions. There has also been an interest in hiding 'dark' or inert satellites in such orbits, either as weapons or as spares for vital communications or other non-weapon systems such as MilStar, with technologies so perfectly reliable that they would only need to be switched on, thereby revealing their positions, when a serious military emergency had arisen. However, current satellites are obliged to make frequent use of their onboard 'housekeeping' systems, to communicate with ground stations and then adjust their orbits for 'station-keeping'.[20] The degree of technical perfection which would eliminate this source of vulnerability cannot even be imagined by space engineers, let alone anticipated.

Moreover, the attacker's asats would not have it all their own way. To start with, any that were themselves mounted on long-stay orbital platforms would be as vulnerable as most other satellites. Those designed to be launched into the attack from Earth, like today's Soviet and US systems, would be too slow to approach and destroy targets in high orbits before the latter could take evasive or retaliatory action. Even if the age of virtually instantaneous 'beam weapons' had arrived, whether based on Earth or on platforms in space, it would be prudent to suppose that the side whose satellites were under attack would have such technologies as well, and could use them against the rising asats or any space-based elements in the attacking force. (Certainly all who argue for US space weapons on the basis of an alleged Soviet 'lead' are obliged to consider this problem seriously, or forfeit all intellectual credibility.) The satellite defender could also seek to 'deter' attack, as US policy already proposes to be able to do, by threatening equal destruction to the attacker's satellites in retaliation.

It is also rather easy to forget some of the attacker's other problems. If for instance he wanted to 'decapitate' the other side's space-borne warning, and command-and-control systems, he would need to mount a completely effective and *simultaneous* strike against a very large number of objects moving in many different planes, and tens of thousands of miles apart, within the 'difficult geometry of the interior of a large spheroid enveloping as its core a smaller and impenetrable spheroid, the earth' (Possony & Rosenzweig, 1955). As Stares laconically remarks, 'for this to be fully co-ordinated with a massive launch of strategic [offensive] missiles would not be easy' (1985: 250).

Not only is it true that to speak separately about weapons for defending and for attacking satellites is to make 'a distinction without a difference', (p. 53). It is also clear that there is a very close link between anti-satellite and anti-missile technologies, such that no effective ABM systems could ever be deployed, even just for force defence, which would not have given a strong boost to parallel applications of the same technology for anti-satellite purposes. The current US F-15 asat programme is a clear case in point:

> [T]he MHV [Miniature Homing Vehicle], which is the heart of the U.S. ASAT system, already displays the symbiotic relationship between ASAT and BMD: the MHV exploits technologies origi-nally developed by the Army for its Homing Overlay project, which explored the feasibility of intercepting ICBMs in midcourse . . . (UCS, 1984: 225)

However, the cooperative overtones of 'symbiotic' may be misleading, when applied to so fundamentally antagonistic a relationship:

> The significance of ASAT for strategic defense lies in the threat it poses against the space platforms of the ABM, in particular against the warning, acquisition, and battle management sensors. On the other hand, the significance of the Strategic Defense Initiative for ASAT is that it will spur technical developments that, inevitably, will be threatening to the critical communication and early warning satellite links on which a ballistic missile defense must rely. This presents an unavoidable dilemma: ASAT threatens ABM, *but ABM developments contribute to ASAT*. (Drell et al., 1984: 62 – emphasis added)

The conclusion must be that, once again, the proposed 'solution' of strategic superiority, this time in space, is highly unlikely to be achieved. It could only come about if one side had been able to deploy an arsenal of completely effective asats *and ABMs* before the other side had got anywhere with its space weapon program-mes at all. The first passage from Gray's book was not completed above (p. 193). It continues:

> the U.S. would probably win – for a while at least. However, even if the weaponization of space is feasible . . . it is far from certain that such a U.S. technological victory could be long sustained. History teaches that military-technological competitions between states tend toward rough equality in qualitative achievement. (1983: 13)

Because it would assuredly be temporary, an apparently decisive lead in space weapons would only be of any value if it were *used* before it was eroded. But the 'winning' side could not threaten or, if defied, actually attack the other's space systems unless it was certain that it also possessed very effective and reliable anti-missile defences through which to nullify any strategic counter-threat the opponent might make, however suicidally, down on Earth. Far from asats, now operating as DSATs, being the way to patch up weaknesses in one's missile defences, the latter are a vital strategic precondition for the former ever being used, either politically or in actual combat.

In short, *if one side had good enough BMD systems to be able to risk space combat, it would not need to engage in it. If it did need to fight in space, to remedy the vulnerability of its defensive systems, the risk would almost certainly be far too great to take.*

But as long as one or both sides in the space arms race remain doggedly obsessed with the fatal chimera of strategic superiority, whether as something to be pursued for the sake of a decisive and bloodless victory, or as something to be denied to the opponent, or as both, the process seems more likely to trigger off a planetary nuclear war than to provide a way of keeping the world in peace.

Robotic Champions?

But what, finally, of Golovine's second possibility, that combat in varying degrees of intensity might be mercifully confined to space, out of mutual fears of annihilation if it were ever allowed to spread down to the planet's surface? A quarter of a century ago this may have seemed plausible, but it is hard to find it so today. There have indeed been frequent and major 'limited wars' in the nuclear age, involving the superpowers or their proxies. But to get something like an analogy to the sort of conflict being postulated as possible in space, one has to imagine numerous, direct and unconcealable clashes between the American and Soviet navies, with serious losses on one or both sides, which were unlikely either to spread onto the land or to escalate towards nuclear war. The picture is unconvincing; for one thing, if this sort of direct US-Soviet warfare, carefully quarantined by unwritten conventions, is both affordable and desirable, what is stopping it now? Why does it need to wait for space weapons before occurring? But if it is improbable on Earth, it seems not a bit more likely to occur in space.

The naval analogy can certainly be criticized in two ways. First, there *are* quite frequent minor clashes at sea (also in the air) between the United States and the Soviet Union, or between their proxies, sometimes even involving loss of life, as well as many other warlike incidents involving other nations, and these do not, in general, lead to outbreaks of full-scale war (Cable, 1981). And on the other hand, space *is* a region in some sense remote from the planet and therefore the only place where an entirely symbolic and sanitized warfare could conceivably now evolve. Unlike navies, its craft are largely unmanned, so that losses in such 'incidents' might be less inflammatory, even, than small-scale naval conflict.

But for one thing a growing part of the world's economic infrastructure is now in space and certain to remain there. The effect of even moderate amounts of space combat on the international banking system, for example, would probably bring untold hardship, and even death, to millions of people. There could also be grave effects on the homœostatic systems of near space and the outer atmosphere from the very deployment, let alone the use of such weapons. Thus even if their close military links to weapons for planetary destruction could somehow be checked or neutralized, and even if conventions were observed whereby it was never 'done' to attack the vulnerable but essential parts of space systems which are in fact down at ground level, the actual physical effects of space weapons would make it impossible to localize and isolate space warfare, as Golovine hypothesized, 'by mutual tacit consent'.

It is also probable that the destabilizing effect of an arms race in asats and other true space weapons would be just as severe even if no attempts were made to develop and deploy effective anti-missile defences. That is, even if the strategic 'balance' continued to consist almost wholly in offensive forces, as it does now, the intensive weaponization of space would drastically lower the threshold to nuclear war. On both sides, fears would be increased yet further that, unless they were to strike first, they would find themselves deprived by enemy action of warning, communications, navigation and intelligence systems, on which the vital military 'edge' in both conventional and nuclear combat could well depend. Only an early mutual decision not to proceed with any further pursuit of some mythically final 'space control' could prevent this happening.

Nowadays, however, disarmament policies and negotiations seem increasingly subordinate to the needs of military space policy, rather than the other way about.

11
Arms Control or Space Control?

*The alternatives are clear: arms control or a shot at developing
defenses. As long as the president sticks to his position, we will have
no arms treaty.*
Former US SALT Ambassador Gerard Smith, autumn 1985

Drop it [SDI]? I ain't gonna do it!
President Reagan, autumn 1985

At the centre of the Strategic Defense Initiative lie two major
contradictions. The first is that efforts to reduce the effectiveness
of the other side's offensive nuclear weapons, by acquiring
anti-missile defences, will in all probability drive the opponents
into seeking to maintain their offensive capability, by increased
qualitative and quantitative efforts. The second is the tension
between the SDI's high degree of dependence on successful arms
control negotiations between the superpowers and the likelihood
that it will in practice prevent such negotiations from producing
any real cuts in missile arsenals:

> Assured survival of the U.S. population appears impossible to
> achieve if the Soviets are determined to deny it to us. This is
> because the technical difficulties of protecting cities against an
> all-out attack can be overcome *only if the attack is limited by
> restraints on the quantity and quality of the attacking forces.* The
> Reagan Administration currently appears to share this assessment.
> (OTA, 1985a: 33 – emphasis added)

At least some advocates of BMD were taking cognizance of this
aspect of the Reagan administration's strategic doctrine from the
outset:

> The objective of damage-limitation mandated by a victory-denial

deterrent should be complemented by deep force level reductions. Active and passive defenses would become increasingly significant as massively heavy attacks become less and less feasible. (Gray & Payne, 1981: 1–75)

Military uses of space are limited by several arms control treaties. Under the Limited Test Ban Treaty of 1963 the United States and the Soviet Union have agreed not to carry out nuclear test explosions in space, such as might one day be required to develop a space-based X-ray laser weapon. And the Outer Space Treaty of 1967 records their decision never to use the Moon or other heavenly bodies for any military purposes at all, and not to place 'any nuclear weapons or any other kinds of weapons of mass destruction' into orbit around Earth. Many international lawyers think that only nuclear weapons that are also mass destruction weapons may be covered by this provision, and therefore question its applicability to the proposed X-ray laser weapon (Wolfrum, 1984; Harndt, 1985). So it is interesting to note that this particular, awkwardly nuclear, strategic defence technology has been suggested more than any other for a pop-up, non-orbital deployment.

Since most technologies envisaged for the SDI would be non-nuclear, and need not involve military activities on heavenly bodies, they are largely unhampered by agreements aimed at preventing the possible spread of *nuclear* weapons into space. Only the ABM Treaty also covers non-nuclear weapons for which there is a foreseeable prospect of deployment. For that reason, the possibility of limiting or preventing the overall arms race in space, by means of further treaties covering asats and other space weapons, now turns primarily on the effective current status and future prospects of that one agreement.

The 1972 ABM Treaty formed an integral part of the SALT negotiating process from 1969 to 1979, both before and after it was signed. It has been generally regarded as the essential basis for the entire process of 'arms control' between the superpowers. And arms control, the achievement of negotiated restraints and partial halts to the arms race, has been seen by many people, including successive governments of the United States and the Soviet Union, as an essential first step towards nuclear and general disarmament. Their view derives from two basic principles. The first is that neither side should seek to gain a decisive advantage over its opponent, either by unilateral actions or at the negotiating table.

The second, that unless the arms race is prevented from taking on new forms, no real start can be made with significant major reductions in existing forces. Of course, new forms of potential warfare could conceivably make some weapons less important, even 'obsolete'. But in so far as such developments were linked to a stubborn refusal to accept the idea of co-existence, they would merely re-shape and accelerate the arms race, instead of slowing it down and seeking political solutions to its fundamental causes. They could not make peace any more secure, and might well make it less so.

Space weapons generally involve arms control issues rather than disarmament ones. For one thing, there are very few space weapons actually in existence today, requiring to be dismantled. Also, until such important arms control issues as the threat of an unchecked arms race in space can be resolved, large-scale negotiated nuclear disarmament seems unlikely to occur. But the distinction should not be exaggerated. Arms control is the attempt to slow or reverse the arms race by prevention, whereas disarmament involves treating those serious symptoms of the disease that have already been allowed to take hold. Neither has a strikingly better record of success than the other.[1] Both share a single view of what would constitute better health in the international body politic.

The claim that the SDI may be self-defeating focuses on the probability of interactions between genuine arms control and disarmament. It is argued that by damaging the more immediate, less superficially 'worthwhile' processes of arms control – in this case limitations on still largely undeployed space weapons – the SDI will prevent just the large-scale disarmament, or 'deep cuts', that would be needed if a strategic BMD deployment were ever to become effective:

Too much is at stake in the present tense state of U.S.–Soviet military and political relations to make it prudent to undermine remaining elements of stability and common understanding before we have something in which we can have more confidence as a replacement. Deterrence (as distinct from strategic superiority and its use as a threatening instrument of political and military policy) has been basic to stability and avoidance of nuclear war to date. An essential guarantee of deterrence, as recognized and defined in the ABM Treaty, has been reciprocal limitation of ABM defenses. . . The role of the [ABM] Treaty therefore needs to be understood and supported, by positive rather than merely passive adherence. (Drell et al., 1984: 87)

There is, however, another point of view. Supporters of the SDI expect it to help bring about large measures of offensive disarmament and diminish the risk of nuclear war. The critics reply that, if effective negotiations are hoped for, it would be safer and simpler to proceed with the reduction of nuclear stockpiles by more direct methods:

> If the President really wants to eliminate nuclear weapons from the face of the earth, why not propose just that to the Soviet Union? . . . If it is desired to limit the damage to the United States should deterrence fail, and reduce the risk of nuclear winter, then at the very least propose reductions to small stockpiles. (Freedman, 1985: 46)

To resolve this debate it is necessary to understand the Treaty itself. But the importance of doing so can best be appreciated after surveying the Reagan administration's pre-1983 policy on the Treaty, and its general approach to arms control in space.

The Reagan Administration and the Treaty, 1981–83

The new administration's lack of enthusiasm for the Treaty was reflected in a change of wording in the first set of *Arms Control Impact Statements* (ACIS), from the Reagan Arms Control and Disarmament agency, by comparison with the last to have been prepared by its predecessor:

> The United States fully supports the ABM Treaty. (US ACDA 1981: 193)

> The United States continues to be a party to the ABM Treaty. (US ACDA 1982: 139)

The shift from 'positive' to 'merely passive adherence' was also made evident from the outset in other ways:

> In its drive to improve the nation's strategic posture, the Reagan administration will consider bulding an antiballistics [sic] missile defense system . . . Caspar Weinberger, the Secretary of Defense-designate, said today.

Mr Weinberger said that extension beyond 1982 of the antiballistics missile treaty . . . was 'not automatic'. . . . the new administration might want to build an antiballistics missile system larger than the two [sic] sites permitted under the present treaty.(*New York Times*, 16 January 1981)[2]

I have difficulty answering the question who would be particularly harmed if the treaty were abrogated . . . I do think that there is a feeling that the ABM Treaty is in effect and if it were suddenly abrogated it might have some destabilizing effect. But I do not really see that that would be a correct assessment. (Weinberger, in S. Foreign Relations, 1981: 34)

Other evidence suggests that, in its first two years, parts of the administration moved quickly and in more practical ways to reject the long-accepted 'restrictive' view of their Treaty obligations. As the commander of the then US Army Ballistic Missile Defense Organization, Major General Tate, explained:

We were limited in the scope of the R&D that we were allowed to carry out, part of it by congressional decree, from 1974 to 1980. For example, we were not allowed to prototype during those years. That restriction has been removed. We are now moving out and demonstrating the kind of hardware that is required if we are going to be able to reach an initial operating capability any time in the near term. (H. Armed Services, 1982: 125)

However, before March 1983 it was commoner for leading members of the administration to interpret the Treaty along traditional lines, and to resent it accordingly. Ever since it was signed, the standard American interpretation of the Treaty had been that given by the legal adviser to the US SALT 1 negotiating team:

[C]onstraints on future ABM systems . . . are an attempt to ensure a long-term, effective limitation on strategic defensive systems in an age of changing technology . . . failure to include a broad proscription against future devices might have led to a significant effort by either or both the U.S. and U.S.S.R. to develop esoteric ABM systems. (Rhinelander, 1974: 128–9)

In the early 1980s US supporters of BMD accepted that the Treaty was a major obstacle:

> Even the development, as well as the deployment, of exotic systems
> such as a laser BAMBI [BAllistic Missile Boost Intercept] system
> are prohibited. (Brennan, 1980: 31)

And not only novel ABM technologies were thwarted by the
Treaty, according to Assistant Defense Secretary James Wade in
March 1981:

> [N]either of the principal BMD concepts for ICBM defense . . .
> – low altitude defense or the overlay defense – can be fully
> developed under the ABM Treaty. (S. Armed Services, 1981: 4113)

Indeed, for the first two Reagan years the US Army BMD
Organization under General Tate was linked with Assistant
Defense Secretary Richard Perle and others in a virtual campaign
to secure some sort of ABM deployment, almost no matter what,
and thereby to initiate a US policy turn away from MAD. This was
seen as requiring an early decision against 'renewing' the Treaty at
its next quinquennial review,[3] due to be conducted in Geneva by
the joint US-Soviet Standing Consultative Commission (SCC) for
the SALT agreements, before the end of 1982:

> With SALT 2 defunct, on prudent estimation, the sanctity of the
> ABM Treaty has diminished dramatically in very recent months.
> (Gray, 1981: 62)

> I am sorry to say that [the ABM Treaty] does not expire. That is one
> of its many defects . . . I would hope that were we to conclude that
> the only way we could defend our own strategic forces was by
> deploying defense, we would not hesitate to renegotiate the treaty
> and failing Soviet acquiescence . . . I would hope that we would
> abrogate the treaty. (Richard Perle, in Paine, 1982)

What lay behind such negative attitudes and practices towards
the Treaty was a political conviction that the United States had
almost no security interests in common with the Soviet Union, and
therefore that there could be no useful agreements between them
about arms limitation and mutual deterrence. Such 'unilateralist'
aspirations for deterrence by US superiority (p. 161) have
determined American policy towards the Soviet Union for many
years, and ever more openly since the late 1970s. Their influence
can be traced in the rise of the 'countervailing strategy', the

acceleration of nuclear war-fighting preparations with Carter's Presidential Directive 59, and the refusal of the US Congress to ratify the SALT 2 agreement.

Another sign of the times in early 1981 was the USAF Academy's conference on military space doctrine, which strongly influenced the Air Force's subsequent Space Plan, and at which:

> The major concern expressed in the legal area [was] the potential for overly rigid laws restricting the development of necessary operational capabilities. (Viotti et al., 1981: 127)

After 23 March 1983 rejections of the ABM Treaty by senior members of the administration and by others throughout the US defence community became franker and more frequent. The Defense Secretary himself revealed that: 'I've never been a proponent of the ABM Treaty' (*ABC Television*, 8 April 1984). As he explained to the Senate Armed Services Committee:

> The idea that we would be perfectly safe if we had no defense at all was one that was quite prevalent. As a matter of fact, it still represents, I suppose, the conventional wisdom.
>
> It was based on the assumption that both sides, Soviet and United States, would stay equal in offensive power and neither would do anything about acquiring defensive power.
>
> That turned out to be a rather tragic mistake.

The retrospective, dissembling support for offensive parity, and the faith that security in the age of nuclear weapons can still be about 'perfect safety', have been noted elsewhere (pp. 75, 189). When asked by Senator Cohen whether he had not implied it was a tragic mistake for the United States to have adopted the ABM Treaty, Mr Weinberger dodged:

> No. I said it was a tragic mistake to take the view that both sides were perfectly safe if neither had any kind of credible deterrent and then not pay attention to the fact that the Soviets were paying no attention to that at all.

But whereas the United States should never depend on Soviet good intentions, the converse situation would be perfectly satisfactory:

If they should achieve an effective system of their own and we did not have one, so that our missiles were rendered impotent and theirs were not, that is not the kind of world I think any of us would want . . .
[But i]f we can get a system which is effective and which we know can render their weapons impotent, we would be back in a situation we were in, for example, when we were the only nation with the nuclear weapon and we did not threaten others with it. (S. Armed Services, 1984: 55, 76, 89)

Outside the Treaty – ATBMs and Asats

Anti-Tactical Ballistic Missiles

Another important aspect of Reagan's early military space policy, besides its negative evaluation of the ABM Treaty, was a willingness to press ahead with missile defence and space weapon programmes in any areas not yet covered by arms control agreements, rather than to seek new agreements to limit or control such areas. Such programmes included asat development, and work on anti-missile defences to counter missiles with shorter 'theatre' ranges, to which the ABM Treaty does not apply.

There are two key arms control issues for the latter type of weapons, anti-tactical ballistic missiles (ATBMs). In the first place, the flight characteristics of longer-range 'tactical' or 'theatre' missiles, such as the SS-20 and the Pershing 2, are similar to those of submarine-launched missiles (SLBMs). Because the latter have normally been placed in the 'strategic' category for arms control purposes, ATBMs capable of countering them may either infringe the Treaty or at least give the impression of doing so (figure 9.1). This would also affect the question of the transfer of ABM systems or components to other states, which is forbidden by the Treaty for those capable of countering '*strategic* ballistic missiles' – an issue made increasingly pertinent by US endeavours to involve allies in the SDI (p. 248). Secondly, even if the development and deployment of ATBMs could be so managed as to avoid any direct breach of the Treaty, they might provide, or appear to provide, a significant capability on which to base later deployments that would do so. Thus they might be a step along the road to developing a nationwide strategic anti-missile capability. But providing such a 'base' for territorial ABM deployments is forbidden by the Treaty.

The United States has voiced repeated concerns over Soviet air-defence systems which might undermine the Treaty by this route. A formal complaint that Soviet air-defence systems were being tested 'in an ABM mode' in concurrent trials of air-defence and ABM systems at the same test range was resolved satisfactorily, at a meeting of the SCC in 1978, with an agreement defining what would or would not be permissible in future. A further draft understanding on the question of concurrent operation of such systems was reached in the SCC in 1982, but Reagan officials thrice backed out of terms they had previously agreed to. Agreement was nonetheless almost achieved in 1982–3, until the US team insisted that if ever, due to special circumstances, any concurrent use of air-defence radars occurred near an ABM test site, notification would have to be given to the other party within ten days, instead of at the next SCC meeting, as the Soviet Union would have been ready to undertake (Longstreth et al., 1985: 59).

It is also relevant to observe that restrictions of this sort are a greater burden to the Soviet Union, because of their long-standing strategic preference for very extensive and constantly modernized anti-aircraft missile systems (SAMs), an option that was for decades out of military favour in the United States.

On the American side, an ATBM role has been envisaged for the Patriot anti-aircraft missile system since its earliest conception over twenty years ago. After the ABM Treaty was signed, the Congress decided that the anti-missile defence software should be removed from the system, lest it infringe the Treaty by developing and testing a BMD capability sufficient to intercept SLBMs.

At the end of the 1970s, however, some US interest in an anti-missile version of Patriot was revived, on the basis that it might provide an alternative means to 'offset' the Soviet Union's SS-20 deployments, rather than 'counter-deployments' of NATO Intermediate-range Nuclear Forces (INF) in the shape of Pershing 2s and Tomahawk cruise missiles. There was also discussion of the inevitable 'softness' of such mobile INF, and of the possibility of using Patriot, which is also mobile, in a complementary BMD role alongside them. (This may have seemed analogous, at the theatre level, to the point-defence BMD systems being proposed in some quarters at the time as part of an MX deployment in the United States.) Pentagon science officials, such as James Wade, reported that initial studies were in hand. Some European interest in this option was expressed, principally from West Germany.

After the President's March 1983 speech there were indications that an ATBM version of Patriot was a strong contender for the European leg of the SDI, to provide a capability for non-nuclear interception of theatre-range ballistic missiles, whether these were conventional or nuclear-armed. Wade and others at the Pentagon continued to support it. The Hoffman Panel, reporting in October 1983, identified ATBM as one of the most promising areas in the entire field for early 'technology demonstrations' and subsequent deployments, which was simply another way of saying that the option had made useful technical and political progress in the late 1970s and early 1980s, before the SDI came along (Hoffman, 1983: 2–3). And specialist press reports spoke of enthusiasm for the idea within the US Army's BMD Organization (since absorbed into the SDIO).

By the end of 1984, however, the Army had concluded an 18-month study into the ATBM question, the results of which were decidedly lukewarm. A small-scale programme was proposed, to begin in 1986, but the chances of eventually producing an effective non-nuclear ATBM version of Patriot were not put high.[4] James Wade, by then Acting Head of all Pentagon research and development, rejected the Army's modest ATBM proposal and told them to think again. The House Armed Services Committee, on the other hand, saw little point in developing ATBMs and instead recommended offensive tactical counter-force systems, 'to negate ground-based tactical missile launchers' (*Defense Week*, 10 December 1984). During 1985 the Army made it clear that it was willing to proceed with ATBM studies on two conditions. First, that a capability to shoot down tactical ballistic and cruise missiles in flight should be seen as a subordinate part of an overall ATM (anti-tactical missile) system that would hinge on offensive 'counterforce strategies' aimed at 'missile launch and control sites'. And second, that the Army should be allowed a relatively free hand to proceed with such studies, independent of the SDIO. The latter condition was finally met in mid-December 1985 (*Defense Week*, 23 December 1985).[5]

Initial deployments of the anti-aircraft version of Patriot began in West Germany in February 1985. The prospect of an anti-missile upgrade or of other US ATBM programmes appears closely tied up with the intra-alliance politics of the SDI as a whole (p. 255), which may become clearer as more details of the SDI's 'European Architecture Study' become available.

Asat Negotiations

Soviet-American discussions about a possible treaty to ban or limit asats took place in the late 1970s, but were broken off by the United States in response to the Soviet intervention in Afghanistan at the end of 1979. The talks were in any case bogged down, for three reasons. The Soviet delegation agreed with many US analysts in seeing the Space Shuttle programme as primarily a military development (Reich, 1984: 40). The Pentagon was as usual reluctant to accept limitations on a feasible weapons technology which it had not finished developing. And the star of those in Washington who believed in a positive military requirement for asats was very much in the ascendant, even before Reagan's election to the Presidency.

In the pre-SDI years of the Reagan administration, however, the question of an asat treaty dominated US policy debates about arms control in space. At Senate hearings in 1982 the then director of the ACDA, Eugene Rostow, echoed that summer's Fact Sheet on Space Policy by spelling out the positive military requirement for weapons capable of 'countering the space components of threats to US forces' and by warning of 'the threat to our national security from advances in Soviet space programmes' (S. Foreign Relations, 1982: 11). Not for the first time, the administration's view was broadly in line with Colin Gray's:

[I]t is critically important that U.S. policy makers . . . disabuse themselves of the notion that outer space will be, or can be, a sanctuary . . . in the event of a general war, the superpowers will fight in and for the control of space as they will fight everywhere else . . . (1983: 105)

The strategic doctrine of space control leaves little scope for arms control. If space could not in fact be a sanctuary, then arms control agreements which claimed to make it so would be worse than useless. But the hard-liners' antipathy or scepticism towards any role for arms control in US strategic policy has a less obvious corollary. It implies that several other, *existing* arms limitation agreements with respect to weapons in space are mistaken. These include the Limited Test Ban Treaty's prohibition on nuclear test explosions in space, and the Outer Space Treaty's ban on any military uses of the Moon and the planets, and on the placing of 'weapons of mass destruction' in Earth orbit. The space-control approach also implies that the legal protection for 'national technical means' of treaty verification, first established by the SALT 1 agreements, is little more than paper-thin.

Congress, however, was unwilling to accept the full logic of that approach, which would mean adopting standards of international behaviour based on little more than the use or threat of brute force. In 1983 it amended the Defense Budget so as to permit testing of the F-15 asat only if the President could give an assurance that his administration was actually negotiating with the Soviet Union for a ban or limitations on such weapons. The programme was in fact delayed for a time. But the Presidential reports on asat policy of March 1984 and August 1985 simply juxtaposed assurances that attempts were being made 'to negotiate with the Soviet Union a mutual and verifiable agreement with the strictest possible limitations on anti-satellite weapons consistent with the national security interests of the United States' (US WH, 1984), alongside stern declarations that no such agreement could ever be regarded as being in the US interest, and that no sufficiently verifiable agreement could be devised anyway.

The administration's critics responded by arguing, firstly, that failure to control asats would be more harmful to US interests than to the Soviet Union, because of America's greater dependence on satellites during peacetime, crisis, or war; secondly, that clear and verifiable asat controls could be agreed now, if they included prohibitions on the easily detectable development and testing phases of such programmes; and thirdly, that any nominal Soviet 'advantage' derived from its existing cumbersome, obsolescent and minimal anti-satellite system was far too small to give the United States any reason not to push for such a treaty (UCS, 1984: 213).

Not long after the President's first negative Report to Congress on the prospects for asat limitations, there were press reports of a secret Senate briefing at which the administration had seized the initiative with claims that there were as many as four Soviet anti-satellite systems, some or all of which were already actually being used against US satellites (*New York Post* & *New York Times*, 13 June 1984). But it is doubtful whether the alleged activity, if it takes place, could reasonably be described as using 'weapons' against US satellites.[6]

With these and other pressures in an election year, the administration succeeded in getting the Congressional restraints upon its asat programme eased far enough for the conditions to be satisfied formally, whilst at the same time proceeding in practice only with the weapons programme, not with the negotiations. At the end of 1984 the focus of public attention moved away from the asat question, for several reasons. The agreement to re-open US-Soviet disarmament talks in a three-part structure, to cover long-range and intermediate-range missiles, and military uses of space, placed the SDI, whose programmes and budgets were already rolling, at the top of the arms control agenda, rather than asats. The Soviet Union seemed also to have moved its interest in a separate asat treaty onto the back burner for a time. And no doubt the initial technical successes of the US asat programme (p. 50) made it harder to put effective brakes on it in Congress. Besides all which:

> [T]here is a . . . more important reason that is not mentioned in the Administration's public position on ASAT arms control: a ban on ASAT would prevent many experiments essential to the SDI research and development effort. (Schneiter, 1985: 219)

The technical and strategic interaction between asat and ABM technologies has already been explained. Dr Keyworth, for example, until December 1985 the President's Science Adviser, openly recognized that a device 'powerful enough to act as an antisatellite weapon' would be a suitable early goal for the laser programme within the SDI (*New York Times*, 7 March 1985). And commentators on the Soviet Union's pre-Summit call for early negotiation of an asat treaty, presented by Ambassador Karpov as something that could be achieved independently of any agreement about the SDI, were quick to point out the problem this would involve. Either an asat treaty would in fact provide a back door to restraining the SDI, if the measures against asat development were effective, or else, if it were loose enough to permit work on corresponding ABM technologies to continue, the treaty would not place any worthwhile restraint on asats either. As independent experts in this area of arms control had long been arguing:

> [I]t is no use banning one of these [categories of] systems and letting the other go ahead. If there were only a ban on ASAT systems, the result would probably be that the technologies which were being developed for ASAT purposes would acquire a new label: we would be told that they were being developed for anti-ballistic missile purposes. It follows that any action against the development or deployment of anti-satellite systems should be accompanied by equivalent action against anti-ballistic missile systems. (Jasani, 1984: 40)

And vice versa, of course. It was no surprise, therefore, that the first 'live' test of the F-15 asat was presented by the US administration as 'demonstrating' the feasibility of similar SDI technologies for missile interception.

Nevertheless, in December 1985 Congress responded to the Soviet overtures by reimposing its ban on full-scale asat testing, despite the fact that two specially-instrumented target satellites had just been placed in orbit. The Pentagon reacted by announcing that it saw the ban as applying only to tests conducted against actual targets, not to shots aimed at mathematical points in space (*Daily Telegraph*, 29 December 1985).

Interpreting the ABM Treaty

Traditional Readings

The Anti-Ballistic Missile Treaty (appendix 1) agreed between the United States and the Soviet Union in May 1972, and amended in 1974, permits deployment of only 100 fixed land-based single-warhead ABM interceptor missiles on each side, within one circular area with a radius of 150 kilometres. This may be centred either on the national capital or at an ICBM silos field.[7] The Treaty also forbids the provision of a 'base' for a broad national anti-missile defence, or the deployment of ABM defence for a territorial region, except to the extent specified within it ((I.2)).[7] As stated by the parties at the time, this arrangement was intended to provide a lasting 'guarantee of deterrence' (in the OTA's phrase) through mutual and permanent abstention from deployment, or advanced development, of ABM systems 'for a defense of the territory of [each side's] country' ((I.1)).

The permitted deployments are possible to monitor from space, but too small to be of strategic value, even as force defences. The core of the Treaty therefore lies in the way it protects each party's 'nuclear deterrent' against any future surprise attempt, by the other, at a strategically devastating 'break-out' from the situation of mutual vulnerability. Its restrictions on potentially large-scale area defence systems, whose deployment would amount to scrapping the Treaty, have always been understood as applying to them *in their development stage*, at which point they can be monitored independently by either party's 'national technical means of verification' (mainly, surveillance satellites) ((XII.1)). (Had each side been free to *develop* new and possibly more effective ABM systems until it was all set to start *deploying* them, both would probably have put major efforts into doing so, and sooner or later a break-out from the Treaty would probably have resulted.)

No ABM systems or components capable of giving broad defensive coverage may be developed, tested or deployed, whether sea-, air-, space-, or mobile land-based ((V.1)). Fixed land-based systems capable of launching numbers of interceptors at once, or in rapid succession, are also banned ((V.2)). Other missile systems and their components may not be given anti-missile capabilities, or be tested 'in an ABM mode' ((VI.a)). And no ABM systems or components may be deployed outside each national territory or transferred to other states ((IX)).

Several provisions were made for the future of the Treaty. The Standing Consultative Commission was established to monitor its implementation and to deal in diplomatic secrecy with any issues of compliance or suggestions for increasing the Treaty's viability, including proposed amendments ((XIII)). The Treaty is reviewed at intervals of five years ((XIV.2)). It is of unlimited duration, but either party may abrogate it unilaterally at six months' notice ((XV)).[8]

The word 'research' is not found in the Treaty or its supporting documents. But not only is it impossible to monitor activity inside laboratories satisfactorily; there is also the fact that modernization and replacement of the permitted, fixed land-based interceptors is allowed ((VII)), thereby legalizing both laboratory research, and subsequent development and testing, for novel BMD technologies in that role. But basic research for this purpose, though legal, may well have applications that extend beyond the fixed, land-based ABM mode – just because it is basic. The line is therefore drawn, by implication, at the more verifiable level of practical develop-ment work, including field-testing, for the kind of banned systems that would be needed for any effective territorial defence. (For instance, the space-based components of boost-phase ABMs.)

Strictly speaking, this does not mean that research into BMD systems other than the fixed, land-based, single-shot variety is positively 'permitted'. What matters is that *all development* of such systems is forbidden, that research only becomes part of develop-ment directed towards one or other possible basing mode at a certain stage, depending on the technology in question, and that only verifiable development can be of any relevance for practical purposes.[9]

Broadly protective defences might be pursued by developing ABMs based on other technologies than the 'launchers', 'interceptors' and 'radars' which the Treaty took as its standard example of an ABM system and its components ((II.1a)). After more than ten years of research into possible military applications of lasers, the parties were well aware at the time that such technologies might some day become available. To prevent its circumvention on this central issue, therefore, the Treaty needed to clarify the applicability of its provisions to any new technologies that might not simply correspond to the examples of components 'currently' employed in the ABM systems of its day.[10]

This issue has become so bound up with the slightly bogus one of what level of research effort is 'permitted by the Treaty', that they cannot really be considered separately. Agreed Statement D was proposed and drafted by the US delegation, accepted after some initial reluctance by the Soviet team, and initialled by Heads of Delegations at the same time as the Treaty was signed by Heads of State. By it the parties agree that:

> in the event ABM systems based on other physical principles and including components capable of substituting for ABM interceptor missiles, ABM launchers, or ABM radars are created in the future, specific limitations on such systems and their components would be subject to discussion . . . and agreement . . .

However, an unfortunate linguistic ambivalence in this statement has caused a problem for the interpretation of the Treaty. The word 'create' is a literal translation of the Russian word which, with its cognates, is used in the main Treaty text to correspond to the English 'develop'. In ordinary English usage, however, to 'create' something implies having completed all that is necessary to bring it into being, whereas to 'develop' it does not imply completion, especially in this context, in which 'development' is associated with 'research' and contrasted with 'testing' and 'deployment'. In short, to 'develop' a military technology is, in English, to try to make it work, whereas to 'create' one is to succeed.[11]

When reading the Treaty alongside this statement in English, therefore, it is possible to find an inconsistency between the Treaty's ban on developing any systems that are not fixed land-based ones ((V.1)), and the statement's reference to the possibility of systems based on novel technology having been, at some future date, 'created'. For such systems are virtually certain either to require or at least to make feasible the forbidden types of basing, as was well understood at the time. This weakness in the Treaty was pointed out as soon as it had been signed (Clemens, 1973: 116). It might be likened to having one notice saying 'Keep Off the Grass' next to another telling people not to wear spiked shoes when crossing the said grass. But since few US arms control experts were then prepared to bet that the Treaty would remain effective for longer than five years, despite its optimistically open-ended character, there was perhaps little interest in making unnecessary professional 'waves' by dissecting all the faults in its wording (Clemens, 1973: 29–33).

The usual, more supportive approach to this problem in the Treaty derives from articles 31 and 32 of the authoritative Vienna Convention on the Law of Treaties, which establish the principle that treaties are to be understood in accordance with their wording, contents, and objective purpose, taken as a whole. Since both parties have agreed to regard the Agreed Statements and Common Understandings as, in effect, additions to the Treaty proper, statement D has to be read within the Treaty – the Treaty cannot be interpreted solely from statement D. On this view, which has been generally accepted until quite recently, statement D was added to the Treaty because the parties were aware that new types of fixed, land-based ABM systems might be devised, under the modernization provision of article VII, for which the *deployment* provisions of article III, specifically referred to in the statement, might come to seem inadequate. Since the statement says nothing to loosen the restrictions of article V against the development of any systems other than fixed, land-based ones, it should be read in the context of those restrictions, not as permitting such a massive exception as virtually to negate them. In short, statement D assumed that technical novelty, over and above modernization, was permitted solely in respect of fixed, land-based systems. And its force was simply that of accepting additional obligations in respect of their deployment, if they departed from 'traditional' ABM technologies.

The verifiable line that needs to be drawn between the sort of basic research which has to be tolerated, and levels of development which are prohibited because they can be independently monitored by either party, was carefully explained at the time by the chief US negotiator in SALT 1, Ambassador Gerard Smith:

The obligation not to develop such systems, devices or warheads would be applicable only to that stage of development which follows laboratory development and testing. The prohibitions on development contained in the ABM Treaty would start at that part of the development process where field testing is initiated on either a prototype or a bread-board model [mock-up]. It was understood by both sides that the prohibition on development applies to activities involved after a component moves from the laboratory development and testing stage to the field testing stage, wherever performed. (S. Armed Services, 1972: 377)

More recently, Smith has elaborated on his understanding of the matter as follows:

I do not think that's a loophole [statement D]. You ought to look at Article I, where the parties agree not to deploy ABM systems . . . That is the heart of the treaty. It was agreed that, yes, you could do research and development on so-called exotic systems . . . but we recognized that space-based systems were especially dangerous because that is moving toward a nation-wide defense system, and, therefore, we put them under more specific constraints in Article Five, so you can't move to development. For [a fixed] land-based [system], however, you could move to development.

[And on the degree of permission extended to 'other physical principles']: The only fair interpretation is that any exotic system is banned. If it is developed and if it is not space-based, then before it is deployed both sides would have to agree that was permitted and that would be an amendment to the treaty. (*Washington Post*, 3 April 1984)

Successive US governments endorsed this view of the Treaty during the 1970s, through their Arms Control and Disarmament Agency (ACDA). Take, for example, the ACDA's *Arms Control Impact Statements* from early 1978:

The ABM Treaty represented a decision on the part of both the United States and the Soviet Union to avoid a massive arms race in ballistic missile defenses that in the end could not prevent destruction of both societies in a nuclear attack. . .

Thus PBWs [particle beam weapons] used for BMD which are fixed land-based could be developed and tested but not deployed without amendment of the ABM Treaty, and the development, testing and deployment of such systems which are other than fixed land-based is prohibited by Article V of the treaty. (US ACDA 1978: 6, 231 note 3)

Until the end of 1985, the Reagan administration had always accepted the legal interpretation established by its predecessors, if not every political gloss they placed upon it. Thus its FY 1985 ACIS repeated that:

> The ABM Treaty prohibition on development, testing and deployment of space-based ABM systems or components applies to directed-energy technology or any other technology used for this purpose. Thus, when such directed-energy programs enter the field-testing phase, they become constrained by these ABM treaty obligations. (US ACDA, 1984: 252)

And the position was restated as recently as April 1985 (US ACDA, 1985: 37).

That this is also the Soviet understanding of the Treaty was recently confirmed by General Secretary Gorbachev:

> What we have in mind [as illegitimate] is not research in fundamental science . . . What we mean is the designing stage . . . And when they start building models or mock-ups or test samples, when they hold field tests, now that is something – when it goes over to the designing stage – that is something that can be verified. (In Fischer, 1985)

Reinterpreting the Treaty for the SDI

The 'Components' Gambit

Until recently, the Reagan administration appeared to be continuing with its pre-SDI position, according to which most significant ABM programmes would eventually breach the Treaty unless an agreement to revise it could be negotiated. This was because they would be directed towards forbidden basing modes, probably associated with novel technologies. But it was argued that the SDI would keep within the limits of the Treaty till about 1990, because it would only be developing technologies that might provide 'adjuncts' or 'sub-components' for ABM systems. As for the forbidden stage of development and testing for full system components and, some day, complete systems, that, it was claimed, was still some way off:[12]

> These near-term technology research projects and tasks are well defined and clearly [Treaty-] compliant. The major technology

220 *Space Weapons*

experiments to be conducted in later years are being planned to be fully compliant. These experiments are designed to demonstrate technical feasibility, they can be established without involving ABM systems or components or devices with their capabilities. Thus, compliant space-based as well as fixed land-based experiments are possible. (US DoD, 1985b: B-2)

This approach is used, in effect, to place all investigation into the 'feasibility' of strategic defence systems under the heading of 'permitted research', no matter how practical and realistic it becomes. Only the final, full-scale engineering development which immediately precedes production and deployment of a weapon system is seen as falling inside the category of banned 'development'. Further extracts from the Pentagon's report can usefully illustrate the legitimation process:

[T]ypes of activity . . . permitted in compliance with the ABM Treaty . . . *Category 2 – 'Field Testing' – of Devices that are not ABM Components or Prototypes of ABM Components* . . .

The . . . Acquisition, Tracking and Pointing (ATP) demonstration program . . . If conducted these experiments will use technologies which are only part of the set of technologies ultimately required for an ABM component . . .

The Ground-Based Laser Uplink experiment . . . The testing mode and capabilities are below the power level and beam quality required for a ground-based laser ABM weapon . . .

The Boost Surveillance and Tracking System . . . will not be a prototype of an ABM component . . . cannot substitute for an ABM component . . .

The Airborne Optical Adjunct (AOA) Experiment will demonstrate . . . technical feasibility . . . The . . . device . . . will not be capable of substituting for an ABM component . . . (US DoD, 1985b: B-4,6,7)

And so on.

However, the Treaty does not define the term 'component'. Launchers, interceptors and radars are certainly listed under that heading ((II.1)), but nothing is said to indicate that these were the *only* admissible types of component, even in 1972. For instance, the battle-management computers of the day, with their necessary software, might not obviously have fallen under one particular category out of the three. But it would have been difficult to see them as anything but ABM system components.

It is sometimes argued in support of the SDI's legality that Agreed Statement D should be read as implying that something is only a 'component' if it can perform the entire mission, within an ABM system, that was previously met by either the launcher, or the interceptor missile, or the radar. But this seems far-fetched. The Treaty clearly envisaged that the discrimination and tracking of targets, together with the guidance of interceptors to destroy them, might require a complex network of radars, each one of which would be a 'component'. No one such component or type of component would fulfil the *entire* role of 'ABM radar'. In the same way, there is no implication in the Treaty that only one type of interceptor missile could be deployed. Indeed, a system combining two different types of interceptor was being constructed on the American side at the time, such that no single missile-type would provide *the* interceptor component. These considerations suggest that something can be a component, in the terms of the Treaty, without having to be capable of executing *all* the functions of one of the three main classes of component listed in Article II. And in any case:

> The fundamental idea behind the definition of a component in article II consists in regarding all such apparatus as may be necessary for the acquisition, tracking and destruction of targets as 'components' of an ABM system. (Fischer, 1985)

The argument that SDI programmes will not test or seek to develop any fully functional ABM systems or their components, only 'experimental' technologies and 'adjuncts' that might eventually be the basis for such things, also fails to respect the overall objective purpose of the Treaty. Article V forbids the development of all systems or components other than the fixed, land-based variety. Development is a process which starts with something which is still far from good enough to perform the task envisaged, and tries to make it good enough, often over many years. If development is banned, then every activity which makes up development is banned – the entire endeavour, after the initial laboratory studies, to achieve the technology in question. Work done on the first day of such development would be just as illegal as work done on the last, many years later. The Treaty does not simply start to bite the moment before the very last connection on the last 'adjunct' in the last 'experimental demonstration' is about to be soldered.

Even if it were supposed that, for some unknown reason (or for some that have already been suggested!), no such programme as the SDI could ever actually 'succeed', that would be irrelevant to the Treaty, which bans the attempt as such, irrespective of whether or not it ever leads to an actual large-scale ABM deployment. It represents a joint US-Soviet decision never seriously to investigate, in respect of any theoretically comprehensive ABM systems sketched out in the laboratory, whether they might or might not actually work. The SDIO, on the other hand, is officially charged by its Charter to proceed in exactly the opposite direction, towards 'The ultimate goal of . . . eliminat[ing] the threat posed by nuclear ballistic missiles' (chapter 6, note 8). As such it has surely already begun to infringe the Treaty's ban on such development, even if for technical, political, or financial reasons it may never bear practical military fruit.

Another way of putting this is to ask whether the Reagan administration still accepts the Treaty's ban on providing a 'base' for a nationwide anti-missile defence, and whether it still stands by the preamble, which includes this declaration:

> [E]ffective measures to limit anti-ballistic missile systems would be a substantial factor in curbing the race in strategic offensive arms and would lead to a decrease in the risk of outbreak of war involving nuclear weapons . . .

If the answer to both questions is affirmative, how can the SDI be justified? And if negative, to either or both, does that not amount to renouncing the Treaty itself, regardless of how permissive a construction can be cunningly extracted from the wording of any of its specific provisions?

Judge Sofaer and the Creation Myth

Suspicions that the US administration was less concerned to uphold the Treaty than to find a legal pretext for the SDI were sharply exacerbated in October 1985. Whereas only the applicability of the standard interpretation to SDI programmes had been queried up to that point,[13] that way of squaring the SDI with the Treaty then seemed to be abandoned in favour of a new one, seen by most experts as posing a far more radical threat to the legal status quo. It was argued that statement D's reference to the possible creation of systems based on new technologies made full-scale development and testing for such systems entirely legitimate, *whether or not they were directed towards one of the forbidden basing modes* ((V.1)). Without forewarning his colleagues, Mr Robert McFarlane, then National Security Adviser to the President, declared that:

> research involving new physical concepts, as well as testing, as well as development, indeed, are approved and authorized by the Treaty. ('Meet the Press', *NBC*, 6 October 1985).

White House officials then confirmed that this had become the 'fixed position' of the administration, after a decision by its Special Arms Control Group, chaired by Mr McFarlane, a few days earlier (*New York Times*, 12 October 1985).

McFarlane's political bombshell – the White House was immediately telephoned by irate allied leaders in Bonn, London and elsewhere – had been six months in the making. In May 1985 two of the Pentagon's most senior and powerful civilian 'hawks', Fred Iklé and Richard Perle, had hired a 35-year–old former Assistant District Attorney, Philip Kunsberg, with no experience in the field of arms control, to take a fresh look at the ABM Treaty and what it did or did not impose by way of limitations on BMD programmes.

Kunsberg's reports, on the treaty text in May and on the (classified) negotiating record in September, were prepared without consulting any of the US negotiating team responsible for drafting the ABM Treaty. He claimed that, contrary to the traditional interpretation, the Treaty did not place any restriction on development and testing for 'exotic' ABM systems, and that possibly their unilateral deployment was not forbidden either.

Once this minimal evaluation of the Treaty's legal import was offered to the rest of the administration, Secretary of State George Shultz placed it in the hands of his Department's new legal adviser, Abraham Sofaer, a former District Court judge with no more arms control experience than Kunsberg. Whilst denigrating the latter's findings as more opinion than analysis, Sofaer was basically supportive of the view of the Treaty which they represented. He argued that the Treaty could be read in two ways, either the traditional restrictive one explained above or, subject to minor reservations, the broader one which the Pentagon had floated. Of the two, said Sofaer, the second was legally the more convincing.

Judge Sofaer's 'broader' reading of the Treaty (1985) turns on two far from newly-discovered issues. Firstly, he argues that the part of article II.1 following the phrase 'currently consisting of' is not a set of *examples* of the sort of thing meant by the definition which precedes them, but a part of the definition itself '[f]or the purpose of this Treaty'. That would mean that the ban on developing ABM systems other than fixed land-based types, in article V.1, does not apply in any way to systems using novel technologies such as infra-red tracking, or directed-energy weapons. Second, he contends that statement D must have been added in part to qualify (his version of) the definition in article II. Seeing the restriction to fixed, land-based systems as holding only for 'traditional' anti-missile interceptor systems, he is then free to argue that statement D's reference to 'exotic' systems possibly having been 'created in the future' gives implicit licence for the development and testing of all such systems, not just fixed, land-based ones.

For Sofaer's argument to work, two quite unreasonable moves have to be countenanced. First, one has to disregard anything in the negotiating records that confirms the repeated statements of members of the US team, to the effect that the two sides deliberately decided to insert the word 'currently' in order to *prevent* the very reading of article II's definition on which Sofaer bases his permissive account of the Treaty. But the judge admitted that he had not been able to review 'every single step in the negotiating process' before preparing his account. Secondly, the Sofaer version of the Treaty has to ignore the explicit reminder within statement D, to the effect that it is addressed primarily to the *deployment* problem for possible novel systems, and not to modifying the Treaty's definition of ABM systems and components or its ban on developing any that were not of the fixed, land-based type ((II.1; V.1)). But in the published version of his explanation of the Treaty (1985), Judge Sofaer has provided no reasons, let alone good reasons, why he and his mentors should be allowed to take such liberties with the plain sense of its provisions.

It should now be clear why ex-Ambassador Smith immediately denounced Mr McFarlane's enormously permissive view of the Treaty as fit to make it 'a dead letter', and therefore tantamount to abrogation. Within the administration there followed a few days of

intense debate. Members of Congress joined in; allied governments continued their anxious representations. Secretary of State
Shultz was widely reported to have threatened resignation if
McFarlane's hard-line position were not modified before his
up-coming appearance at the North Atlantic Assembly and
subsequent consultation with NATO Foreign Secretaries. A
compromise position seems finally to have been suggested by
Ambassador Nitze, whose personal views on the McFarlane line
have not been made public, and then established by the
President's intervention. But even this 'compromise' represented a
further serious erosion in US respect for the Treaty. McFarlane's
broad and destructive reading was endorsed in principle, but as a
gracious act of supererogatory virtue it was decided to adhere to
the old, stricter interpretation for the time being, even though the
administration had now agreed that it was legally invalid. In Mr
Shultz's words:

> It is our view, based on a careful analysis of the Treaty text, and the
> negotiating record, that a broader interpretation of our authority is
> fully justified. This is, however, a moot point. Our Strategic
> Defense Initiative research program has been structured and, as the
> President has reaffirmed on Friday, will continue to be conducted,
> in accordance with a restrictive interpretation of the Treaty's
> obligations. (*New York Times*, 15 October 1985)

But as one 'Pentagon official deeply involved in the arms
negotiations' told the *Washington Post* (15 October 1985), 'it was
not clear to him that his side had lost the argument, pointing to the
fact that Shultz did not say in the announcement how long the
administration will continue to abide by the "restrictive interpretation" of the treaty' (see p. 274).

Soviet Proposals for Arms Control in Space

The Soviet Union has been fairly consistent in taking initiatives for
arms control in space, since well before the SDI came on the
scene. The breakdown of the asat negotiations in 1979 (p. 211)
initiated a worsening climate for bilateral talks, through the last
year of the Carter administration and the first two years of its
successor. With US arms control policy temporarily 'without form,
and void', the Russians turned to the United Nations, which in any
case they use for such initiatives more often than do the
Americans.

In August 1981 the Soviet representative presented a draft treaty 'On the Prohibition of the Stationing of Weapons of Any Kind in Outer Space' to the General Assembly. The treaty was in essence an extension of the Outer Space Treaty's existing ban on the orbiting of 'weapons of mass destruction' to cover 'weapons of any kind'. However, its provisions would not have banned ground-based asats such as the Soviet F-LV and the planned US F-15 systems. The draft contained a legal weakness. It implied that attacks on the space objects of other parties would be legitimate if those objects either were or contained weapons. The lack of any definition of 'weapon', however, or of any provision for arbitration as to what could or could not be deemed a weapon, left open the possibility that any government could seek to justify a destructive act in space, simply by declaring that it had decided in good faith that the object attacked either was or might be a weapon. Since some space-traversing objects are already being designed to function as weapons by direct collision, it is difficult to lay down an unambiguous criterion for deciding what is and is not a space weapon. Beside this apparent flaw in the text, there was also a reference to the military potential of 'reusable manned space vehicles' which would not have commended the draft to the Americans, just as their Shuttle programme was beginning to look like a major success. It was no surprise when US opposition prevented the 1981 Soviet draft treaty from being adopted as the basis for discussion at the Committee on Disarmament.

Two years later the Soviet Union tabled a new, more effectively worded draft treaty 'On the Prohibition of the Use or Threat of Force in Outer Space and from Space Against the Earth'. Instead of simply banning 'weapons' from space, it proposed a ban on all 'use or threat of force' either against any space objects, or by means of space objects if the latter were themselves 'instruments of destruction'. (The insertion of this phrase means the draft treaty would not exclude the many threats and uses of force in practice today, which utilize military satellites to 'enhance' destruction by Earth-based weapons rather than to cause it directly.) The draft lists detailed measures to secure this goal, including prohibitions on the testing, deployment or use in space of weapons for the destruction of objects in space or anywhere else. But the proposed undertaking 'not to test or use manned spacecraft for military, including anti-satellite, purposes' would be impossible to verify with certainty, and might lead to endless wrangles about what, if any, human activities in space could be guaranteed to be of absolutely no military value. Another important clause would require the destruction of existing asats, implying a relative 'sacrifice' by the Soviet Union, which had some asats deployed, however slight their military value, whereas the United States did not.

A more general problem for the Soviet 1983 draft is its central focus on the concept of 'the use of force', an expression from the UN Charter which lawyers have already struggled in vain to apply effectively to international disputes for over forty years. Many legal experts feel that arms control in space will need to be far more tightly focused and precisely worded if progress is ever to be made.

No political moves by the Soviet Union at the United Nations, on the issue of arms control in space or anywhere else, could overcome the coldness between the two superpowers a few months before the new US cruise and Pershing 2 missile deployments began in Western Europe. The missiles duly arrived and the Russians, wisely or not, walked out of the bilateral talks on strategic and medium-range missiles that winter.

As the 1984 presidential election drew near, however, the Soviet Union decided to try once more for talks on space weapons, the aspect of the developing strategic situation over which, perhaps, the least bad feeling and mistrust had been stirred up in recent years. This time the proposal, made on 19 June 1984, was for bilateral talks on 'preventing the militarization of space', an expression which was all too easily misunderstood in English, with implications for removing all military satellites that seemed to come almost thirty years too late (p. 37). But four days later National Security Adviser McFarlane replied – in a sort of affirmative.

Instead of simply accepting the Soviet invitation in principle, the United States sought to turn it to advantage by proposing that if talks were to resume their subject matter should not be confined to space matters, but should be broadened to take in the topics of offensive missile reductions as well. These had been shelved for several months, since the Soviet walk-out on the missiles issue, so the counter-proposal was unlikely to be accepted. There ensued an unseemly exchange of accusations, as each side sought to blame the other for the non-occurrence of the meeting, suggested for September 1984 in Vienna. What is certain is that the meeting did not in fact take place. But at the end of the year a new three-part bilateral negotiations package was agreed to, in which some degree of linkage between the three topics of long- and medium-range missiles, and space weapons, was after all established:

> The sides agree that the subject of the negotiations will be a complex of questions concerning space and nuclear arms, both strategic and intermediate range, with all the questions considered and resolved in their inter relationship. (Shultz–Gromyko communiqué, *New York Times*, 9 January 1985)

The Soviet Union did not relinquish its efforts at the United Nations in 1985. Instead, it gave them a more positive tone, in the form of its 16 August proposal on 'international co-operation in the peaceful exploitation of outer space *under conditions of its non-militarization*' (emphasis added).

Every aspect of Soviet policy suggests a genuine feeling of urgency about preventing a new round of the technological arms race from getting under way in space. The one question that remains is whether they place as much importance on banning ground-based space weapons as on banning those that might be deployed in space itself. Though this question should not be neglected, it is not in itself an obstacle to progress in East–West negotiations for arms control in space, whether at the UN or the bilateral talks in Geneva.

The November 1985 Summit

General Secretary Gorbachev and President Reagan held their two-day Summit meeting in Geneva on 20 and 21 November 1985. It was the first such meeting between rulers of the two superpowers for six years. Three formal agreements between the Soviet Union and the United States were either signed or endorsed at the meeting, providing for improved educational and cultural exchanges, for possible direct commercial flights between the two countries, and for one additional consulate in each other's territory. In the arms control and disarmament field, however, little was achieved beyond a recognition of certain negotiating principles, which were already binding on the two parties, at least in part.

The two sides recognized that any nuclear war 'could have catastrophic consequences' for the world, and that 'a nuclear war cannot be won and must never be fought.' The twenty-year-old obligation of both superpowers to negotiate seriously and urgently for real disarmament, under the 1968 Non-Proliferation Treaty, was referred to specifically in the communiqué, and seemed to underlie the two leaders' agreement 'to accelerate the work at [the bilateral] negotiations.' There was also a slight echo of the 1961 'McCloy-Zorin Agreement', in their pledge not to seek to achieve military superiority over one another.[14] But whereas the earlier agreement only binds the superpowers to abjure the pursuit of 'military advantage' *by means of* 'measures of general and complete disarmament' (§5), their undertaking at Geneva was completely general in scope.

On matters of detail, the Summit added little to positions already reached between the two sides elsewhere. The idea of a 50 per cent cut in nuclear weapons was formally recognized as one target for their ongoing bilateral negotiations, and the possibility of an interim agreement on limiting INF deployments was welcomed. Naturally, the overall goal that had been set for those negotiations ten unproductive months earlier, of preventing any arms race in space and ending it on earth, was reiterated.

Neither side had anything new to say about the issue of space weapons. For Mr Gorbachev, this remained the central question, and progress on it had been zero:

> SDI does involve taking weapons into outer space. Try to imagine what sort of a confrontation could arise if the nuclear charge were suddenly to break off from some sort of space ensemble . . . There's an attempt being made, say, to destroy it. And all sorts of computers will be at work. Political leaders at that point would have no way of exercising any control over it . . . We are prepared to engage in radical cuts in nuclear weapons provided that the door to unleashing an arms race in outer space be firmly slammed shut. (*The Guardian*, 22 November 1985)

> The Americans weren't very happy with our reasoning [on SDI], but frankly we failed to see theirs . . . I said to the President, you are not dealing with simple folk, and if you as President are determined to stand by SDI, then you realize that I in my position will have to do something about it. (*Daily Telegraph*, 22 November 1985)

President Reagan, on the other hand, had stood firmly by his support for SDI, as shown by his final statement to the press:

> Will we join together in sharply reducing offensive nuclear arms and moving to non-nuclear defensive strengths for systems to make this a safer world? The people of America, the Soviet Union and

throughout the world are ready to answer, Yes. (*The Guardian*, 22 November 1985)

And US Foreign Secretary Shultz later confirmed that:

The President feels as strongly as ever that the research programme [SDI] is essential. He insists on that. There was no give on that. (*The Guardian*, 22 November 1985)

Throughout the Summit the Soviet leader continued to insist on his country's established position, that no agreement on cuts in strategic-range offensive nuclear weapons could possibly be reached unless the SDI were restricted to the Treaty's permitted level of laboratory research, with no field-testing of any kind. This was seen by commentators both as a Soviet response to the possibility of a politically significant US 'advantage' of some sort, and as a move designed to put pressure on a President seeking some suitably peaceful achievement in his closing years of office. At the proposed Summits of 1986 and 1987, Mr Gorbachev may see an advantage in there being an agreement-in-readiness on reducing offensive systems, as an effective lever to use against the SDI with US public opinion and the NATO allies. He may also be working hard on preparing the Soviet negotiating position vis-a-vis the *next* President.

Willingness to talk of possible agreements is of course an established part of the political 'game of disarmament'. But so far there has been little evidence to support the US administration's contention that the SDI will do wonders for Soviet readiness to reach actual agreements on cutting offensive nuclear forces. And unfortunately for the administration, the political circumstances are such that the SDI may well be required to pass its test on the arms control side long before it could ever deliver significant anti-missile defences. The problem therefore remains that whereas the American side appears interested in deep cuts with and for the sake of building strategic defences, the Soviet Union, whatever may or may not be its interest in them for their own sake, is clearly determined to use the prospect of offensive cuts as its primary lever with which to prevent any major developments in BMD.

President Reagan's initial reluctance to breach the informal SALT 2 ceilings on offensive nuclear weapons, after the unratified agreement 'expired' at the end of 1985, was one small indicator that space and nuclear arms control may not be completely done for yet. The first round of the bilateral Geneva talks in 1986, however, showed little of the would-be constructive and collaborative approach to which lip-service was paid at the November summit. The negotiators, and their governments, remained as concerned as before to defend irreconcilable positions, based on perceived 'national interest', and to attack the standpoint of the other side as power-seeking and hypocritical.

Apart from any signs of progress or lack of it at Geneva or at other East-West negotiations, an important clue to the way policy is evolving in Washington has been the ACDA's vigorous endorsement of the 'McFarlane Version' of the ABM Treaty (p. 274), notwithstanding Mr McFarlane's resignation in December 1985. With the next Treaty review conference due at the end of 1987, the ACDA's reminder of the possibility of US abrogation of the treaty, in the same document (US ACDA, 1986b), must also be regarded as ominous. On the Soviet side, any resumption of its asat testing programme would be a disappointing development, even though this might be an 'instinctive' reaction to the Pentagon's insistence on persisting with the F-15 asat programme.

12
A Tour of the Battlefields

The battlefield is elsewhere – the battlefield is in Allied capitals, notably in Europe.
Larry Speakes, White House Press Secretary, October 1985

The political repercussions of the threatened onset of a space weapons race in general, and of the SDI in particular, have been and continue to be extensive. They are also more immediate and therefore, perhaps, more important than either the military or the arms control aspects of the question, narrowly defined.

To begin with the United States, the SDI has produced an even more intensely divided debate than did the earlier Reagan decision to pursue an ability to 'prevail' in nuclear war by means of a major build-up in offensive weapons. The basic strategic questions are the same in both cases. Should people in the United States be content to live without special political privileges, as citizens within a world community? Or should they, on the other hand, attempt to revive their country's earlier, go-it-alone tradition, hoping to replace the isolationism of former centuries, founded on geographical remoteness and economic wealth, with an imperial supremacy as the world's only true superpower, able to 'command the portals of space' (Wallop, 1979) against all others?

US opinion polls have shown some positive response to the SDI's professed goal of rendering nuclear missiles, above all those targeted on the United States, 'impotent and obsolete'. But polls also suggest that a majority would be willing to forego any lead in space weapons in return for a reasonable disarmament treaty. The arms control question may be only a corollary of the deeper one of what role and status in the world US citizens now aspire to. But it suggests, at least, a preference for a political fix in the hand over half-a-dozen technological fixes that might conceivably come along in the second term of President Bush or A. N. Other.

Much of the US public debate has been of high quality. But

232

there remains some unclarity about it, since many participants appear to find it more convenient, for domestic political reasons, to focus on technical or economic aspects of the SDI, rather than on its central strategic 'philosophy'. Scientists and arms control professionals have been to the fore amongst the administration's critics. Politicians have been more concerned with damage-limitation exercises in relations between Congress and the administration, and with protecting their election prospects by accepting as much as possible of the SDI's 'research only' rationale, and with it contracts for aerospace corporations and military establishments across the country (appendix 2).

Some leading American opponents of the SDI suggest there would be a better chance of improving US policy over space weapons, if only the Soviet Union would keep out of the discussion and, especially, desist from publicly criticizing the US government on the matter. This is a short-sighted and defeatist proposition, based on the idea that most ordinary Americans are now, and must forever be, incapable of thinking the issues out freely and fairly, regardless of whether or not their conclusions happen to resemble notions publicly espoused by the Soviet Union. If that were so, then the central question of US foreign policy, the struggle between a responsible globalism (not one which endorses high-handed military or political interventions in other countries) and an irresponsible isolationism, must always be decided in favour of the latter. The administration's critics have worked very hard and achieved a great deal. But a truly compre-hensive and popular opposition, not just to the SDI but to what the SDI means and stands for, has yet to emerge under nationally prominent leaders, from the Democratic Party or anywhere else. Over the next two years the mid-term congressional elections, followed by the presidential campaigns, will be of major interest in this regard.

Relations between the United States and her allies are affected not only by the SDI and other space weapon programmes and policies as such, but also by the babel of disunity to which they have given rise within American society, and even within the administration. There is another important factor to be consi-dered, namely what effect the SDI is having on the underlying framework of US-Soviet relations.

Stephen Meyer (1985) has argued persuasively that the Soviet Union is more likely to view the SDI as technically feasible, in some version at least, than some of its Western critics have been.

Meyer also contends that Soviet leaders may well doubt the programme's short-term military relevance, but that, rather like the West Europeans, they may nevertheless see its economic and technological challenges as an immediate problem, and not only for reasons of international military and space prestige. Soviet dependence on imported Western technologies in some areas, above all computing, requires them to stay 'in touch' as far as possible, but they are already finding this difficult to achieve. Meyer explains that, because of the marked dependence of Soviet military procurement and production on planning certainty and predictability, the SDI will also create fears that:

> Shifting the arms race to an SDI environment would . . . greatly reduce the efficiency and effectiveness of the Soviet military economy. (1985: 277)

It has been a paramount goal of the Soviet Union for at least two decades to achieve and then maintain strategic nuclear (and 'para-nuclear') parity with the United States (Holloway, 1983: 43–58). The SDI is bound to be perceived in Moscow as a major strategic threat, and as one which cannot be wholly countered by measures to weaken the potential capabilities of US BMD systems. There is therefore a reasonable doubt as to whether the Soviet Union would bargain away its militarily significant space programmes in return for anything less than a total US disarmament in this field, together with reliable political assurances of the permanence of that state of affairs.

The SDI also threatens to revive in Soviet perceptions the spectre of a possible US first-strike attack. The massive expansion of the US arsenal of strategic nuclear warheads which is planned for the 1990s, with the MX and the Trident D-5 missile deployments, might enable US planners to target them in a two-to-one pattern against Soviet missile silos and other 'time-urgent' targets, even if Soviet forces, also, continue to grow very rapidly. Furthermore, the Soviet Union has a very high proportion indeed (about 80 per cent) of its nuclear missile submarines vulnerably in port at any one time, and has few long-range bombers in her strategic nuclear forces (table 7.1).

Within one or two decades, the Soviet Union might therefore perceive herself as capable of little more than a heavily 'thinned' retaliation after a would-be disarming US first strike. Precisely because of the unyielding technical problems for an effective

SDI-type shield against an all-out Soviet attack, a future US BMD system could seem to make far more sense within such an aggressive scenario. Mr Weinberger's recent allusion to the undesirability of allowing the Soviet Union to retain a second-strike capability (p. 61) only serves to fuel such anxieties.

It is of course highly improbable that even the most fanatically anti-Soviet strategist now advising the Reagan administration would seriously counsel an unprovoked 'preventive' attack on the Soviet Union, or that, even if he did so, his advice would be heeded. But the view that such a capability is essential to sustain the 'peacetime' effectiveness of US nuclear forces is widely held by people with influence in Washington today. And Soviet policy makers will no more be able to think of such an American combination of offensive and defensive forces as a purely token, symbolic threat than would their Western counterparts.

An alternative long-term scenario sometimes put forward for the military space aspect of US-Soviet relations has been that the United States may some day 'share' the technology of a highly capable BMD system with the Soviet Union, so that both could benefit from it equally. Though frequently ignored or dismissed in serious strategic circles – Freedman, for instance, mentions it briefly without comment (1985), and Meyer (1985) omits it altogether – this possibility deserves at least some consideration.

The idea of the United States simply passing over its 'Star Wars' defensive systems to its major military rival has only ever been advanced in this straightforward form by President Reagan himself, as a personal speculation about what *might* happen in the future:[1]

> [A future President] could offer to give that same defensive weapon to prove to them that there was no longer any need for keeping these missiles. (Press Conference, 29 March 1983)

Defense Secretary Weinberger, however, understanding perhaps that BMD would not be separate from all his country's other military strengths, had already framed a more cautious version of the same idea:

> I would hope and assume that the Soviets, with all the work they have done and are doing in this field, would develop about the same time as we did the same kind of effective defense . . . ('Meet The Press', *NBC News*, 27 March 1983)

For over two years the President scarcely reverted to this part of his 'vision' at all. Meanwhile officials hastened to explain that what might be shared with the Soviet Union would not be the closely kept secrets of US military technology, to which even NATO allies have scant access, but the 'concept' of a new emphasis on strategic defences. Concepts, of course, come cheap:

> We have offered to begin discussions in the upcoming Geneva talks with the Soviets as to how we might together make a transition to a more stable and reliable relationship based on an increasing mix of defensive systems. (Nitze, 1985)

Stephen Meyer's insights can be applied here too. They suggest that, due to the greater wealth and technological achievements of the United States, the 'cooperative effort' proposed by Ambassador Nitze would probably be perceived in the Soviet Union as an attempt to subordinate her within an overall US scheme for planetary management. As such, it could be rejected, with some justification, as little more than a different approach to ensuring a Soviet 'defeat' in the global East-West struggle.

Both in March 1983 and when he refloated the idea of 'sharing' the SDI's results, in an interview given shortly before the Summit ('World At One', *BBC*, 30 October 1985), President Reagan made it clear that even in his fondest dreams the future US President was not expected to make the Soviet Union a firm offer until the United States had actually begun 'deploying this weapon'. Only at that point, when the way had been opened to removing the Soviet threat to the United States, but not necessarily its converse, would any 'sharing' be on the cards.

But this first, marginally more generous aspect of President Reagan's idea has not been the whole story. The sentence quoted above from March 1983 was followed immediately by this:

> Or with that defense he could say to them: 'I am willing to do away with all my missiles. You do away with all of yours.'

And in the 1985 interview the President stressed that 'there would have to be the reduction of offensive weapons . . . [and] a switch to defense instead of offense . . .'

Fred Iklé, US Under Secretary of Defense for Policy, put a further gloss on this side of the President's idea in a statement to the Senate Appropriations Committee on 21 February 1985. It was, he explained, 'unlikely' that any sharing could take place until after the Soviet Union had removed most if not all of her offensive nuclear missiles (*New York Times*, 22 February 1985). But even if, most improbably, the United States had unilaterally dismantled many of her own offensive missiles before an agreement with the Soviet Union had been reached, the possession of a 'perfect' shield would leave her holding all the strategic aces. In both respects, the American 'offer' to share BMD has been tacitly premised on the prior subordination and, in effect, the political defeat of the Soviet Union.[2] Which probably explains the President's unconscious assumption (above) that a simple US promise could secure a major Soviet nuclear 'sacrifice' in return.

From a European perspective, the 'sharing' version of the SDI's strategic approach is a striking example of the recurrent US aspiration for technological fixes, through which to escape the irksome responsibility of finding and laboriously implementing real political solutions to problems in East-West relations. There is another aspect to it too. It is a particularly clear illustration of the way in which, shared or not, the SDI concept has been promoted as a game for two players only. As the conservative British columnist Peregrine Worsthorne has pointed out, the SDI 'springs from . . . a view of East–West relations that virtually ignores the interest of both Western and Eastern Europe' (*Sunday Telegraph*, 17 November 1985). Prudent managers in Washington may have been moved to advise against overdoing the talk about sharing 'Star Wars' with Moscow at least as much by this political consideration as out of any embarrassment over its intimation of technological naivety at senior levels of government.

Western writers on strategic aspects of BMD have frequently drawn attention to the technological, economic and military roles that US allies would be expected to play (Possony, 1963). Most member governments of NATO, however, were apparently ill-prepared by their military 'experts' for the US President's March 1983 announcement of a new drive for strategic defences. Perhaps, also, their political commitment to the permanence of 'nuclear deterrence', which had allegedly kept the peace (of Europe) for nearly forty years, prevented them from understanding any such warnings as they may privately have received.

Canada was different. It is a country rendered sensitive to the issue through the intimacy of its joint defence arrangements with the United States in NORAD, and its Department of National

Defence had taken the measure of Reagan's interest in BMD very early on (Ranger, 1981), though how far such insights permeated in Canadian government circles is uncertain. Founded in 1957, the joint US-Canadian North American Air Defense Command (NORAD) added the BMEWS early-warning radars in Greenland, Britain and Alaska to its responsibilities in the early 1960s. From that time, Canadian governments required the insertion of a special clause in the bilateral renewal agreements (1968, 1973, 1975, 1980) stating that they did 'not involve in any way a Canadian commitment to participate in active ballistic missile defence.' It was a measure of the difference the Reagan team intended to make, when they at once renegotiated the agreement (in 1981) to remove this caveat and to change the organization's title from 'Air' to 'Aerospace Defense'. An embarrassed Canadian government was reduced to the belated and unconvincing explanation that the clause 'had become unnecessary and perhaps offensive in view of the prohibitions embodied in the 1972 ABM Treaty' (Polanyi, 1985).

Weinberger's March 1985 invitation (p. 248) for official participation in the SDI led to a more thorough, focused and democratic debate in Canada than, perhaps, in any other US ally. In August 1985 a special all-party committee of both Canadian Houses of Parliament reported as follows:

> The majority of the Committee is concerned about the implications of ballistic missile defence on international stability and on the future of Canada's involvement in the arms control process . . . The majority of the Committee recommends that the government remain firmly committed to the letter and spirit of the Anti-Ballistic Missile Treaty of 1972 as essential to the maintenance of order and stability. (In Polanyi, 1985)

A fortnight later Prime Minister Mulroney announced that Canada would not be taking up the offer of a government-to-government arrangement, since it would be inappropriate to Canadian 'policies and priorities' for her to do so. Individual firms, however, might bid for SDI contracts as they chose, if they thought they could obtain them without government help under the restrictive conditions that would doubtless be laid down.

That is not, however, the whole of the story. In May 1985 the *New York Times* revealed that a new plan was being put together, under the direction of Fred Iklé, to integrate all parts of the US offensive nuclear force posture with existing air defences and projected anti-missile and anti-satellite capabilities. True to its new title, NORAD is being increasingly drawn towards 'space defence' roles as well as its traditional air defence one, and finds itself coordinated alongside the US Space and Strategic Air Commands under the new US Space Defense Operations Centre (SPADOC) in Colorado:

> The money and effort being expended on SPADOC is still another example of NORAD's growing involvement in space. It seems unlikely that air defense, space defense and missile warning can continue for long as separate functions; they will most likely be integrated into a single aerospace defense system within a very few years. Whether NORAD will be absorbed into the unified Space Command or will remain separate is open to question; but in any case, the organizations will be highly interdependent and will operate from what is becoming a massive aerospace complex. (Gumble, 1985: 90)

The Pentagon's new plans had been 'discussed with Canadian leaders' according to the *New York Times* (29 May 1985). But the process of 'integrating the NORAD Space Surveillance Center [SPADATS HQ] into Spadoc' (*Aviation Week & Space Technology*, 9 December 1985) is likely to cause further controversy, especially around the NORAD agreement's next renewal, due in 1986.

Declining to join in the SDI at government level may have seemed a clear and enlightened decision to many Canadians. But it has to be evaluated in context. The Canadian commitment to NORAD and NATO is shown by her acceptance of the new North Warning System radar development, with its anti-cruise missile capability, and by her current purchase of 138 CF-18 fighters from the United States, together with a US-aided expansion of the northerly line of air bases. Then there is the general cooperation, through NORAD, with an overall US posture that is increasingly premised on some kind of 'projected antimissile shield', in the words of the *New York Times*, no matter how nebulous that may still be in technical terms.[3] Nor should one overlook Weinberger's casual reference to the possibility that future ABM systems 'might be [based] here', that is, in Canada (*New York Times*, 19 March 1985).

The terms of the 1986 NORAD renewal will show what real distance, if any, Canada can put between herself and the space weapon plans of the United States. Washington may have tolerated symbolic dissent so far, by way of appeasing an ally's self-esteem. But practical obstruction of the administration's new strategic 'vision' might be another matter, and seems unlikely to be permitted within the United States' 'front yard'.

Canada's European partners in NATO should be particularly interested in the outcome of the test posed, by the NORAD renewal, for the Canadian policy of 'firm commitment' to the ABM Treaty. For their own basic policies in response to the SDI have been similar to Canada's in some respects, even if they have been implemented differently. Britain, West Germany and Italy are possibly the most interesting cases, because of the short-term accommodations with the SDI which they have tried to reach despite, or because of, their underlying negativity towards it. France, the other main 'space power' of Western Europe, has like Denmark and Greece declared herself flatly opposed to the whole enterprise, and declined to participate accordingly. But because of its positive formulations, French policy has been a unique element in the overall pattern of disquiet about the SDI in NATO today, an anxiety which concerns not merely the military problems it may bring in the distant future, but also the political ones which it has created in the present.

It is helpful, indeed, to distinguish short-term interests and policies from medium- to long-term ones, in NATO responses to the SDI. The general pattern has been that the latter predominated in the first two years from March 1983, but that during 1985 the former assumed a higher profile.

US allies in Europe were not consulted before the March 1983 speech, and their public reactions were slow to take shape. This was not only because European ministries were ill-prepared, and needed time to work their ideas out. Widespread political upheavals over the imminent deployment of new NATO theatre nuclear weapons – ground-launched cruise (Tomahawk) and Pershing 2 missiles – meant that the focus of the nuclear policy debate between NATO governments and their domestic anti-nuclear opponents was elsewhere at the time. It also meant that governments needed to believe, or to affect to believe, that there was nothing disturbingly new and radical about US nuclear strategic principles under Reagan.

A presidential speech that seemed both to join the anti-nuclear movements in their dissent from 'deterrence', and to declare a determination to achieve some kind of defence-based strategic superiority, was thus definitely unwelcome. Reagan's unforeseen move threatened to destroy 'the precarious domestic consensus on the NATO double decision [missiles deployment and INF negotiations]' explained Christoph Bertram, former director of the London International Institute for Strategic Studies (*Die Zeit*, 1 April 1983). Furthermore, what the European NATO governments may have lacked by way of up-to-the-minute analysis of the issue was more than made up for by their long-established policies, combining technical and strategic scepticism towards BMD with strong support for maintenance of the ABM Treaty regime of 'mutual assured vulnerability', until and unless major progress should have been made with detente and disarmament (Yost, 1982).

Initial West European reluctance to make too much of the speech was perhaps encouraged by the formation of the Fletcher and Hoffman panels (p. 331), which allowed for a 'wait-and-see' response, at least until the depth and likely political staying power of the new Reagan policy could be properly evaluated. And when, in the autumn, the panels' reports showed the SDI to be still very much on course, European attention was even more taken up with the 'missiles crisis' than it had been in March.

The first proper US briefing for the NATO allies on the SDI was given at the Brussels headquarters of the alliance in February 1984. It was followed in April by a meeting at Cesmé, Turkey, of NATO's Nuclear Planning Group (NPG) of Defence Ministers and senior aides. At both meetings the Europeans were sharply critical : doubtful whether such defences could provide any worthwhile protection for their countries against intermediate- and shorter-range ballistic missiles, and against the other nuclear capabilities of Warsaw Pact forces; concerned about their possible negative effects on the arms control process; fearful that US BMD would result in 'zones of differing security' (a German euphemism) within the alliance; and anxious about the SDI's destabilizing effects, not merely on the arms race but also during any major East-West crisis, should one break out during or after the deployment of any future strategic defences. West Germany's Defence Minister, Manfred Wörner, was especially frank:

> My impression is the Europeans were broadly united in their critical questions . . . I can't see that it [SDI] would provide greater protection or stability. I can only hope it would give an incentive for arms control. (*Washington Post*, 4 April 1984)

Much of the early European criticism of the SDI came in fact from conservative voices, basing themselves on what they had assumed was a clear and positive evaluation of what nuclear deterrence was supposed to be.[4] For example, the first editorial in the London *Times*, later to feature as one of the SDI's leading European sympathizers, was flatly negative (25 March 1983). And in 1984 the differences over the SDI within West Germany's ruling coalition were over how, not whether, it was to be opposed. Whilst the Christian Democrat Defence Minister was joined by Bavaria's Minister-President Strauss (Christian Social Union) in outright opposition to it *as an autonomous US development*, the Free Democrat Foreign Minister, Hans Dietrich Genscher, was pressing for an independent West European initiative for arms control in space. Meanwhile Chancellor Kohl cast desperately about for some diplomatic veil under which to handle his goverment's dissent from what seemed increasingly likely to become the centrepiece of American strategic aspirations (Brauch, 1985a 5–6).[5]

It was therefore no surprise when European NATO governments responded favourably to the Soviet proposal in June 1984 for new bilateral talks on arms control in space, and so informed their American ally. The proposal itself, and Prime Minister Thatcher's speech to the European Atlantic Group on 12 July, expressing support for the ABM Treaty and hopes for further arms control measures to limit military activities in space, should be seen in that context.[6] In the same month, however, Under Secretary Fred Iklé visited NATO capitals with warnings from the administration of a possible isolationist backlash in American public opinion, if European criticisms of the SDI were not promptly muted. For the time being, particularly in West Germany, his advice appeared to take effect. At the October NPG meeting in Strega, Italy, the SDI was not a major issue.

In September, Christoph Bertram had advised that West Europeans would only be able to influence the Reagan administration over SDI 'if they arrive at a clear, unambiguous and closely coordinated assessment. Diplomatic soft-pedalling may be interpreted in Washington as a reflection of support' (*Die Zeit*, 21 September 1984). Subsequent events suggest, however, that any apparent reduction in official European objections to the SDI that autumn was part of a wider political strategy. European NATO leaders had certainly engaged in high-level exchanges over their side of the SDI's 'alliance management' problem. They appear to have concerted a purely temporary intermission of criticisms of the idea, out of regard for the administration's sensitivity during the President's election campaign, in which the anti-nuclear pretensions of 'his' Initiative had been allocated a large role. But they were content to do so for the good reason that they had already formed a joint plan of action for later in the year, in the highly probable event of Reagan's re-election.

This view of events in the last part of 1984 and the early months of 1985 can only be hypothesis. But it offers an explanation of why the transatlantic disagreement over SDI suddenly went critical *after* President Reagan had secured his second term of office. The crisis began, oddly enough, with what looked like a solution. In December 1984 Mrs Thatcher took the unusual step of joining the predicted future Soviet leader, Mr Gorbachev, in a common declaration of their opposition to any arms race in space, just before setting off on an international trip intended to conclude with discussions with President Reagan on that and other issues.

Whatever may have been the immediate reaction in Washington (Thompson, 1985: 107–8), Thatcher's visit ended in an arrangement that would set a pattern for attempts at intra-alliance 'damage limitation'. In return for repeated disingenuous endorsement of the SDI as a sensible and prudent investment in mere research, a set of four points, known as the Camp David agreement, was approved by the two leaders. With their vitally ambivalent terms emphasized, these were, first, that the overall aim of SDI was not to achieve superiority over the Soviet Union but to maintain a *balance* between the two sides. Second, that the SDI itself would remain within the *limits* of the ABM Treaty and that any subsequent activities, beyond those limits, would have to be conducted within another legal framework, to be negotiated with the Soviet Union after prior consultations within NATO. Third, that the overall aim of the SDI would be to enhance *deterrence* rather than to undermine or replace it. And last, that the US–Soviet disarmament negotiations, about to be renewed, would aim to ensure the security of both sides at lower levels of offensive weapons.

The Camp David points avoided direct reference to two important intra-alliance issues which had been mentioned in several other, similar 'lists' (such as Bertram's September article), namely the possible damage to alliance unity from any BMD which 'worked' either only for the United States, or far better for the United States than for Western Europe, and the strain which such an effort was thought likely to place on scarce alliance resources, especially if the Europeans were expected somehow to join in. Nevertheless, such matters were widely understood to be included on the likely agenda for the intra-alliance negotiations which the Prime Minister insisted would remain a precondition for any alliance-wide consent to actual deployment. And it seems improbable that the British leader had not obtained, and did not in due course lay before the President, the endorsement of her principal NATO colleagues for the political bargain she was endeavouring to secure. In any case, its terms were squarely in line with both Italian and West German policy on space weapons, and largely with the French position also (Lenzer, 1985: §§ 84–121).

But Camp David failed to end NATO's troubles with the SDI. Though the principles it had sought to enshrine were emphasized from time to time thereafter by British leaders, and by similar declarations from their European allies, it transpired that powerful figures within the US administration were simply not prepared to pretend that if, or as they usually seemed to see it, *when* some technically effective BMD system had been devised, the question of its actual deployment would still be completely open for an alliance-wide discussion and decision, to be followed by similarly unpredictable negotiations with the Soviet Union.

The other half of the Camp David bargain, European support for 'research' as a 'prudent hedge' against some allegedly feasible Soviet break-out from the ABM Treaty, was felt in Washington to be a different matter. In February 1985 Weinberger returned to the theme of possible European participation in the research which 'everyone supported', whilst passing through London to an important conference convened in Munich by the journal *Europäische Wehrkunde*.

At this meeting of top NATO defence and foreign policy officials, advisers and journalists, the US Defense Secretary was supported by his Assistant, Richard Perle, and by arms talks Ambassador Paul Nitze. He was joined by his French and German colleagues, by Chancellor Kohl, and by the British technology Minister, Geoffrey Pattie. The conference proved extremely difficult for the SDI. France's Charles Hernu feared that BMD would only lead to further proliferation of offensive weapons, and restated his government's call for a five-year moratorium on laser weapon deployments – hardly any great sacrifice at present technological levels – and for strict limitations to be agreed on asats. Pattie expressed fears that conventional defence would be sacrificed to BMD, while the latter would prove of little military value. Egon Bahr, a leading German SPD member, agreed that an alternative to the inadequate security of 'nuclear deterrence' was needed, but thought it should be found in stronger but strictly defensive conventional forces, rather than in a strategic anti-nuclear shield. Without taking any final position, Chancellor Kohl made considerable play with the need to avoid any increase in strategic instability if new defensive technologies were to be deployed, to which President Reagan had of course referred in his original speech (p. 79), and spoke of the need for any defensive system to be a unitary asset for the entire alliance, and for its development and possible deployment to be a matter for close and continuing consultation between the various governments.

On the American side, Weinberger repeated his assurance that the technical thrust of the SDI would be just as much towards missile defence in Europe as to defences for the US mainland. Nitze produced assurances, later to be repeated in a more public setting (1985) and since known as 'the Nitze criteria', to the effect that BMD deployments would not be initiated unless they were judged to be survivable (proof against direct counter-attack) and 'cost effective at the margin', such that a given quantity of reliable anti-missile defensive capability should be cheaper to develop and deploy than could be any equivalent amount of off-setting offensive proliferation by the opponent. Nitze also laid stress on the fact that offensive deterrence would be maintained 'for some time', until some fairly remote future of perfect BMD systems, at which point only would 'the global elimination of nuclear weapons' become feasible (in Brauch, 1985a).

An important feature of the Wehrkunde Conference was the growing interest being shown by NATO Europeans in technical and economic aspects of the SDI, both as potential threat and as opportunity. Chancellor Kohl pointed out that West Germany could not allow herself to be excluded from any overall technological innovations that might result.[7] Pattie complained at existing US barriers to technology transfer, implying that they augured ill for any real cooperation around the SDI. It seemed that if the SDI refused to go away, the Europeans were determined to ensure they were not immediately damaged by allowing it to produce differential benefits for the US economy at the expense of their own. (This development also contained the seeds of a divergence of priorities for new military technology, notably between Britain and West Germany.) In the midst of transatlantic disagreements acute enough to reveal the fragility of the Camp David agreement, the makings of a more pragmatic but less united arrangement began to appear.

However, the President's February 1985 Report to Congress (US WH, 1985b) on Soviet treaty violations roused further European anxiety that arms control was about to be abandoned in favour of space control. For all her 'declaratory' enthusiasm for the re-elected President and his entire BMD research programme, Mrs Thatcher, who is nothing if not a stayer, above all on issues she regards as having been fairly and squarely settled, may well have formed a graver private assessment of the situation on her visit to Washington that month. What seems certain is that her Foreign Secretary's pointedly unenthusiastic speech about the SDI, given in London on 15 March (Howe, 1985), could not have been delivered without prior clearance with his chief.

Whilst affirming his confidence that 'present and planned US and NATO [offensive] nuclear forces provide a stable balance of deterrence and will continue to do so', Sir Geoffrey Howe was a great deal less sure about 'active defences'. They might enhance deterrence; alternatively, 'their development may set us on a road that diminishes security.' The ABM Treaty was explained as permitting research, but only research, in the relevant fields, and was described as 'a political and military keystone in the still shaky arch of [Western] security.' The SDI, by contrast, was characterized as 'subject to uncertainty', and as research which must not be allowed to acquire 'an unstoppable momentum of its own', thereby leading to deployment without regard to such 'complex and difficult' questions about its strategic implications as the following, most of which the reader has encountered earlier in this book:

> Could the process of moving towards a greater emphasis on active defences be managed without generating dangerous uncertainty?

> [W]ould the supposed technology actually work? And would it . . . provide defences that . . . were survivable and cost-effective? . . .
> [H]ow would protection be extended against the non-ballistic nuclear threat . . . ?

> And what would be the effect on all the other elements of our defences . . .?[8]

Nor was Sir Geoffrey unwilling to sound a political warning about the possible consequences for NATO of any strategic inequalities resulting from partial BMD systems:

> Finally, as members of the Atlantic alliance, we must consider the potential consequences for this unique relationship. We must be sure that the US nuclear guarantee to Europe would indeed be enhanced . . . from [the] very inception [of BMD] . . . All the allies must continue . . . to share the same sense that the security of NATO territory is indivisible.

Beside this lightly veiled claim to a European veto on US BMD deployment, the Foreign Secretary included several other 'musts' on his list, such as the requirements that 'visions . . . must be clearly stated . . . [and] we must be especially on our guard against raising hopes that it may be impossible to fulfil.' There was a great deal more. The Camp David points were spelled out. Asat limitations today were judged vastly preferable to technically doubtful BMD the day after tomorrow. And it was asked whether the effect of Soviet counter-measures and competing BMD deployments might not be that NATO would 'lose on the swings whatever might be gained on the roundabouts.' Reagan's claims about Soviet non-compliance with arms control agreements became a US commitment 'to reverse the erosion of the ABM Treaty'. And American arms control policy was welcomed as something on which NATO's unremitting seriousness should 'be seen and heard' without giving the impression that 'we have something else in mind.' And so on, at considerable length and with little more than cosmetic remarks in favour of SDI research, as long as it was only research.

It was, in short, a skilful and thorough demolition job, falling just short of outright apostasy, as indeed it appeared to US Assistant Secretary of Defense Richard Perle, whose public response to it, given in London a few days later, was so extraordinarily derogatory that press officers on both sides had to work overtime at quelling rumours of a major dispute between the two governments (*The Times*, 19/20 March 1985).

Howe actually did little more than put together doubts and criticisms of the SDI that had already come from many quarters within NATO. But the effect of his speech must be judged from its context, in the aftermath both of the Munich *conversazione* and of Thatcher's carefully worded pledges of solidarity, restricted always to *her* definition of what the SDI should be. As such, it can only have confirmed the Reagan administration in a course of action upon which it had already decided, with a view to lancing the slowly swelling boil of dissent, and perhaps worse, that had been gathering about the SDI issue within the alliance.

There had as yet been no formal NATO-wide endorsement for the long-term goals of the SDI, as contrasted with its immediate, 'research only' aspect, and none seemed likely to be given.[9] The traditional US response in such circumstances has been to adopt a bilateral, country-by-country approach, using whatever leverage may be available. In the circumstances, the weakest point in the position of European NATO governments was clearly their economic anxiety about being 'left out' of a major technological revolution. A secondary lever was the interest of some governments, notably the West German but also in some ways the French, in developing some sort of European anti-missile capabilities, a theme which came to be referred to, misleadingly, as the 'European Defence Initiative' or EDI. But above all, the economic, 'new technology' dimension was one in which the United States stood greatly to gain from the Europeans' inability to take Bertram's platitudinous back-seat advice, about the strength to be gained from unity.

On 26 March 1985, US Defense Secretary Weinberger wrote to all his NATO colleagues, as well as to Australia, Israel, and Japan, with a request for them to let him know within sixty days whether or not they were interested in joining in the SDI research programme, and if so what they thought they had to offer. The short deadline brought forth a good deal of protest. It also had two other effects. The Australian government immediately declined the offer, on the grounds that they were not prepared to do business on such terms. Besides following suit, the French had already decided to complicate matters with an alternative offer of their own.

French space policy has been slightly distinct from that of her principal European partners in recent years in two respects. The opposition to US space weapon programmes has been crisper, with France acting independently at the UN Committee on Disarmament in support of a partial asat ban, in June 1984, and with Mitterand referring to the SDI as a case of 'over-armament'.

But French space policy has also had a strong military element, firmly linked to maintenance of 'the defensive potential of present means . . . without becoming involved in a new arms race' (Lenzer, 1985: § 90). Hence France has made overtures to West Germany for joint development of military intelligence satellites, and has also been a proponent of similar capabilities either for Western Europe as a whole, or for the entire international community of the United Nations. She has also pressed hard for some years for joint European technology development efforts, in space as well as other fields, aimed at 'making [Western] Europe appear as a great technological power' (Lenzer, 1985: § 93). The French government therefore saw the economic anxieties of their NATO colleagues, towards the SDI as a technological thrust, as an excellent opportunity to renew their efforts in this direction. In April 1985 they sent out initial proposals for a number of international joint technology programmes in the fields of computing, telecommunications, robotics, materials, and biotechnologies. The overall endeavour was to be known as 'Eureka'.[10]

The detailed history of Eureka and the economic issues that it raises lie beyond the scope of this book. It was clearly intended to offer Western and neutral European states an alternative, in-house route towards 'the third industrial revolution', and one that was both more systematic and less openly military than such US-regulated options as might become available within the SDI. By the end of 1985 it had begun to acquire a reasonable momentum of its own, with a nominal enrolment of eighteen Western and neutral European states, although the Dutch government had been the only one prepared to join the French in actually putting up any seed-money.

Responses to Weinberger's offer were slow and vague by comparison, and his sixty day deadline had to be dropped after three weeks. Besides France, the other clear refusals inside Europe came from Denmark, Greece and Norway. But it cannot really be said that Eureka has enabled any state to refuse SDI participation that would not otherwise have done so. Some companies and research institutions in the relevant fields have shown signs of interest in both the SDI and Eureka. Were the same organization to receive similar contracts, a 'reverse spin-off' of supposedly 'civil' Eureka-sponsored technology into military applications could scarcely be avoided. In view of French interests both in high-tech conventional and offensive nuclear military systems, including penetration aids to overcome any future Soviet BMD systems, and also in non-weapon military space programmes, it seems most unlikely that Eureka was ever intended to produce purely non-military results.[11]

However, Eureka may well have fuelled precisely those West European economic anxieties on which the Reagan administration had decided to play, in seeking a stop-gap solution to its dispute with the rest of NATO over SDI.

By mid-1985 the less evangelistic, more pragmatic US approach to its dissenting allies began to make some headway. The Thatcher government, arguably the most awkward of them all till then, appeared to imagine it had secured a useful bargaining position. Its Defence Minister, Michael Heseltine, initiated negotiations with the Pentagon on the terms under which British firms and government research establishments might bid for SDI research contracts, and over whether some guaranteed share of the SDI's five-year budget could be assigned for Britain. As the months went by, sections of the British press gave increasingly confident details of the coming bonanza. Thus despite other reports to the effect that no contracts would come to British firms except on the merit of their bids, the defence correspondent of the *Daily Telegraph* referred to a '£1450 million programme which currently would involve major work for 20 British companies' (25 October 1985). Only a few weeks later, any such expectations were decisively and predictably quashed by a Congressional ruling that there could be no 'set-asides' of any sums within the SDI budget for the purpose of awarding contracts to allied bidders.

By autumn, another factor also became prominent. The British government was vexed by the apparent lack of high-tech contracts for British companies in the proposed new BMEWS phased-array radar construction at Fylingdales Moor. Their chagrin was compounded by the subsequent decision by the other half of their 'special relationship' to purchase the French 'Rita' communications system for the US Army, instead of the British 'Ptarmigan' contender, despite a special lobbying effort on behalf of the latter by the Prime Minister herself.

A UK-US Memorandum of Understanding (MOU) on British participation in the SDI was finally signed by Weinberger and Heseltine, after Thatcher had personally approved it, on 6 December 1985. The MOU will not be published; indeed, it is reputed to contain a provision that it shall remain secret 'in perpetuity', and there have been reports that Thatcher refused a personal appeal from Chancellor Kohl for a sight of it. (An incident which, if it occurred, demonstrates succinctly the 'divide-and-rule' aspect of the Reagan administration's decision to play the economic card on this issue.)[12]

In typical British fashion the Memorandum was agreed when – more than two years after the President's 1983 speech – there had still been no formal debate on the SDI in the House of Commons. Besides the lack of any US commitment to an approximate spending figure, and likewise of any overall programme for SDI work in Britain, what has been positively announced about the agreement is that it establishes rules under which British companies may bid for and possibly secure contracts from the SDIO, and sets up an office within the British Ministry of Defence to coordinate all such negotiations and to handle their security aspects. It is not certain how far US negotiators succeeded in achieving their demand for the right to inspect the books of participating UK establishments, as a check against onward 'technology transfer' to undesirable parties. In March 1986, the planned visit of a US team to Britain for this purpose, which had not been cleared with the British SDI office, was cancelled amidst some confusion.

Eighteen technology areas have been chosen as especially promising, from the SDI's point of view (table 12.1). Though Heseltine claimed to have made certain that one-way technology and personnel flows, from Britain to the United States, could not occur, it is significant that this nominally two-sided venture has been defined, not on a fifty-fifty basis, but by areas of special *British* expertise alone.

Seven technology areas were also identified in the negotiations as candidates for two-way information exchange (table 12.2).

The lists have several interesting features. From the strategic standpoint, the Architecture Study, concerned with the overall form that anti-missile defences in or for Europe should take, may be seen as an opportunity to influence American SDI decisions. About £6.5 million is to be transferred from the Pentagon directly to the British SDI Programme Office for this purpose. In general there is a certain balance between areas of relatively 'pure' advanced technology, and specifically military applications, such as terminal interception or countermeasures.

Cooperation on penetration aids is thought likely to involve the Aldermaston Weapons Research Establishment, on the basis of its recent experience with designing the 'Chevaline'

Table 12.1 Technology Areas for UK Participation in SDI

1 European Architecture Study	10 Terminal interceptor research
2 Laser/particle beam/RF lethality vulnerability/ hardening	11 Laser radar/vibrometry/ imaging
3 Electro-magnetic gun and effects research	12 Countermeasures
4 Ion sources	13 Software security
5 Optical computers, switches and limiters	14 Electronic materials
6 Advanced thyratrons	15 Phase conjugation
7 Non-electronic materials	16 Battle management/command control and communications
8 Sensors	17 Signal processing
9 Terminal radar research	18 Space research

Source: House of Commons Defence Committee, Public Sessions on the SDI, December 1985.

Table 12.2 Candidate SDI Technology Areas for Information Exchange

1 Command, control and communications (C3) and battle-management research
2 Laser and optics
3 Advanced computing
4 Surveillance, target acquisition, identification and tracking
5 Non-nuclear electro-magnetic pulse (EMP) and radio frequency (RF) weapon technology
6 Space technology
7 Special materials

Source: as table 12.1.

warhead for Britain's Polaris nuclear submarines. As with other areas in the package, it is possible to doubt whether they have in fact come up with anything different or better than what has been achieved in the Pentagon's ASMS programme already, and to doubt even more whether much of value will be received from ASMS in return. On the other hand, the British penetration aids are believed by some experts to have been obliged to be 'cleverer' than the American, both because they had to be designed to ensure the success of a relatively 'small' attack, in the unlikely event of Britain ever 'going it alone', and because they had a much smaller total throw-weight to work with. There is of course no way to confirm such rumours, which may be no more than offshoots of the bargaining process between the two governments. And although the British government has declared, on the basis of expert advice, that it is satisfied that future British Trident warheads will be able to penetrate any probable Soviet defences up to the year 2015 at least, this may reflect a low evaluation of Soviet BMD, at least as much as a high one of British penetration aids.

Meanwhile, the US-UK negotiations and subsequent agreement led to considerable disquiet within British scientific circles. Antipathy to the SDI's long-term goals covers a great part of the political spectrum in Britain, as it does elsewhere in Western Europe. But those most directly affected by, or familiar with, the implications of British participation were also concerned with other aspects. There was much scepticism on the technical feasibility question, above all from experts in computing.[13] And beside the familiar 'brain drain' anxiety, that major US aerospace firms are more interested in acquiring small groups of talented individuals than in sharing the contracts pool with their European competitors, there was considerable disquiet over issues of academic freedom and, on the commercial side, of intellectual property rights. In this connection, there was frequent mention of a cooperative exchange of information between French and US pharmaceutical companies researching into treatments for AIDS that is widely held to have led to the French being obliged to pay royalties to their US partners for access to their own intellectual property as a result of American patent law (*New Scientist*, 15 May 1986). Anxieties over such issues are unlikely to be appeased by an agreement to which most of the interested parties are to have no access. Thus the absence of even some bland public version of the text may well have been a political blunder. On the other hand, the British government holds most of the managerial or economic cards with respect to British producers of advanced military technology, and a reasonable crop of SDI contract bids is likely, with time, to emerge.

By March 1986 contracts worth about £8 million had been signed with British establishments. They included two £142,000 allocations linking optical computing teams at Ferranti and at Heriot-Watt University in a package centred on the University of Dayton, Ohio, which had in essence been agreed without benefit of the MOU or the British SDI office. Ferranti also received a contract to develop laser mirrors; Marconi, for battle-management systems; Plessey, for another computer system; and English Electric Valve, for electronic switches. The only large contract was for about £7 million over five years, awarded for neutral particle beam work at the UK Atomic Energy Authority's Culham Laboratory. Another award for a similar amount, for a scaled-up version of the Rutherford Appleton laboratory's ultraviolet excimer 'Sprite' laser, was at an advanced stage of negotiation.

It would not, however, be correct to say that the British government had allowed its more fundamental demands about the limitations that should be observed by the SDI to be shelved, in return for the dubious benefits of research participation. Only a week before the November 1985 Summit, Britain's Ambassador to the United States gave an important speech containing a strong reaffirmation of the Camp David agreement, and of his government's commitment to genuine negotiations rather than to pursuit of one-sided military advantage for its own sake, in what US hawks now frankly admit to seeing as 'a winner-take-all contest' (Swift, 1985). That speech seems unlikely merely to have been a move in the MOU negotiating process.

The political value of the MOU to both sides was above all that it enabled the Thatcher government to represent its disquiet as satisfaction, its hindrance as support, and US waywardness as a calm, considered and cooperative alliance venture. The disagreements over the SDI between Britain, together with her fellow NATO allies, and 'their' superpower are still real; but so too is the paper that has been used to cover over the cracks. Furthermore, the paper is so newly hung that it may not begin to tear for some time.

The British agreement was not just important in itself, but even more so as the first of its kind, and therefore likely to help the US administration achieve a favourable outcome in similar negotiations with their main partner in NATO, West Germany. Most of the issues and perceptions there have resembled those in Britain and need no repetition. But in West Germany opposition has come not only from the scientific community, and perhaps in stronger and more organized forms than, so far, in Britain. There have also been strongly negative responses to the SDI from some major German companies, such as Bosch, Leitz and Siemens. Others, such as Dornier, expressed an interest, initially with the proviso that a government-to-government agreement would be needed to protect their rights within such arrangements.[14] The Kohl government, however, proceeded with considerable caution, due largely to the anxieties about the SDI felt by the second largest coalition party, the Free Democrats. Defence Minister Wörner and CSU leader Strauss transformed their earlier misgivings on SDI as an *American* development into at least a verbal enthusiasm, linked to demands for full consultation, a sharing of America's military high technology with her allies, and effective theatre missile defences (ATBMs). In Wörner's opinion, 'all this can be looked after better by participating than by being a mere spectator' (*Daily Telegraph*, 25 November 1985). The interest in some sort of 'European Defence Initiative', recently expressed by leading members of the SPD such as Egon Bahr, should perhaps be seen in the light of the idea that a technological fix can be discovered that will provide a suitably 'strong but defensive' posture for a loyal pro-NATO Left to advocate as the reasonable and moderate alternative to 'Star Wars'.

In general, the West Germans are more directly concerned than the British about the possibility of a Soviet lead developing in *conventional* theatre-range counter-force capabilities, alongside a defensive capability to deny, in part, a NATO resort to its corresponding tactic of Follow-On Forces Attack (Wörner, 1986; Yost, 1985: 15). Thus if BMD in Europe could be presented as limited to ATBMs, and if the terms for joining in the SDI were not too inequitable, some degree of German participation could be expected to develop.

Nonetheless, until the end of 1985 the Kohl government gave the impression of deliberately stalling, in the hope that something would 'come up', perhaps by way of offensive arms reductions, to remove the incentive for large-scale BMD altogether. Strauss urged Kohl to give the SDI his clear backing. Kohl's eventual response, in December, was a decision to open discussions about participation. To patch over the divisions within the Bonn Cabinet, however, it was stated that cooperation with the US effort would only be possible on condition that BMD research should lead to 'cooperative solutions' between the United States and the Soviet Union *about all subsequent (post-1989) development and deployment.* Since the Reagan administration had repeatedly declared that the Soviet Union could not have a veto over eventual US deployment of BMD systems, the West German stipulation appeared to be strictly unworkable. The other element in Kohl's decision was more likely to have practical effects. Martin Bangemann, Economics Minister and leader of the Free Democratic party, was to head the negotiating team.

The two-part US-West German agreement that was eventually signed on 27 March 1986 was thus remarkable for the junior partner's insistence on what Bangemann called its 'strictly private and civilian nature' (*The Guardian*, 29 March 1986). Full texts of both secret accords, were published three weeks later by the Cologne *Express* (18 April).

Though, on the experience of recent years, European NATO governments have little reason to hope for much from Geneva, there can be no doubt that their commitment to the ongoing negotiations is fairly substantial. For one thing, they see domestic patience with the nuclear arms race wearing thin, as perhaps their own does too. If 'multilateralism' cannot produce some really effective results before long, the 'unilateralists', temporarily rebuffed by the recent missile deployments, are only too likely to make major gains with public opinion. Thus the pre-Summit protests from European leaders, over US moves to undermine the ABM Treaty from within, were signs that the alliance's political problem over disarmament has certainly not been removed by the modest amount of 'SDI cooperation' so far obtained.

In the winter of 1985/6 agreements began to seem slightly more likely on 'confidence-building measures', at the European Security Conference in Stockholm, and on 'mutual force reductions', at the long-stalled NATO–Warsaw Pact talks in Vienna. If such progress is achieved, it would be welcomed more for its possible political significance than as contributing much in itself to European security. The acid test for West European confidence in US alliance leadership would remain the ability of the latter to bring about significant nuclear weapons reductions in Europe. Even the vague indications of Soviet willingness to consider an INF agreement at Geneva whilst other issues, including the SDI, were still unresolved, may not necessarily be seen as evidence that recent US strategic policy, including the SDI, has increased the likelihood of achieving actual results in this field. On the contrary, they could well be, or appear to be, part of a Soviet strategy of increasing pressures on the United States to make concessions on the SDI, by being as reasonable as possible, short of actually going through with any nuclear disarmament.

Another area of potential friction within NATO is provided by the space policies of France, Italy, Britain and West Germany, together with their common interests in the European Space Agency (ESA). The setting up of national space centres in Italy and Britain is one demonstration that, together with Canada and Japan, these countries now have sizeable and permanent national interests in space, and therefore in the manner and extent to which it continues to be used for military purposes.

As for ESA, there was little European response to the news that Spacelab, a 'civilian' space module originally provided to NASA by ESA under their cooperation agreement, was to have been used in 1987 for laser pointing and tracking experiments within the SDI. The ESA–NASA agreement states that the modules ESA handed over may only be used 'for peaceful purposes'. On this basis, however, 'NASA believes the flight is acceptable . . . [and] has heard of no problems from ESA in this regard' (*Aviation Week & Space Technology*, 19 August 1985).

The anti-tactical missile issue, reflected in the British SDI cooperation agreement (items 1, 9 and 10 in table 12.1), is certain to enter the political limelight in Europe during 1986. The question of whether an expensive East-West competition in new military technologies can or cannot be avoided is of course far larger than the SDI/EDI question, and has been steadily gathering momentum for several years. In that respect, the recent Euro-missiles controversy may have been the last time around for the relatively simple strategic and political debates of the 1960s and 1970s. Highly automated, flexible and strategically ambivalent weapon systems may be setting new and dangerous patterns for the future:

> [T]echnology can also create strains. An incredibly rapid pace of technological change confronts Alliance decision-makers with an enormous range of possible options. The stakes and the risks associated with their decisions are high. So are the costs – a major difficulty at a time when the Alliance's economic resources are limited. Inevitably that combination produces strains.
> These strains are exacerbated by unease within the Alliance that the dream made possible by new technology could become a nightmare in which the opponent exploits the same technologies . . . (Abshire, 1985: 1)

But anxieties about the technical and military implications of an all-out offence-defence competition between theatre nuclear forces in Europe may be relatively manageable, by comparison with their possible repercussions at the political level, if they hamper the pursuit of detente and of sensible arms limitations (Holst, 1969; Brauch, 1985a).

Besides the ATBM question proper, there is also the more remote but highly sensitive matter of possible European military contributions to a future BMD system *for the United States*. The British and West German views on the need to avoid 'zones of differing security' have already been cited. But if such an actual or perceived defensive US privilege comes to require some form of European basing for some of its elements, the issue would certainly grow more acute. Not only would there be local resentment at providing facilities which might raise the already enormous human costs to one side of the alliance, should war ever come, whilst lowering them for the other. Fuel would also be added to the flames of West or Pan-European independence sentiments by the impression that, once again, major decisions about security arrangements in Europe would have been largely pre-empted by the very few Americans who were able to take part in the US decision, in the first months of 1983.

It is perhaps too early to speak of a full-blown 'SDI crisis' in NATO as a successor to the 'INF crisis' of the early 1980s. Nor is it possible yet to judge whether the major shift in US strategic nuclear policy, which the administration and its supporters, at least, consider the SDI to represent, involves any fundamental political changes within the alliance, and whether such changes, if they are being produced by these events, will be negatively or positively perceived by most people in European NATO countries. The military space element, however, does place new strains upon the transatlantic relationship. In the past, the autonomy of the US government over its own nuclear forces together with their ancillary hardware in space has never been in question. A BMD system supposed to provide a major physical element in the national security of each ally, however, might well create demands for joint access to at least the basic priorities written into its software. And if the United States begins to move weapons out into the great 'common' of space, just as its West European and other allies are starting to go into business with their own civil and non-weapon military spacecraft, some fairly acute conflicts of policy and interest are likely to result.

As to responses to the SDI from other parts of the world, the

Australian decision not to take part in the programme has already been mentioned. Prime Minister Hawke subsequently endeavoured to reverse or undermine it, but as of May 1986 his efforts had had little effect.

The fact that Mr Weinberger's invitation was sent to the Japanese government did not even rate a mention in the region's principal business magazine, the *Far Eastern Economic Review*. There have since been delegations from Tokyo to Washington which stated politely that the Japanese 'understood' the US position. But the two governments are in some disarray over a 1983 agreement on sharing military technology already, with US officials hinting that Japan is holding out on them in certain respects. Whilst such difficulties continue, progress on *effective* Japanese participation in the SDI seems highly improbable. Short of major political changes in Japan, however, there may be a cosmetic agreement in being before the end of 1986, despite the Nakasone government's still unrealized proposal for a Eureka-like $5 billion 'Human Frontier' initiative for advanced *civil* technology development, which was no sooner announced than it began to encounter major domestic political obstacles.

The very origins of the SDI were touched upon by one indirect account of probable reasons for both the US interest in such cooperation, and the gentle Japanese snub that it has so far received:

> The immediate US concern over possible Japanese inroads in supercomputers was evidenced in reports of scientists from Los Alamos and Lawrence Livermore National laboratories after their visits to Japanese computer centres in early 1982. Their conclusions that the Japanese could eventually surpass the Americans was [sic] substantiated in the following 15 months by successive announcements of progressively superior machines to be placed on the market by 1985. . . A super-sensitive security-related technology which the US had gone so far as to specifically deny to France for nuclear development in the late 1960s was clearly out of control. . . [W]hat in a techno-economic context is vital to the security of Japan is seen in the US as a threat to national supremacy in crucial military technology. (*Far Eastern Economic Review*, 31 January 1985)

Such evidence of a pre-SDI US concern for a threat *from its allies* goes far to support the 'French hypothesis' about what the SDI is really all about (Thompson, 1985: 119–20). As expressed by a former French Foreign Minister, EEC Commissioner Claude

Cheysson, this is that:

> In the name of the threat which they pretend hangs over the United States and Europe, it will be possible to inject considerable sums into scientific and technological research. The Americans wish in this way to recover their leadership in certain areas of the high technology of tomorrow. (*Libération*, 3 May 1985)

But clearly M. Cheysson was being far too trusting, or too diplomatic, to do justice to the anxiety of America's Japanese competitors, which has been not merely that the United States would intensify its own legitimate investment in the technological competition, by whatever bogus political device, but also that there would be an attempt to gain an illicit access for American capital to Japanese intellectual assets, under cover of the security scare about a Soviet 'space gap'.

The theory might also point to the fact that Britain, the weakest of America's major competitors in the relevant fields, was the first to give in to the transatlantic 'Star Wars' onslaught. Japan, however, is the most powerful commercial and technological rival in the world to the United States. Of all the major Western economies, Japan's has been least penetrated by US capital; her authorities have sought to limit the role of the yen as a currency on the international market; and the world's largest aerospace and other arms multinationals seldom include major Japanese participants.

By May 1986 the Japanese disinterest meant that, apart from rumoured deals in Argentina, the only major non-European participation in SDI had come from Israel, where a government-to-government agreement was signed, and several research and development contracts were clearly imminent.

As for the rest of humanity, uninvited by Mr Weinberger, unrepresented at the Geneva Summit or in the councils of NATO and the Warsaw Pact, their opinions on the question of space weapons may in part be represented by large votes at the United Nations in support of Soviet and other proposals for the 'demilitarization of space', as well as by support for the 'Five Continent Peace Initiative'.[15] The polite message sent by the latter to the Geneva Summit was perfectly clear about the need to prevent any extension of the arms race into space:

> We recognize as a positive development that during the present

year your governments have initiated in Geneva negotiations covering both space and nuclear arms to be considered in their interrelationship. We are concerned that such negotiations have not yet produced results. We feel, however, that various recent proposals and developments seem to offer new hope that both deep cuts in the arsenals of nuclear weapons and effective measures for the prevention of an arms race in outer space will now be seriously considered . . . (*New York Times*, 30 October 1985)

So much of the strategic competition between the United States and the Soviet Union is about their would-be political influence, prestige, and access to economic resources in the rest of the world that it would be unwise for either superpower to adopt a simple, 'How many ICBMs has the Pope?' attitude to such representations. Nevertheless, world public opinion is by its nature neither single-minded nor swift to take effect on such an issue. In recent years, this has been made all too plain by the lack of tangible results from the two UN Special Sessions on Disarmament, in 1978 and 1982, at which the governments of all UN member states declared, amongst much else, that 'to prevent an arms race in outer space, further measures should be taken and appropriate international negotiations be held' (in Huzzard & Meredith, 1985: 68). Regrettably, the immediate result of such widely held beliefs may be to intensify the efforts of the two powers currently making the greatest military use of space to excuse and legitimize their various activities, rather than to bring them to an end.

13
Wasting Space

Everything that happens to us – to the country, and to the human race, is strictly a matter of application. It's up to us to use properly the tools we have at hand . . .

[I]f we choose to use this vehicle to take us out into the solar system, I think we're moving the right way; in the only direction that's sensible. I know enough about this business to look on the future of the Thor not with a constant fear, but with hope. In my estimation, this missile – a weapon I helped to create, a weapon that can blot out the lives of millions of people – is going to become the true workhorse of our early explorations of the immediate solar system.

And that, goddammit, is a pretty wonderful thing to feel. It's something I like working with.

It does wonders for my sleep, too.

Thor missile engineer, night of 16/17 August 1958, Starlite Motel,
near Cape Canaveral

In all the long history of space weapons, from the early speculations of the 1920s to the laser laboratories at Livermore and Sary Shagan in the mid-1980s, there has never been such an excellent time for humanity to make up its mind about them as now. For that, at least, the current US administration should take some kind of credit. The very confusions and inconsistencies surrounding the Strategic Defense Initiative have been a part of the admirable frankness with which, for the past three years, it has been announced and conducted. If the decisions about space weapons that must now be taken by people in all countries sometimes amount to participating in the American political process at one remove, that is both a result of the latter's relative accessibility for outsiders, and a sign of where, in a practical sense, the issue must in large part be settled.[1]

Three years after the political birth of the Strategic Defense Initiative, what is to be made of its effects? Is it bringing the world perceptibly closer to the President's hopes of 'eliminating the threat posed by strategic nuclear missiles . . . [which] would pave the way for arms control measures to eliminate the weapons themselves' (Reagan, 1983)? Is it helping to secure a real detente between East and West, represented, perhaps, by the recent Geneva Summit? And technically, does it look like fulfilling George Keyworth's June 1985 prediction?

> Before the president leaves office [January 1989], we're going to be able to demonstrate technology that convinces the Soviets that we can – if we choose – develop a weapon to shoot down their entire ICBM fleet as it tries to enter space. (1985a: 2)

On the other side, what of Soviet space policy and programmes, both as they stood before the SDI came along, and as responses to the US initiative? What constructive options does the Soviet Union have, and is there any evidence that they may be taken?

The technical question about the SDI is probably the easiest to answer. And since it *is* a technical question the answer applies also to any Soviet attempt at developing full-scale population defences against nuclear missiles, whether or not such a programme is now going on in secret as the US administration claims. In a way Dr Keyworth answered it himself in London two weeks after making the boast just quoted, when he produced yet another assessment, as was his generous wont, this time for the European Atlantic Group (p. 000). He explained that, after all, 'achieving perfect defenses' was 'an unnecessarily stringent requirement'. And that 'the President's ultimate goal of impotence and obsolescence for nuclear weapons isn't based on the need for such perfection' (1985b: 5 – see p. 176 above).

Since its formation in 1984 the SDIO has frequently played down, even abjured, any talk of perfect or near-perfect population defence becoming a reliable prospect by the end of its five-year programme. It has stressed, instead, the short-term attainability of limited technical goals, such as ATBMs for deployment in Europe, or force defence by means of terminal and late mid-course interceptor missiles based in or near the United States. (This version of its objectives is sometimes described as dropping 'Star Wars 1' in favour of 'Star Wars 2'.) An even clearer sign that it now regards itself as running into some heavy technical going was given in its April 1985 Report to Congress (US DoD 1985b), which opened a Pentagon campaign to revise the State Department's strict criteria for any BMD system worth placing before US allies, promulgated by Ambassador Nitze at the start of the year (p. 245). These were, of course, firstly that any proposed BMD system should have overcome the problem posed by the possibility of active countermeasures, in which Soviet scientists have already shown some interest (Sagdeyev & Kokoshin, 1985), and be able to survive direct attacks upon itself. And secondly, that it should be 'cost effective at the margin'. This is just as much a key concept for force defence as for population defence. In both cases, the idea is to present the opponent with an unacceptable 'price' within the peacetime nuclear arms race, so that (in the slightly more feasible force-defence scenario) the fearful necessity of having to charge him an actual 'entry price', by requiring him to use so many offensive warheads to destroy each defended missile silo, will never actually arise.[2]

At the beginning of March, even before the report appeared, Dr Fred Hoffman a prominent SDI supporter, who had chaired the panel of outside experts in 1983, presented an authoritative analysis of the SDI's strategic rationale which paid little attention to the survivability criterion. Amidst much about the military value of imperfect BMD that was not contentious, in its own terms, but was not conclusive either, came the following, a supposedly definitive refutation of the critique of partial defences that they tend to undermine crisis stability:

> The grain of relevance in the argument is its identification of the problems presented by vulnerable offensive forces. It then superimposes partially effective defenses on the vulnerable offensive forces and concludes that defenses are destabilizing. (Hoffman, 1985)

Deliberately or not, Hoffman entirely missed the point of his opponents' argument, with which he certainly should have been familiar:

> The argument is made that defenses increase crisis stability by reducing the prospect of a successful first strike. *Such an argument fails if the defense is itself vulnerable to preemption.* Indeed, vulnerable defenses are particularly destabilizing during a crisis since enormous leverage can be obtained if only a fraction of them can be neutralized or destroyed. (Wilkening, 1984: 163 – emphasis added)

The Pentagon's report itself appeared to some commentators to be downgrading the 'survivability' requirement. Though survivability was listed as an important matter, there was at least an implication that it could not be an absolute requirement, and the evasive tautology 'adequate survivability' occurred at least once. Another expression, 'reasonable survivability', is believed to have been used in unofficial press briefings from administration members. But if the first Nitze criterion were weakened in any way, its point would be lost. After ruling out the meaningless 'survivable, provided the enemy does not attack it in unreasonably effective ways', a watered-down version of it might be that a BMD system could be deployed, provided it appeared more or less survivable to a scrutiny conducted without too much 'unreasonable' rigour. This approach may be related to the notion of an 'Initial Operating Capability', the Pentagon's traditional euphemism for deploying things first and trying to bring them up to adequate performance levels afterwards. In other words, a BMD system's 'initial operating survivability' would not have to be an actual capability to defeat attacks aimed at itself, merely a 'reasonably' good imitation of one. But whatever selfish gains were envisaged by the corporations who might build such a baroque fake, or by the generals who might command it, it is clear that defences which could not protect themselves from being bypassed or destroyed would be useless for protecting anything else. Dr Kissinger, for one, now sees the State Department as working against the SDI, above all in the person of Paul Nitze, one of its senior policy-makers:

> And now, with the SDI, he has stated criteria that are unfulfillable, so that – operationally – he is working more in the direction of limiting or destroying SDI. (In Charlton, 1985)

If Kissinger is right, a Pentagon campaign against the State Department over these issues seems inevitable, however discreetly it may be conducted.

The April 1985 report paid little heed to Nitze's second criterion, that defences have to be cheaper than any possible offensive build-up to counteract them by the other side. Its view appeared to be that such calculations would make no sense for many years to come. The idea of an economic relationship between building defences and the opponent's ability to proliferate offensive systems was not even mentioned in the relevant sections (US DoD, 1985b: C-24-5; 11–14).

The Pentagon's motive for side-stepping the second Nitze criterion is evident from the analysis by Dr Paul Rogers (1986), referred to above (p. 167). Dr Rogers, it will be recalled, found that by the year 1995, the earliest conceivable date for even a partial US BMD deployment, total Soviet strategic warheads could have risen from their present estimated number of roughly 9000 to as many as 31,000. (The Pentagon itself, of course, considers that just such a Soviet build-up has in fact begun.)

Matters became somewhat disordered in the summer of 1985, with the National Security Council issuing a National Security Decision Directive (NSDD 172) over the President's signature at the end of May, to the effect that the Nitze criteria still stood, and that enhancement of deterrence by protecting offensive missiles was certainly a major part of the SDI's objectives for many years to come. Only four days later, Dr Keyworth struck back with the 'absolutist' statement noted above (p. 263), in which he empha-sized the centrality of boost-phase interception systems (not required for silo defences) and stressed that rendering 'the ICBM useless as an offensive weapon' was now 'only a matter of time' (1985a: 2). At the end of October Weinberger returned to the attack on the Nitze criteria, with a clumsy declaration to the Senate Foreign Relations Committee that the second one did not mean very much, but that the kind of defences he, Weinberger, was after – evidently still of the 'thoroughly reliable and total' variety – would pass it anyway:

> Well, I have to say, senator, that I really do not know what cost-effective at the margins means. It is one of those nice phrases that rolls around easily off the tongue and people nod rather approvingly because it sounds rather profound. I have the greatest admiration for Ambassador Nitze, but I do not know specifically what he has in mind with that. If he means is it less expensive to build strategic defenses than continually to engage in trying to add offensive systems, I would say the syllogism proves itself. It is clearly less expensive because the defense can, in effect, ultimately, if it is as effective as we hope it is, make it quite apparent that further offensive systems are not useful.

To which Committee member Senator Gore responded:

> Any decision to discard this criterion would strip the program and the concept of its last shred of intellectual legitimacy. It would only stimulate a race to deploy offensive countermeasures. This was the

realization that led us to the ABM Treaty in the first place. If they do this, they're saying, 'Damn logic, damn reasoned debate, full speed ahead!' (*International Herald Tribune*, 18 December 1985)

Besides the recent setbacks to the X-ray laser programme, and the SDIO's decision to respond by doubling its budget (p. 189), two further, more general features of the SDI's recent history indicate continued serious self-doubts about its own feasibility. The steadily increasing complexity of the prevailing rough 'architectural' blueprint for an eventual system, from three, through four to seven or more layers, may have something to do with a desire to let every major contending research centre feel sure of a seat at some future contracts banquet. But it also reflects an increasing recognition that the efficiency of any individual defensive technology is likely to be very low, against what the offence could do by way of penetrative countermeasures.[3] Talk of a new 'vertical layering' approach to BMD does rather little to disguise the basic fact that what President Reagan and Dr Keyworth have repeatedly referred to as a single entity, with such expressions as 'a weapon' and 'this weapon', would in fact be an extremely complex and poorly integrated panoply. Even setting aside the question of its enormous costs, this suggests that it will have all the usual disadvantages of military over-elaboration (Kaldor, 1982), as some at least of those who are developing it are aware. As one key computer scientist within the Lawrence Livermore Laboratory's weapons development team expressed it:

> I get a little worried about the complexity of the defensive systems we are talking about, which have to function autonomously. This isn't too difficult a technology for technocrats like us. But with the maintenance staff you get in the volunteer Navy, I worry about a system that smart. (In Broad, 1985: 82–3)

The second aspect of the SDI's current self-evaluation that should influence the way others see it is the emerging pattern of priorities within its overall weapons development programme. This pattern (buzzword 'temporal layering'[4]) suggests that the programme's original bias, towards longer-term, possibly more effective but also more speculative options, is growing even stronger, perhaps due to improved understanding of the technical problems it faces, including possible countermeasures. The result is a marked emphasis on less mature, more scientifically demanding ('advanced') options, such as free-electron lasers, at the expense of more mature but less militarily feasible ones, such as infra-red chemical lasers. But whilst the former, 'gold' class of mid-term technologies is planned to receive the lion's share (65 per cent) of the programme's budget for the next two years, the bottom-division 'silver' class will get 25 per cent of the money in return for all the politically valuable 'demonstrations' they can provide, regardless of whether they are still thought capable of being developed in a form that would overcome their basic unsuitability for actual deployment.[5]

This approach to the organization's combination of technical and political problems was outlined by its Chief Scientist Dr Yonas in December 1985, to participants in the Laser '85 conference at Las Vegas. It is a fairly transparent bid to cope with the difficulty that no *really* adequate BMD technology, by any criteria, has much chance of being demonstrated within the next two or three years, despite repeated promises to the contrary. What it means, for example, is that during 1986 and 1987 the EMRLD excimer laser programme will be 're-configured' from its current role of 'baseline technology development' to provide a demonstration laser at the White Sands Missile Range. Hard or soft, the targets for such a 'weapon' are not to be found in Soviet missile silos, but on Capitol Hill. Doubtless by 1988, if not before, some suitably encouraging news will be beamed from White Sands to Washington, in the hope that hands on the appropriate committees will be duly raised.

The questions of what the SDI and other space weapons programmes are likely to cost the United States, and of how they come to be brought forward at all in the current state of its economy, are too large to receive a thorough answer here. But they certainly cannot, in the real world, be ignored. Indeed, the struggles between the administration and Congress over what is to be done to redress the enormous US budget deficit now seem more likely than ever to affect the medium-term prospects for such programmes. The passing of the Gramm–Rudman–Hollings Balanced Budget Act in December 1985 requires an immediate reduction of some $40 billion in the domestic US budget deficit of over $200 billion, and further similar cuts thereafter.[6]

No US decision-maker can think of the SDI purely in terms of its present rising but still 'modest' share of the federal budget. The enormous expenditure which further development and eventual deployment would require must always be borne in mind (appendix 2). The programme should therefore be seen against the general background of United States economic decline, with the dollar substantially down against other major currencies in 1985, and with the United States transformed, if perhaps not permanently, from a creditor to a debtor nation. Of course, worsening prospects for the defence industry as a whole do not necessarily endanger the SDI. Thanks to its exceptionally high political profile, it may come to seem more and more like a life-raft to which every major aerospace corporation will attempt to cling, at a time when others, such as General Motors, are seeking to enlarge their share of the military economy in order to take shelter from the even less hopeful situation on the civil side.

The outcome for the SDI and its successor programmes will doubtless be affected by the large sums given to members of key congressional committees, by aerospace corporations anxious to protect their high profits and extraordinarily favourable tax status from the slight but growing risk of presidential reform. (The money reaches legislators indirectly, of course, through things called Political Action Committees.) As George Ball, a distinguished former diplomat and official under several past administrations, recently explained:

> Perhaps the most effective support for Star Wars is now being generated not by ideology but by good free-enterprise greed. Firms in the hypertrophic defense industry, along with their thousands of technicians, are manifesting a deep patriotic enthusiasm for Star Wars. Since they are experienced in lobbying and wield heavy influence with members of Congress who have defense plants in their constituencies, they are creating formidable momentum for the project. Whether or not it would contribute to the security of the nation, it offers them security. Thus an investment analyst for the industry published a newsletter about the President's space program entitled 'Money From Heaven', while another analyst wrote: 'For the US aerospace industry the redirection of the strategic arms competition towards defense can hardly come soon enough.' (*New York Review of Books*, 11 April 1985)

Meanwhile the town of Aurora, Colorado, sixty miles from the Pentagon's new $1.4 billion Space Defense Operation Centre, has begun calling itself 'Star Wars City', and expects to become the largest city in its state within ten years. Concern is growing amongst economic experts at the contrast between such islands of

artificial prosperity, based on military subsidies and institu-
tionalized inefficiencies, and the underlying trends of a vast
foreign debt (estimated in January 1985 as \$7.2 trillion – three
times its 1975 level), of an overvalued dollar, of falling industrial
employment and productivity, and of shortages of industrial
investment, as savings are attracted elsewhere by the high interest
rates in the finance markets which are produced by the funding
arrangements necessary to sustain the deficit. No wonder the
connection between the US military budget and the worsening
state of the US economy is beginning to be noticed around the
world. As General Secretary Gorbachev put it when meeting a
visiting Congressional delegation in April 1985: 'Maybe we should
wait. Maybe you'll want to talk when things get worse' (*Washington Post*, 18 April 1985).

The SDI seems unlikely either to help or to be helped by the US
economic crisis, in which the proportion of people in lower income
groups continues to rise, 40 million Americans live below the
official poverty line, and 20 million are estimated to be chronically
short of food. But such historical periods, which combine declining
standards and expectations for the many with shallow and
short-lived prosperity for an irresponsible few, many of them in
uniform, are notoriously prone, not merely to eventual crisis and
even war, but also to a credulous acceptance of phoney solutions
for major problems. And not just those concocted in the finance
markets either. It is not far-fetched, therefore, to suggest that a
disturbing cultural connection exists between 'Star Wars' and the
shaky state of the US economy which might, absurdly, work *in
favour* of the SDI.

The 'I want it! I want it!' tone with which the 'vision' of BMD
has recently been addressed in some quarters is nothing new
(chapter 10, note 15). It is perhaps not untypical of a society whose
millionaires can sometimes be persuaded to invest in that ultimate
technological fix, the deep-freeze Californian catacombs whose
occupants are supposed to have secured an exclusive lien on
earthly resurrection. And it is of course just one more example of
what Herbert York long since dubbed 'the fallacy of the last step',
meaning the mistake of supposing that there is a winning post in
the arms race, some unanswerable force deployment which must
leave the other side permanently beaten into strategic inferiority
and hence political subordination.

But what matters in the late 1980s is the close fit between two
psychological patterns. There is the turning away, by leaders, by

many 'experts', and by a substantial, even perhaps a decisive, section of the US electorate, from the increasingly gloomy facts of their economic situation, towards fantasies such as that Gramm–Rudman–Hollings itself provides a solution, or that the rip-off exploits of corporate merger-mongers have something to do with real economic growth. And there is the subject of this book, the 'Star Wars' make-believe, which imagines that 'our' side might acquire some new, unbeatable West-Coast wonder-weapon, whilst the opponent's forces remain spell-bound, conveniently frozen at their present numerical and technical levels.

> Space weapons are attractive because of their glamor and their 'magical' qualities. To judge by the fairy tales of yesteryear and their contemporary counterpart science fiction, people have always been fascinated with the concept of the little magical device that bestows immediate almost supernatural powers on its owner: the good fairy's magic wand, Aladdin's lamp and, more recently, Flash Gordon's death ray, the phaser of Star Trek and the beam weapons of Star Wars. What Congressman would dare not vote for funds for such magical weapons that in the public's mind would protect the country from the 'evil Empire'? (Tsipis, 1984)

But the technical, political and economic saga of the SDI as a weapons technology programme is necessarily far from its conclusion, and many millions of gallons of water will doubtless flow through the cumbersome cooling systems of its static lasers before that day is reached. It is both more urgent and more possible, however, to evaluate the immediate effects of the space weapons competition on East-West relations. The Reagan administration and other supporters of the SDI believe that thwarting Congressional attempts to restrain their asat programme and starting a new drive for strategic defences have done nothing but good in this area, by demonstrating the administration's concern for an alternative to MAD, and thus taking the wind from the sails of the American anti-nuclear peace movements. This, they say, has begun to restore the internal political 'strength' required for the conduct of US foreign policy. The Soviet government has allegedly been obliged to seek a new, more constructive relationship with its US counterpart, as evidenced by the Geneva Summit. And this in turn has given a new lead, supposedly, to a Western alliance increasingly troubled by the lack of progress with disarmament since the end of the 1970s, and by the grim prospect of an unending competition in offensive nuclear weapons as the only apparent alternative.

Even without contesting such claims directly, it is possible to rephrase the question in a form the SDI lobby might wish to answer differently. If space weapon programmes are good for East-West relations, why have they themselves objected so strongly to those they attribute to the Soviet Union? Why has a single radar station, however arguably illegal, been made into such a Cold War issue? If the Soviet Union had responded to the SDI with an indignation proportional to the Pentagon's anger over the far smaller matter of Abalakova, what scope do they think would now remain for diplomacy?

It is possible, of course, that one side's view of what is good for a two-sided relationship is correct, even if it is disputed by the other party for a short time. This possibility becomes increasingly remote as time passes and indeed, in this case, is likely to prove self-defeating. However, supporters of the SDI appear to see the matter as follows: If the United States can demonstrate her strength, and her determination to retain it, the Soviet Union may be forced to abandon the arms race and to negotiate from a position of weakness; however, if that does not come about, the aggressive purposes of the 'evil Empire' will at least have been exposed. It seems probable that US hawks are more interested in the second of these propositions than the first.

Many of the young scientists in the X-ray laser and other 'Star Wars' programmes at the Lawrence Livermore Laboratory were screened for their basic political 'soundness' as well as for intellectual brilliance by the right-wing Hertz Foundation, whose board includes Edward Teller and Livermore's X-ray laser boss Dr Lowell Wood, and has in the past contained such people as J. Edgar Hoover and Curtis LeMay. The scientists whom Teller and Wood have gathered round them have little time for the view that the SDI should benefit the Soviet Union as much as their own country. Here is a representative sample:

'I'd like to see them try to escalate and spend their entire budget and see their country go to ruin. I only hope it would go that way.'

'If we can pull away from the Soviet Union and leave them in the dust along with Afghanistan and India, technologically, then in essence we've won.'

'I disagree that you want things that are to the advantage of both powers. We're only playing one side of this game.'

'I don't see how you could work on nuclear weapons if you didn't care about the Soviets taking over. As for me, I have to be anti-Soviet first. I don't give a shit what they do to their own people.' (In Broad, 1985: 57, 63, 82, 144)[7]

However, research degrees in physics may not be the best qualifications for understanding either the Soviet Union or the East-West relationship:

> We found we were playing [war games] against defense contractor personnel and others who knew nothing about Soviet doctrine. It took our whole team, the Red Team, less than 20 minutes to agree that our first counter to 'star wars' would be to increase offensive missile numbers. Their team, the Blue Team, said 'No, that is not how the Soviets think'. Every step we took surprised them. (Soviet affairs specialist, *International Herald Tribune*, 20 December 1985)

The claim that the Geneva Summit's 'agreement to continue talking' was in large part due to the effect of the SDI in 'bringing the Soviets to their senses' is surprising, when made by people who at other times point freely to the ulterior uses that the Soviet Union either has or may believe itself to have for such diplomatic opportunities. Perhaps the SDI did have something to do with the resumption of bilateral arms control talks in 1985, and thus indirectly with the Summit. But the nature of any such link is more important than its existence. It is only possible to cite it as evidence of an improvement in relations if the talks stop being used solely as an inexpensive medium for the struggle between the superpowers, as they were throughout 1985, and start to produce some genuine results. Unfortunately the first round of negotiations in 1986 showed no improvement.

In the interval between the writing and the publication of this book, further bilateral negotiations may show even more clearly whether or not the Summit made any real difference to the prospects for arms control. If hope has anything to do with reason, it is always reasonable to go on *hoping* for progress in disarmament. In chapter 11, however, it was suggested that the Reagan strategic nuclear policy provides Soviet leaders with excellent motives for either being or seeming to be on the verge of signing agreements for widespread reductions in offensive weapons, if at all possible, whilst all the time refusing actually to do so as long as no concessions are forthcoming on SDI. If that assessment has any validity, it would be unreasonable to *expect* much in the way of actually dismantled nuclear weapons from the 'Star Wars' epoch.

This judgement is supported by the continuing friction, to put it no stronger, between the SDI and the ABM Treaty, and indeed, by the US government's own account, between Soviet space weapon programmes and the Treaty also. It would be a little more consistent, however, for the Reagan administration to say the same thing in both cases. Either the early stages of developing effective space weapons do amount, in both the United States and the Soviet Union, to infringing the Treaty by starting to provide a 'base' for BMD, or they do not. Either *all* new early-warning radars not on the periphery of the United States or the Soviet Union are in breach of the Treaty, or they are not. Of the two options, the first seems more convincing in both cases.

In view of the difficulties the ABM Treaty is now in, the only sensible line by which to judge whether one or other party is observing it effectively is to assess *the degree of effort being spent on rescuing it from imminent disaster*. Merely standing back and firing off accusations at the other party, thereby incidentally making the prospect of an effective restoration of the Treaty still more remote, ought not to count with world opinion as a way of meeting the obligations it imposes. For these include promoting 'the objectives and implementation of the provisions of this Treaty' by, amongst other things, considering 'as appropriate, possible proposals for further increasing the viability of this Treaty' ((XIII)).

Appendix B of the Pentagon's April 1985 report, dealing with the SDI's compliance with the ABM Treaty, met with objections from the 'arms control lobby' to its use of the components gambit (p. 219) to loosen legal restraints upon the SDI. The objectors are very unlikely to have made any mistakes about the legal position. However, they often seemed to overlook the fact that, even if this was a piece of legal trickery, it was by no means a new one, but had been going on for some time. The 'McFarlane Version' of the Treaty, on the other hand, is a more serious legal and political development precisely because it has never received serious consideration in the past, let alone been formally endorsed by the highest political authorities in the United States (p. 225). The political controversy in Washington was exacerbated by the administrations decision to hold these more permissive 'boundaries of treaty interpretation' in reserve, as a threat that would not be implemented 'as long as the program receives the [budgetary] support [from Congress] needed to implement its plan' (US ACDA, 1986b).

The 1986 ACIS thus confirmed that there had been a radical break with long-established US policy on the ABM treaty, creating a short-term outlook for arms control can only be assessed as bleak. As well as the Soviet Union, America's allies in NATO and elsewhere will strongly resent such a prospect. The bargain which some of her closest partners in Western Europe appear to have reached with the United States over the SDI emphatically does not include support for the programme's long-term ends. Nor does it include any license for further erosion of the ABM Treaty, any more than it does a willingness to wait indefinitely for further real progress with bilateral nuclear arms control, the last substantial result from which the unratified SALT 2 agreement, came in 1979. All around the world, from Ottawa and Tokyo, from Canberra and Bonn, and even from London, governments friendly to the United States have spoken on this matter with a single voice. Their patience with a bloc leader which remains deaf to such complaints cannot be without limit. And even if their representations can continue unheeded for a while longer, before the deterioration in their relations with the United States becomes irreparable, calims that the SDI has benefited NATO and will continue to do so are likely to be as counterproductive as they are unconvincing.

US leaders may nonetheless continue to push within NATO and elsewhere the strategic reform for which, above all, they value the SDI, whilst their increasingly resentful allies wait for them to be out of office or out of pocket, or both. To the degree that either half of the partnership finds the other's stance less and less tolerable, West European firms and governments will struggle with increased determination to protect themselves from US technological raiding, but if possible to gain something tangible from the programme. As the process unfolds, Professor Chomsky's insight of a few years ago (1981), to the effect that NATO is essentially an arrangement to conceal and to repress the natural antagonism between the United States and Western Europe, and by implication also the natural community of interests between the United States and the Soviet Union, may seem increasingly convincing to many Europeans. The upheavals that may then transpire could be at least as great as those surrounding the so-called 'Euro-missiles crisis' of the early 1980s. They will have their dangerous side; but they will also have their benefits and their opportunities, long wished for by many Europeans in all 'zones'.

Disarmament, and above all nuclear disarmament, is not a process to be evaluated in terms of force levels alone, desirable as it is for these to be reduced. What matters even more is the reduction of the probability of war, by means of an increasing demilitarization of the conduct of planetary human social life, which for the time being revolves around 'international relations'. Human endeavours in space so far, however, may be assessed as an opportunity to assist with this process that has been spectacularly missed.

Most experts on the military uses of space subscribe to the thesis that there is a so-called 'arms control dilemma' in this field:

> It is obvious that there is a dilemma. Military spacecraft have contributed considerably to the formulation of strategies for actually fighting nuclear war rather than just deterring it. On the face of it it would seem a reasonable idea to ban all military satellites. But would it? These same satellites can be used to monitor treaties safeguarding international security; they can be used to maintain military communications so that war by miscalculation is less likely – but some form of treaty is needed to protect satellites for peaceful and stable use. There must also be other ways of reducing the destabilizing effects of the military use of space. (Jasani and Lee, 1984: 93)

This dilemma, or contradiction, is a far more fundamental problem for disarmament than the damage done by East-West struggles for propaganda or military advantage around any recent or current negotiations, or by the SDI's attack on the ABM Treaty, or by any possible Soviet infringement of this or that arms control provision. It is not enough simply to ascribe the shortcomings or absence of arms control in space to the misjudgements or tardiness of particular governments, as does Stares:

> In hindsight, the early part of the 1980s will most probably be viewed as a fundamental watershed in the militarization of space. During this period the chance for a significant antisatellite arms control agreement was lost – possibly for ever. The Reagan administration squandered an opportunity to take advantage of unprecedented Soviet flexibility on this issue. By the time it had begun to reconsider US policy, the Soviet position had hardened. . . tests of the US ASAT system and the commitment to the SDI will only complicate further negotiations on this topic, no doubt leaving future administrations to regret bitterly this lost opportunity. (Stares, 1985: 235)

To qualify the position taken by Stares does not imply a willingness to exonerate either side in the military space competition. Whilst the United States is clearly responsible for much of the world's current problem with space weapons, the Soviet Union cannot be judged entirely blameless in the matter, for all the flexibility and positive aspects of her negotiating positions in recent years. The persistence with which militarily valueless ABMs and asats were developed and deployed was a needless source of destabilizing anxiety to the West. The paucity of official information about Soviet space programmes has enabled Western 'hawks' to peddle worst-case scenarios with impunity. And positive Soviet moves at the United Nations have too often courted dismissal as primarily a propaganda show, because they seemed to be out-of-step with a security policy based firmly on military strength and on a preoccupation with the bilateral relationship between the superpowers. The opportunity now exists to proceed at least with informal negotiations on a treaty for comprehensive arms control in space, whether or not they have been blocked by the United States at the UN Committee on Disarmament. If necessary, a start could be made on constructing the basic legal and institutional framework for collective human activity in space, leaving the United States to come to its senses and rejoin the rest of the planet later. But there are few signs that Soviet leaders will actually adopt such a course of action, instead of merely seeking political advantage from the failure of the Americans to do so.

It is possible, indeed, that the Soviet Union may be no more prepared than the United States to part with the freedom to pursue purely national military interests in space. Though Soviet military space priorities may well be different, and for the time being less inherently dangerous, this remains a fundamentally reactionary direction for space policy. If it is indeed shared by the two opposing superpowers, that might be the most threatening aspect of the whole situation.

Such a harsh judgement can be explained by considering the kind of legal and physical controls that might be needed, ideally, to ensure that future space systems would no longer spur on the arms race, but instead would help to supervise its steady deceleration. A ban on the development and testing of any weapons designed to function in, from or against objects in space, as proposed in the 1983 Soviet draft treaty (p. 226), would only be the start.

It was argued above (pp. 44–5) that the development of asats is driven by the increasing capacity of military satellites to serve as 'force enhancers' during actual combat. (A related problem is the multiple overlap between non-weapon military satellites and corresponding 'civil' space systems, for instance for communications, terrain surveys, and meteorology.) As long as such devices are placed in orbit, there would be strong military motives to oppose an effective ban on asats. And each side might have to regard as potential asats almost any of its opponent's space objects, especially those carrying a human crew, that could come within range of its own military space assets.

One partial solution might be for spacefaring nations to agree to stop using their ever more capable data-processing technologies to build satellites with functions that can not merely assist military preparations in peacetime but also threaten to make a crucial difference in actual war-fighting, should it ever come about. An agreed halt on these lines might in turn diminish the motivation for acquiring space weapons which those technical capabilities increasingly provide. But even such negative agreements, if drafted in verifiable form, would have to reach much further than anything yet seen in arms control. They would require a large degree of active and direct international supervision over national space systems. It might be necessary for reconnaissance satellites to be checked by an international agency, and certified free of any 'real time' transmission capabilities. Civil communication satellites, at least, might have to be certified as lacking any capacity for the sophisticated evasion of intercepts by such techniques as rapid and randomly programmed frequency shifts, 'squirt' transmission, or super-high frequencies. Navigation satellites might need to be operated as a single, shared and publicly available system, with no special capabilities reserved to the military forces of only one side. And so on.

It quickly becomes obvious that encroachments on 'sovereignty' of that order are unlikely to develop painlessly from the present institutional basis for arms control. If simply postulated by themselves, in utopian fashion, they may seem hardly worth discussing. But the exercise does illuminate the basic political problem with space technologies, whether military or not. Machines that have obviously global functions seem ill-adapted to being used and controlled by insecure nation-states for their partial and therefore humanly inadequate purposes. Nevertheless, it was nation-states that built and launched them, and which have made the international 'law' of space which directs that they remain, for the most part, under national sovereignty.

At the same time, the inappropriateness of imposing national or corporate limitations onto portions of the solar system, be they regions of near-Earth space, or celestial bodies like the Moon or planets, has been legally acknowledged through an enactment of the United Nations, the flawed but still surviving collective political assembly of our species.

In the long run, this fundamental dilemma of space technology can only be resolved in one or other of two ways. Either the limited political authority and partisan priorities which characterize present space systems will continue to call the tune, in which case they will continue to be dominated by military requirements. As such, they are likely to provide the trigger, or a volatile catalyst, for a planetary holocaust. Or else their human creators can grow towards a more cooperative planetary society, one that matches their increasingly universalizing technologies, whilst accepting and valuing the human diversity which could then be preserved, communicated and more effectively enjoyed.

Some may argue that, just as in the past, the most effective and most probable agent for unifying divided and feuding human groups would be the political or military exercise of superior force. That would be to resolve the contradiction along lines surmised by Golovine a quarter of a century ago:

> The winner of the space war would . . . be in a position to dictate terms not only to the losing side but to all its allies and 'uncommitted' nations. In fact, a World Government would be created, but bearing little resemblance to the Utopian dreams of those who seem to be constitutionally unable to comprehend the grim realities of twentieth-century Power Politics (1962: 119).

For reasons given above (p. 192), this no longer seems as practicable an option as it did to Golovine. The alternative route, of a gradual, peaceful and increasingly democratic reconciliation between societies now in conflict may well give rise to anxieties, and worse, about the major adjustments and sacrifices that would have to be made. Such fears may even have contributed to the rightward movement of public opinion over recent years in many Western countries, and hence to some of the problems discussed above. Nevertheless, the fantasies of politicians who denigrate disarmament and cry up the myth of a decisive Western military superiority in its place are no basis for a sensible strategic policy.

This is not a plea for the revival and restoration of old-fashioned 'arms control'. In the 1970s it was partly the inadequacies of that process that brought matters to their present pass. Something more radical is needed, based on large-scale public participation. Space technologies themselves offer, perhaps, an opportunity for democratic and internationalist alternatives to absolute national sovereignties. For instance, what amateur enthusiasts can already achieve by way of access to non-encrypted satellite telemetry could be made available to every television set.

To understand what harm is being done to present and future generations by the worsening military competition in space, it is necessary to understand the opportunities that confront humanity as we begin to move around the solar system. Over and above the vast and more obvious material possibilities are two which can only be measured by what they might add to our values, our imagination, our self-knowledge, in short, to our collective wisdom. The greater one, already appreciated by many millions of people, is that somehow, some day, our human civilization may come into beneficial contact with beings like ourselves, yet rewardingly different from ourselves.[8]

The importance of that opportunity is equalled only by its probable remoteness from our own time. It is conceivable that future human generations might inhabit many star systems in our small corner of our own galaxy, over thousands of years, without bringing it appreciably closer. They would have to face the moral and spiritual problems of such a frustration, if it assailed them, in their own fashion.

But the important business that our generation and those to follow it in the twenty-first century have with space lies with the opportunity we have already made for ourselves with space technology. And that is, the chance it gives us to grow up. What humanity is wasting, or allowing to be wasted in space today, is above all else the scope it provides for a major advance in the long historical process of our self-development towards greater wisdom and awareness of ourselves as a planetary human community, and indeed as part of a planetary ecology embracing also the millions of other living beings and their supporting physical systems. People may differ as to whether or not they choose to see that process as itself a part of the preparation of our species for the greater opportunity, should it ever come about, of that almost unimaginable encounter. What should be incontestable, however, is that such an increase in maturity is desperately necessary in the here and now, for our general well-being and in particular for our security.

This small occasion of our first venturing into space follows many previous but even more limited steps towards the same end, that were taken more or less successfully in the human past, from monotheism and coinage, through printing and circumnavigation, to modern telecommunications and data processing. Unlike all these, however, our arrival on the edge of space presents us with a chance not merely to know or to practise, but also to see and to feel, almost directly, the growing unity of human society. We may not yet have reached the point, for instance, at which a simple picture of the planet seen as a whole from space is displayed in every place of public human assembly, whether that room or building serves a religious, a commercial, a recreational, or a political function. But at least it has already become possible to think along such lines, and the process has begun in a small way, though still frustrated by petty military rivalries.

We do still have that lesser chance, the chance of our lifetimes rather than of any that may come after us. We may be wasting it, out of widespread ignorance and officially managed deception, but we have not yet finally lost it. And that is where a critical understanding of the motives and intentions behind space weapons comes in. Paradoxically, the disastrous loss of the Shuttle *Challenger*, which occurred just before this book was completed, may have done much to help people in this respect. As one British columnist explained:

> The urgency of the accelerating [Shuttle] programme scheduled for 1986 comes from its function in the Strategic Defence Initiative. This means that for all the elevating rhetoric about the questing spirit, the fearless probe into the unknown, the determination to carry on 'for their sake', the series of 'missions' also involves the launch into space of conflicts and contests that are all too familiar and terrestrial. The spirit of adventure is one thing; but its outcome, in this context, is unlikely to be half so noble and altruistic as the comment on this unhappy event would have us believe. Thus, the conquest of the cosmos is a logical extension of struggles in a world whose very furthest corners have been penetrated by ideological wars. Indeed, mere dominion over the planet is no longer enough; nothing less than infinite space becomes the object of victory. (*The Guardian*, 3 February 1986)

Jeremy Seabrook's references to 'the conquest of the cosmos' and to 'infinite space' may have been exaggerations, begotten by the very rhetoric that was being called in question.[9] But his point about the distortion of human aspirations and ideals in current space programmes is no less valid for that. If there was any doubt in the matter, the negative results that have already been produced in world affairs by the Strategic Defense Initiative are surely a tangible and by now decisive proof.

Space will not solve our problems just because we can reach it. The ability to 'touch the face of God' is not a matter of altitude, as it seems to have appeared to the fundamentalist imagination of the American President and, one suspects, to many of his fellow-citizens. There is only one way we can start to 'humanize space' as the dead schoolteacher-astronaut Christa McAuliffe wanted. And that is by using space more effectively to humanize ourselves. Space weapons can only delay that process or, if we fail to renounce them, bring it to an unnecessary, untimely and unjustifiable conclusion.

Appendix 1 The ABM Treaty

Treaty Between the United States of America and the Union of Soviet Socialist Republics on the Limitation of Anti-Ballistic Missile Systems

Signed at Moscow May 26, 1972
Ratification advised by U.S. Senate August 3, 1972
Ratified by U.S. President September 30, 1972
Proclaimed by U.S. President October 3, 1972
Instruments of ratification exchanged October 3, 1972
Entered into force October 3, 1972

The United States of America and the Union of Soviet Socialist Republics, hereinafter referred to as the Parties,

Proceeding from the premise that nuclear war would have devastating consequences for all mankind,

Considering that effective measures to limit anti-ballistic missile systems would be a substantial factor in curbing the race in strategic offensive arms and would lead to a decrease in the risk of outbreak of war involving nuclear weapons,

Proceeding from the premise that the limitation of anti-ballistic missile systems, as well as certain agreed measures with respect to the limitation of strategic offensive arms, would contribute to the creation of more favorable conditions for further negotiations on limiting strategic arms,

Mindful of their obligations under Article VI of the Treaty on the Non-Proliferation of Nuclear Weapons,

Declaring their intention to achieve at the earliest possible date the cessation of the nuclear arms race and to take effective measures toward reductions in strategic arms, nuclear disarmament, and general and complete disarmament,

Desiring to contribute to the relaxation of international tension and the strengthening of trust between States,

Have agreed as follows:

Article I

1. Each party undertakes to limit anti-ballistic missile (ABM) systems and to adopt other measures in accordance with the provisions of this Treaty.

2. Each Party undertakes not to deploy ABM systems for a defense of the territory of its country and not to provide a base for such a defense, and not to deploy ABM systems for defense of an individual region except as provided for in Article III of this Treaty.

Article II

1. For the purpose of this Treaty an ABM system is a system to counter strategic ballistic missiles or their elements in flight trajectory, currently consisting of:

   (a) ABM interceptor missiles, which are interceptor missiles constructed and deployed for an ABM role, or of a type tested in an ABM mode;

   (b) ABM launchers, which are launchers constructed and deployed for launching ABM interceptor missiles; and

   (c) ABM radars, which are radars constructed and deployed for an ABM role, or of a type tested in an ABM mode.

2. The ABM system components listed in paragraph 1 of this Article include those which are:

   (a) operational;
   (b) under construction;
   (c) undergoing testing;
   (d) undergoing overhaul, repair or conversion; or
   (e) mothballed.

Article III

Each Party undertakes not to deploy ABM systems or their components except that:

   (a) within one ABM system deployment area having a radius of one hundred and fifty kilometers and centered on the Party's national capital, a Party may deploy: (1) no more

283

than one hundred ABM launchers and no more than one hundred ABM interceptor missiles at launch sites, and (2) ABM radars within no more than six ABM radar complexes, the area of each complex being circular and having a diameter of no more than three kilometers; and

(b) within one ABM system deployment area having a radius of one hundred and fifty kilometers and containing ICBM silo launchers, a Party may deploy: (1) no more than one hundred ABM launchers and no more than one hundred ABM interceptor missiles at launch sites, (2) two large phased-array ABM radars comparable in potential to corresponding ABM radars operational or under construction on the date of signature of the Treaty in an ABM system deployment area containing ICBM silo launchers, and (3) no more than eighteen ABM radars each having a potential less than the potential of the smaller of the above-mentioned two large phased-array ABM radars.

Article IV

The limitations provided for in Article III shall not apply to ABM systems or their components used for development or testing, and located within current or additionally agreed test ranges. Each Party may have no more than a total of fifteen ABM launchers at test ranges.

Article V

1. Each Party undertakes not to develop, test, or deploy ABM systems or components which are sea-based, air-based, space-based, or mobile land-based.

2. Each Party undertakes not to develop, test, or deploy ABM launchers for launching more than one ABM interceptor missile at a time from each launcher, not to modify deployed launchers to provide them with such a capability, not to develop, test, or deploy automatic or semi-automatic or other similar systems for rapid reload of ABM launchers.

Article VI

To enhance assurance of the effectiveness of the limitations on ABM systems and their components provided by the Treaty, each Party undertakes:

(a) not to give missiles, launchers, or radars, other than ABM interceptor missiles, ABM launchers, or ABM radars, capabilities to counter strategic ballistic missiles or their elements in flight trajectory, and not to test them in an ABM mode; and

(b) not to deploy in the future radars for early warning of strategic ballistic missile attack except at locations along the periphery of its national territory and oriented outward.

Article VII

Subject to the provisions of this Treaty, modernization and replacement of ABM systems or their components may be carried out.

Article VIII

ABM systems or their components in excess of the numbers or outside the areas specified in this Treaty, as well as ABM systems or their components prohibited by this Treaty, shall be destroyed or dismantled under agreed procedures within the shortest possible agreed period of time.

Article IX

To assure the viability and effectiveness of this Treaty, each Party undertakes not to transfer to other States, and not to deploy outside its national territory, ABM systems or their components limited by this Treaty.

Article X

Each Party undertakes not to assume any international obligations which would conflict with this Treaty.

Article XI

The Parties undertake to continue active negotiations for limitations on strategic offensive arms.

Article XII

1. For the purpose of providing assurance of compliance with the provisions of this Treaty, each Party shall use national technical means of verification at its disposal in a manner consistent with generally recognized principles of international law.

2. Each Party undertakes not to interfere with the national technical means of verification of the other Party operating in accordance with paragraph 1 of this Article.

3. Each Party undertakes not to use deliberate concealment measures which impede verification by national technical means of compliance with the provisions of this Treaty. This obligation shall not require changes in current construction, assembly, conversion, or overhaul practices.

Article XIII

1. To promote the objectives and implementation of the provisions of this Treaty, the Parties shall establish promptly a Standing Consultative Commission, within the framework of which they will:

(a) consider questions concerning compliance with the obligations assumed and related situations which may be considered ambiguous;

(b) provide on a voluntary basis such information as either Party considers necessary to assure confidence in compliance with the obligations assumed;

(c) consider questions involving unintended interference with national technical means of verification;

(d) consider possible changes in the strategic situation which have a bearing on the provisions of this Treaty;

(e) agree upon procedures and dates for destruction or dismantling of ABM systems or their components in cases provided for by the provisions of this Treaty;

(f) consider, as appropriate, possible proposals for further increasing the viability of this Treaty; including proposals for amendments in accordance with the provisions of this Treaty;

(g) consider, as appropriate, proposals for further measures aimed at limiting strategic arms.

2. The Parties through consultation shall establish, and may amend as appropriate, Regulations for the Standing Consultative Commission governing procedures, composition and other relevant matters.

Article XIV

1. Each Party may propose amendments to this Treaty. Agreed amendments shall enter into force in accordance with the procedures governing the entry into force of this Treaty.

2. Five years after entry into force of this Treaty, and at five-year intervals thereafter, the Parties shall together conduct a review of this Treaty.

Article XV

1. This Treaty shall be of unlimited duration.

2. Each Party shall, in exercising its national sovereignty, have the right to withdraw from this Treaty if it decides that extraordinary events related to the subject matter of this Treaty have jeopardized its supreme interests. It shall give notice of its decision to the other Party six months prior to withdrawal from the Treaty. Such notice shall include a statement of the extraordinary events the notifying Party regards as having jeopardized its supreme interests.

Article XVI

1. This Treaty shall be subject to ratification in accordance with the constitutional procedures of each Party. The Treaty shall enter into force on the day of the exchange of instruments of ratification.

2. This Treaty shall be registered pursuant to Article 102 of the Charter of the United Nations.

DONE at Moscow on May 26, 1972, in two copies, each in the English and Russian languages, both texts being equally authentic.

FOR THE UNITED STATES OF AMERICA	FOR THE UNION OF SOVIET SOCIALIST REPUBLICS
RICHARD NIXON	L. I. BREZHNEV
President of the United States of America	*General Secretary of the Central Committee of the CPSU*

Agreed Statements, Common Understandings, and Unilateral Statements Regarding the Treaty Between the United States of America and the Union of Soviet Socialist Republics on the Limitation of Anti-Ballistic Missiles

1. Agreed Statements

The document set forth below was agreed upon and initialed by the Heads of the

Delegations on May 26, 1972 (letter designations added);

AGREED STATEMENTS REGARDING THE TREATY BETWEEN THE UNITED STATES OF AMERICA AND THE UNION OF SOVIET SOCIALIST REPUBLICS ON THE LIMITATION OF ANTI-BALLISTIC MISSILE SYTEMS

[A]

The Parties understand that, in addition to the ABM radars which may be deployed in accordance with subparagraph (a) of Article III of the Treaty, those non-phased- array ABM radars operational on the date of signature of the Treaty within the ABM system deployment area for defense of the national capital may be retained.

[B]

The Parties understand that the potential (the product of mean emitted power in watts and antenna area in square meters) of the smaller of the two large phased-array ABM radars referred to in subparagraph (b) of Article III of the Treaty is considered for purposes of the Treaty to be three million.

[C]

The Parties understand that the center of the ABM system deployment area centered on the national capital and the center of the ABM system deployment area containing ICBM silo launchers for each Party shall be separated by no less than thirteen hundred kilometers.

[D]

In order to insure fulfillment of the obligation not to deploy ABM systems and their components except as provided in Article III of the Treaty, the Parties agree that in the event ABM systems based on other physical principles and including components capable of substituting for ABM interceptor missiles, ABM launchers, or ABM radars are created in the future, specific limitations on such systems and their components would be subject to discussion in accordance with Article XIII and agreement in accordance with Article XIV of the Treaty.

[E]

The Parties understand that Article V of the Treaty includes obligations not to develop, test or deploy ABM interceptor missiles for the delivery by each ABM interceptor missile of more than one independently guided warhead.

[F]

The Parties agree not to deploy phased-array radars having a potential (the product of mean emitted power in watts and antenna area in square meters) exceeding three million, except as provided for in Articles III, IV and VI of the Treaty, or except for the purposes of tracking objects in outer space or for use as national technical means of verification.

[G]

The Parties understand that Article IX of the Treaty includes the obligation of the US and the USSR not to provide to other States technical descriptions or blue prints specially worked out for the construction of ABM systems and their components limited by the Treaty.

2. Common Understandings

Common understanding of the Parties on the following matters was reached during the negotiations:

A. Location of ICBM Defenses

The U.S. Delegation made the following statement on May 26, 1972:

Article III of the ABM Treaty provides for each side one ABM system deployment area centered on its national capital and one ABM system deployment area containing ICBM silo launchers. The two sides have registered agreement on the following statement: "The Parties understand that the center of the ABM system deployment area centered on the national capital and the center of the ABM system deployment area containing ICBM silo launchers for each Party shall be separated by no less than thirteen hundred kilometers." In this connection, the U.S. side notes that its

ABM system deployment area for defense of ICBM silo launchers, located west of the Mississippi River, will be centered in the Grand Forks ICBM silo launcher deployment area. (See Agreed Statement [C].)

B. ABM Test Ranges

The U.S. Delegation made the following statement on April 26, 1972:

Article IV of the ABM Treaty provides that "the limitations provided for in Article III shall not apply to ABM systems or their components used for development or testing, and located within current or additionally agreed test ranges." We believe it would be useful to assure that there is no misunderstanding as to current ABM test ranges. It is our understanding that ABM test ranges encompass the area within which ABM components are located for test purposes. The current U.S. ABM test ranges are at White Sands, New Mexico, and at Kwajalein Atoll, and the current Soviet ABM test range is near Sary Shagan in Kazakhstan. We consider that non-phased array radars of types used for range safety or instrumentation purposes may be located outside of ABM test ranges. We interpret the reference in Article IV to "additionally agreed test ranges" to mean that ABM components will not be located at any other test ranges without prior agreement between our Governments that there will be such additional ABM test ranges.

On May 5, 1972, the Soviet Delegation stated that there was a common understanding on what ABM test ranges were, that the use of the types of non-ABM radars for range safety or instrumentation was not limited under the Treaty, that the reference in Article IV to "additionally agreed" test ranges was sufficiently clear, and that national means permitted identifying current test ranges.

C. Mobile ABM Systems

On January 29, 1972, the U.S. Delegation made the following statement:

Article V(1) of the Joint Draft Text of the ABM Treaty includes an undertaking not to develop, test, or deploy mobile land-based ABM systems and their components. On May 5, 1971, the U.S. side indicated that, in its view, a prohibition on deployment of mobile ABM systems and components would rule out the deployment of ABM launchers and radars which were not permanent fixed types. At that time, we asked for the Soviet view of this interpretation. Does the Soviet side agree with the U.S. side's interpretation put forward on May 5, 1971?

On April 13, 1972, the Soviet Delegation said there is a general common understanding on this matter.

D. Standing Consultative Commission

Ambassador Smith made the following statement on May 22, 1972:

The United States proposes that the sides agree that, with regard to initial implementation of the ABM Treaty's Article XIII on the Standing Consultative Commission (SCC) and of the consultation Articles to the Interim Agreement on offensive arms and the Accidents Agreement,[1] agreement establishing the SCC will be worked out early in the follow-on SALT negotiations; until that is completed, the following arrangements will prevail: when SALT is in session, any consultation desired by either side under these Articles can be carried out by the two SALT Delegations; when SALT is not in session, *ad hoc* arrangements for any desired consultations under these Articles may be made through diplomatic channels.

Minister Semenov replied that, on an *ad referendum* basis, he could agree that the U.S. statement corresponded to the Soviet understanding.

E. Standstill

On May 6, 1972, Minister Semenov made the following statement:

In an effort to accommodate the wishes of the U.S. side, the Soviet Delegation is prepared to proceed on the basis that the two sides will in fact observe the obligations of both the Interim Agreement and the ABM Treaty beginning from the date of signature of these two documents.

In reply, the U.S. Delegation made the following statement on May 20, 1972:

---

[1] See Article 7 of Agreement to Reduce the Risk of Outbreak of Nuclear War Between the United States of America and the Union of Soviet Socialist Republics, signed Sept. 30, 1971.

The U.S. agrees in principle with the Soviet statement made on May 6 concerning observance of obligations beginning from date of signature but we would like to make clear our understanding that this means that, pending ratification and acceptance, neither side would take any action prohibited by the agreements after they had entered into force. This understanding would continue to apply in the absence of notification by either signatory of its intention not to proceed with ratification or approval.

The Soviet Delegation indicated agreement with the U.S. statement.

3. Unilateral Statements

The following noteworthy unilateral statements were made during the negotiations by the United States Delegation:

A. Withdrawal from the ABM Treaty

On May 9, 1972, Ambassador Smith made the following statement:

The U.S. Delegation has stressed the importance the U.S. Government attaches to achieving agreement on more complete limitations on strategic offensive arms, following agreement on an ABM Treaty and on an Interim Agreement on certain measures with respect to the limitation of strategic offensive arms. The U.S. Delegation believes that an objective of the follow-on negotiations should be to constrain and reduce on a long-term basis threats to the survivability of our respective strategic retaliatory forces. The USSR Delegation has also indicated that the objectives of SALT would remain unfulfilled without the achievement of an agreement providing for more complete limitations on strategic offensive arms. Both sides recognize that the initial agreements would be steps toward the achievement of more complete limitations on strategic arms. If an agreement providing for more complete strategic offensive arms limitations were not achieved within five years, U.S. supreme interests could be jeopardized. Should that occur, it would constitute a basis for withdrawal from the ABM Treaty. The U.S. does not wish to see such a situation occur, nor do we believe that the USSR does. It is because we wish to prevent such a situation that we emphasize the importance the U.S. Government attaches to achievement of more complete limitations on strategic offensive arms. The U.S. Executive will inform the Congress, in connection with Congressional consideration of the ABM Treaty and the Interim Agreement, of this statement of the U.S. position.

B. Tested in ABM Mode

On April 7, 1972, the U.S. Delegation made the following statement:

Article II of the Joint Text Draft uses the term "tested in an ABM mode," in defining ABM components, and Article VI includes certain obligations concerning such testing. We believe that the sides should have a common understanding of this phrase. First, we would note that the testing provisions of the ABM Treaty are intended to apply to testing which occurs after the date of signature of the Treaty, and not to any testing which may have occurred in the past. Next, we would amplify the remarks we have made on this subject during the previous Helsinki phase by setting forth the objectives which govern the U.S. view on the subject, namely, while prohibiting testing of non-ABM components for ABM purposes: not to prevent testing of ABM components, and not to prevent testing of non-ABM components for non-ABM purposes. To clarify our interpretation of "tested in an ABM mode," we note that we would consider a launcher, missile or radar to be "tested in an ABM mode" if, for example, any of the following events occur: (1) a launcher is used to launch an ABM interceptor missile, (2) an interceptor missile is flight tested against a target vehicle which has a flight trajectory with characteristics of a strategic ballistic missile flight trajectory, or is flight tested in conjunction with the test of an ABM interceptor missile or an ABM radar at the same test range, or is flight tested to an altitude inconsistent with interception of targets against which air defenses are deployed, (3) a radar makes measurements on a cooperative target vehicle of the kind referred to in item (2) above during the reentry portion of its trajectory or makes measurements in conjunction with the test of an ABM interceptor missile or an ABM radar at the same test range. Radars used for purposes such as range safety or instrumentation would be exempt from application of these criteria.

C. No-Transfer Article of ABM Treaty

On April 18, 1972, the U.S. Delegation made the following statement:

In regard to this Article [IX], I have a brief and I believe self-explanatory statement to make. The U.S. side wishes to make clear that the provisions of this Article do not set a precedent for whatever provision may be considered for a Treaty on Limiting Strategic Offensive Arms. The question of transfer of strategic offensive arms is a far more complex issue, which may require a different solution.

D. No Increase in Defense of Early Warning Radars

On July 28, 1970, the U.S. Delegation made the following statement:

Since Hen House radars [Soviet ballistic missile early warning radars] can detect and track ballistic missile warheads at great distances, they have a significant ABM potential. Accordingly, the U.S. would regard any increase in the defenses of such radars by surface-to-air missiles as inconsistent with an agreement.

Appendix 2 A Guide to the SDI Programme and Organisation

Ed Reiss

A: The Major Programme Elements

The Strategic Defense Initiative Program (SDIP) comprises five major programme elements (PEs). Each is defined by its stated aim (US DoD 1984b):

> PE 63220C, *Surveillance, Acquisition, Tracking and Kill Assessment* (SATKA), aims to develop and demonstrate the capabilities required to detect, track and discriminate objects in all phases of the ballistic missile trajectory.

> PE 63221C, *Directed Energy Weapons* (DEW), aims to develop and demonstrate the technologies for ground-based and space-based laser weapons systems, space-based neutral particle beams, and nuclear-bomb-pumped systems for ballistic missile defence.

> PE 63222C, *Kinetic Energy Weapons* (KEW), aims to develop and demonstrate technology for kinetic energy weapons, that is, for interceptor missiles and hypervelocity gun systems which rely on 'non-nuclear kill' to destroy the target.

> PE 63223C, *Systems Concepts and Battle Management/C^3*, aims to develop survival-enhanced information processors, real-time software systems, advanced communications technology, battle-management computational algorithms, and policy-responsive weapons-control-and-release capabilities.

> PE 63224C, *Survivability, Lethality, and Key Support Technology*, aims to develop and demonstrate technologies for space-system survivability, space prime power and conditioning, and space transportation/logistics.

B: Budget Summary

Table AP2.1 summarizes the SDI Budget. Fiscal Year 1986 (FY 1986) began on 1 October 1985. However, the complex Congressional decision process for the Defense Budget took even longer than usual to conclude this time around. By December 1985 the total DoD Budget for the current year had been set at $297.4 billion, of which $2.75 billion were for the SDI. This was a cut of almost $1 billion from the sum of $3.72 billion requested by the SDIO, but the latter had been deliberately inflated, in anticipation of just such an outcome. Looked at another way, the SDIO's allocation for FY 1986 is almost double (+ 97 per cent) what it received in FY 1985.

Figures in the tables below are given as follows. Those in roman (normal) type represent definite past or present amounts that have been authorized by Congress, namely, those for the major programme elements and for the overall SDIP up to and including FY 1986. Within the FY 1985 and FY 1986 columns individual project totals are close official estimates, adding up to the authorized PE totals. Their status as estimates has been marked by placing them in *italic* type. The same has been done for *all* figures in the FY 1987 column, which represents only the initial SDI budget *requests*. These were sent to Congress in February 1986, but are unlikely ever to be fully endorsed. Another category of figures is shown by placing *italics inside brackets*. These are, firstly, estimates of the breakdown of spending *within* certain major projects, taken from the SDIO's April 1985 Report to Congress (US DoD, 1985b). They do not usually add up to the official current estimate (*italic*, no brackets) for the project in question, but are given as indicators of the balance of effort within the project in question before the major revision of the SDI which was introduced in late 1985 (below). A comparison of such earlier figures with the subsequently decided project total can also illustrate the extent of recent changes within the SDI. Though reductions have been the norm, occasional very large increases, as in DEW Project 0004 and Systems Project 3002, should not be overlooked. Secondly, since the FY 1984 SDI break-down figures derive from the same document, and do not add up to the most recent official estimate for SDI or SDI-equivalent expenditures in that period (see table AP2.1 note c, and section G below), they have been rendered in the same style.

In November 1985, once the approximate size of its actual FY

1986 budget had become clear, the SDIO announced major changes in several projects. Within the SATKA Program Element, the space-based imaging sensor Projects 0003 and 0004 were redirected towards possible airborne systems for tactical BMD, for instance in Europe. Project 0006 was heavily downgraded, by dropping the goal of accurate space-based tracking of boosters altogether, and confined to marginal improvements in early-warning satellites. Project 0007 will continue to work for accurate mid-course tracking, but without discrimination of decoys from warheads. The discrimination task will now be attempted by several of the directed energy technologies in the DEW Program Element 63221C, whose feasibility as anti-missile *weapons* is now in doubt. They include the space-based chemical lasers of Project 0001, the neutral particle beams of Project 0003, and, in a secondary application, the X-ray laser of Project 0004, which is now seen primarily as a potential asat (see table AP2.3).

Table AP2.1   SDIO Budget Summary[a] *($ in millions)*

| | Appropriations | | | Request |
	FY 1984[b]	FY 1985	FY 1986	FY 1987
Surveillance, Acquisition, Tracking & Kill Assessment	*(366)*	537	858	*1262*
Directed-Energy Weapons Technology	*(313)*	377	844	*1615*
Kinetic Energy Weapons Technology	*(196)*	265	596	*1002*
Systems Concepts/Battle Management	*(10)*	111	226	*462*
Survivability, Lethality, & Key Technologies	*(23.5)*	126	221	*454*
SDIO Management (PE 65898C)	*(0.5)*	8	9	*17*
Subtotals, Department of Defense	*(909)*	1423	2754	*4812*
Department of Energy	N.A.	200	282	*602*
Totals	*(909)*[c]	1623	3036	*5414*

Notes:
a Some rounding occurs.
b This column is based on figures in US DoD 1985b.
c An official estimate given in Congressional testimony in 1985 was that the Department of Defense spent just over $1 billion in FY 1984 on projects duly transferred to the SDIO (S.Appropriations, 1985: 474).

Sources: Pike, 1986b; US Department of Defense, 1985b.

The overall tendency away from space-based BMD system components has led to renewed emphasis on potential kinetic energy weapons, where most projects are for late mid-course or terminal phase interception techniques. KEW Project 0010 however, is a kinetic kill technology now being accorded high priority for the boost phase.

C: SDI Programme Elements, Projects, Tasks, Budgets

The main SDI programme elements are sub-divided into numerous projects, listed in the following tables (AP2.2 – AP2.6) together with their past and projected budgets:

Table AP2.2   PE 63220C: Surveillance, Acquisition, Tracking & Kill Assessment ($ in millions)

	FY 1984	FY 1985	FY 1986	FY 1987
Project 0001 – Radar Discrimination Technology & Data Base				
Cobra Judy		*(17.10)*	*(14.40)*	*(31.31)*
Post-Boost Vehicle Data Collection		*(2.70)*	*(46.30)*	*(53.03)*
Discrimination Development & Radar Technology		*(10.10)*	*(13.40)*	*(14.12)*
Subtotals	*(24.10)*	25.00	19.00	48.00
Project 0002 – Optical Discrimination Technology & Data Base				
Infra-red (IR) Exoatmospheric/ High Endoatmospheric Signature Data		*(95.60)*	*(113.80)*	*(115.91)*
Laser Image Data		*(3.20)*	*(28.90)*	*(30.85)*
IR Background Studies		*(34.90)*	*(56.00)*	*(45.55)*
Subtotals	*(50.50)*	128.00	116.00	110.00
Project 0003 – Imaging Radar Technology				
Large Array Technology		*(9.40)*	*(27.80)*	*(66.46)*
Near-Term Imaging Demonstration		*(5.90)*	*(8.00)*	*(4.64)*
Satellite/Aircraft Imaging Radar			*(10.00)*	*(51.86)*
Subtotals	*(6.30)*	14.00	35.00	50.00
Project 0004 – Imaging Laser Technology (Optical)				
Large Optics Technology		*(8.40)*	*(23.00)*	*(22.27)*
Laser Imaging Technology		*(10.90)*	*(71.00)*	*(99.96)*
Imaging Laser Measurements		*(9.00)*	*(28.00)*	*(38.80)*

Table AP2.2 contd

Early Demonstration of Angle-Only Tracking			*(5.00)*	*(27.72)*
Subtotals	*(5.60)*	28.30	47.00	64.00
Project 0005 – IR Sensor Technology				
Optics Technology		*(16.90)*	*(54.40)*	*(55.01)*
IR Focal Plane Development		*(20.40)*	*(64.00)*	*(66.70)*
Space Cryocoolers		*(20.50)*	*(33.00)*	*(36.12)*
Subtotals	*(86.80)*	53.00	80.00	135.00
Project 0006 – Boost Surveillance and Tracking System Experiment				
	(28.30)	38.00	100.00	65.00
Project 0007 – Space Surveillance and Tracking System Experiment				
	(35.60)	44.00	82.00	170.00
Project 0008 – Optical Surveillance Experiment				
	(28.80)	110.00	131.00	167.98
Project 0009 – Terminal Imaging Radar Experiment				
	(100.50)	5.00	20.00	79.00
Project 0010 – Space-Based Imaging Experiment				
				5.00
Project 0011 – Common Technology & Architecture				
SATKA Systems Studies		*(17.15)*	*(49.00)*	*(43.70)*
Imaging Algorithm Development		*(3.00)*	*(9.50)*	*(14.71)*
Radiation Hardened Circuits		*(33.40)*	*(99.80)*	*(114.89)*
Real-Time Signal Processing		*(29.40)*	*(91.90)*	*(99.20)*
Subtotals		60.00	109.00	200.00
Project 0012 – Program Management SATKA				
		24.00	25.00	25.00
Project 0013 – SATKA Systems Studies				
		3.00	25.00	60.00
Project 0014 – Director SDIO SATKA				
		3.00	70.00	83.00
Totals	*(366.50)*	537.00	858.00	1262.00

Sources: as for table AP2.1.

Table AP2.3   PE 63221C: Directed-Energy Weapons Technology ($ in millions)

	FY 1984	FY 1985	FY 1986	FY 1987
Project 0001 – Space-Based Laser Concepts				
Laser Devices		(5.52)	(18.00)	(41.10)
Beam Control		(4.10)	(18.30)	(34.30)
Large Optics		(4.40)	(11.10)	(21.20)
Aiming, Tracking & Pointing (ATP)		(8.18)	(37.10)	(69.10)
Major Experiments		(120.40)	(179.80)	(234.00)
Space ATP Support			(80.00)	(80.00)
Concept & Development Definition		(7.00)	(20.60)	(28.00)
Innovative Science & Technology		(4.80)	(7.00)	TBD
Other Technology		(8.30)	–	–
Subtotals	(157.00)	153.00	144.00	406.00
Project 0002 – Ground-Based Laser Concepts				
Laser Devices		(82.05)	(110.60)	(95.00)
Ground Segment Beam Control		(66.12)	(99.00)	(103.00)
Space Segment		(2.00)	(32.00)	(67.00)
Major Experiments		(4.13)	(149.24)	(150.00)
Concept & Development Definition		(5.00)	(19.70)	(25.20)
Innovative Science & Technology		(5.20)	(10.50)	TBD
Other Technology		(13.65)	(10.50)	(15.00)
Subtotals	(133.10)	184.00	293.00	725.00
Project 0003 – Space-Based Particle Beam Concepts				
Neutral Particle Beam (NPB)		(20.60)	(74.00)	(89.00)
Antigone		(7.00)	(29.00)	(31.00)
Advanced NPB Concepts			(2.00)	(15.00)
Integrated NPB Experiments			(13.60)	(35.00)
Concept & Development Definition		(2.90)	(7.80)	(14.00)
Innovative Science & Technology		(1.00)	(7.00)	TBD
Other Technology		(0.65)	–	–
Subtotals	(13.90)	33.00	145.00	225.00

Table AP2.3 contd

Project 0004 – Nuclear-Driven Directed Energy Concepts

ATP Technology & Demonstration		*(2.30)*	*(19.10)*	*(22.00)*
Laboratory Experiments		*(0.20)*	–	*(23.88)*
Concept & Development Definition		*(0.50)*	*(1.00)*	*(2.80)*
Innovative Science & Technology		*(0.30)*	*(8.50)*	TBD
Other Technology		*(0.10)*	–	–
Subtotals	*(8.50)*	13.00	112.00	176.00
Project 0005 – DEW Concept Development		13.00	20.00	28.00
Project 0006 – DEW SDIO Director		−20.00	130.00	55.00
Totals	*(312.50)*	376.00	844.00	1615.00

Sources: as for table AP2.1.

Table AP2.4   PE 63222C: Kinetic Energy Weapons Technology ($ in millions)

	FY 1984	FY 1985	FY 1986	FY 1987
Project 0001 – Endoatmospheric Nonnuclear Kill Technology	*(111.10)*	72.00	48.00	85.00
Project 0002 – Exoatmospheric Nonnuclear Kill Technology	*(0.10)*	39.00	59.00	115.00
Project 0003 – Subsystem Engineering and Analysis	*(2.40)*	4.00	7.00	15.00
Project 0004 – Hypervelocity Launcher Technology	*(6.60)*	18.00	40.00	40.00
Project 0005 – Novel Concepts	*(2.30)*	9.00	17.00	30.00
Project 0006 – Endoatmospheric Nonnuclear Kill Test Bed (HEDI)	*(36.80)*	19.00	71.00	150.00
Project 0007 – Exoatmospheric Nonnuclear Kill Test Bed (ERIS)	*(24.20)*	10.00	74.00	160.00
Project 0008 – SLBM Boostphase Engagement (new project)		?	?	?

Table AP2.4 contd

Project 0009 – Sagittar Space Rail-gun

	(6.60)	3.50	23.00	35.00

Project 0010 – Space-based Kinetic Kill Vehicle

	(5.50)	25.00	90.00	180.00

Project 0011 – Terminal Technologies Integration

	(0.10)	10.00	20.00	91.80

Project 0012 – Program Management Army

		36.00	40.00	55.00

Project 0013 – Director SDIO

		21.00	101.00	35.00

Project 0014 – Army Materials & Mechanics Research Center

			6.00	10.00
Totals	(195.70)	265.00	596.00	1002.00

Sources: as for table AP2.1.

Table AP2.5   PE 63223C: Systems Concepts/Battle Management ($ in millions)

	FY 1984	FY 1985	FY 1986	FY 1987
Project 3001 – Battle Management/Command, Control and Communications (BM/C³)				
BM/C^3 Architecture		(9.00)	(38.00)	(43.45)
Processor Technology		(14.50)	(42.80)	(48.60)
Software Technology		(13.70)	(44.86)	(50.90)
Communication Technology		(8.80)	(19.40)	(21.90)
Subtotals		29.00	85.00	177.00
Project 3002 – SDI Systems Architectures				
Threat Analysis		(5.80)	(10.68)	(11.69)
Systems Architecture Studies		(20.00)	(35.20)	(24.29)
Pilot Architecture Study		(3.10)	(2.00)	(2.00)
Program Integration		(2.50)	(5.00)	(5.00)
Functional Analyses & Modeling		(6.00)	(16.10)	(24.80)
Systems Concepts & Simulation		(15.60)	(29.26)	(39.90)
Subtotals		82.00	142.00	285.00
Totals	(10.00)	111.00	226.00	462.00

Sources: as for table AP2.1.

Table AP2.6   PE 63224C: Survivability, Lethality and Key Technologies ($ in millions)

	FY 1984	FY 1985	FY 1986	FY 1987
Project 0010 – System Survivability				
Survivability Assessment		*(1.50)*	*(3.45)*	*(4.90)*
Survivability Analysis		*(3.60)*	*(5.00)*	*(6.00)*
Threat Refinement		*(3.00)*	*(3.25)*	*(5.10)*
Countermeasure Development		*(28.10)*	*(60.45)*	*(72.70)*
Subtotals	*(9.20)*	*33.00*	*28.00*	*94.00*
Project 0011 – Lethality and Target Hardening				
Thermal Lasers		*(23.70)*	*(22.25)*	*(24.50)*
Impulse Lasers		*(8.10)*	*(14.95)*	*(12.30)*
X-ray Lasers		*(8.20)*	*(20.35)*	*(28.50)*
Particle Beams		*(7.00)*	*(8.25)*	*(9.90)*
Kinetic Energy		*(10.00)*	*(23.55)*	*(26.50)*
High Power Microwaves		*(6.30*	*(14.15)*	*(19.90)*
Subtotals	*(11.30)*	*63.30*	*97.00*	*140.00*
Project 0012 – Space Power and Power Conversion				
Multimegawatt Management		*(1.00)*	*(3.20)*	*(3.90)*
Multimegawatt Industry Concept		*(1.00)*	*(3.20)*	*(3.90)*
Multimegawatt Technology			*(38.20)*	*(45.40)*
SP–100 (100 kWe Class)		*(8.00)*	*(16.00)*	*(19.70)*
Subtotals	*(2.00)*	*19.00*	*53.00*	*103.00*
Project 0013 – Space Logistics				
	(1.00)	*(1.50)*	*(23.00)*	*(60.00)*
Project 0014 – Materials				
			5.00	*38.00*
Project 0015 – Other Support				
		9.00	*15.00*	*20.00*
Totals	*(23.50)*	*126.00*	*221.00*	*454.00*

Sources: as for table AP2.1.

D: Omissions and Exclusions from the SDIP

The tables above list projects making up the main programme elements of the SDIP. But it is equally important to consider what is *not* in the SDIP, since in several senses these exclusions also shape and define the programme. Perhaps the most striking such omission is the anti-satellite programme, or 'Space Defense System', to give it its official name. The budget request for this programme in FY 1987 is $322 million.

Table AP2.7 lists programmes not officially within the SDIP, but closely related to its work.

Table AP2.7   Programmes not in the SDI ($ in millions)

Program Element	Name	FY 1984[a]	FY 1985	FY 1986[b]
64406F	Space Defense System	202.70	133.00	*149.90*
63226E	Air Defense Surveillance Warning (Teal Ruby)	32.20	31.00	*25.00*
63401F	Advanced Spacecraft Technology (Satellite Power & Survivability)		6.90	*9.70*
63605F	Advanced Radiation Technology	46.70	5.00	*19.70*
62707E	Particle Beam Technology	30.90	17.40	*91.50*
62307A	Laser Weapons Technology	20.00	21.10	*21.40*
63424F	Missile Surveillance Technology		3.00	*11.60*
65806A	DoD High Energy Laser Facility	37.10	32.80	*20.20*
Totals		369.60	250.20	*279.00*

Notes:

a As there is no problem here of reconciling FY 1984 programme figures with a total which they do not match, these have been left in roman type.

b These are budget request figures only, and are therefore given in italic.

Source: US Department of Defense, 1985b.

E: The SDI Organization

Figure AP2.1 shows the administrative structure of the Strategic Defense Initiative Organization as of 1 April 1985. The SDIO adheres to an ideal of 'centralized direction' and 'decentralized execution' of its programme. It is structured as a set of directorates, each answering to the Office of the Director, Lieutenant General Abrahamson. It has a staff of some eighty people, working closely with R&D units in the US Army, Air

Force, Navy, Defense Advanced Research Projects Agency (DARPA) and the Department of Energy (DoE).

SDIO officials stress that it aims to be flexible. They speak of 'ad hocracy rather than bureaucracy', to suggest that things get done in an improvisatory and flexible fashion. People dealing with the SDIO sometimes see elements of inefficiency in its 'ad hocracy'. And despite Dr Yonas's acceptance of double duty as Chief Scientist and as General Abrahamson's Deputy, there have been complaints about the absence of any permanent Deputy Director. In view of General Abrahamson's frequent extended absences, on promotional trips to Congress, the media, defence corporations, and allied capitals around the world, this would seem to be an under-emphasized executive position.

F: The 'Infrastructure' of the SDI

The SDIO staff, of course, are only a small fraction of those working on SDI projects in the United States. Most actual work for the SDI is delegated to larger, more significant, but less public organizations. The independent, New York-based Council on Economic Priorities has estimated that there were about 5000 scientists, engineers and technical workers employed on SDI-related work in 1984. It predicts that this number will have risen to 18,600 by 1987 (*Washington Post*, 20 October 1985). About 450 organizations, including over seventy universities or other academic bodies, have received SDI contracts, of which more than 1000 were executed in 1985.

The *US Army* Strategic Defense Command (USASDC), formerly known as the Ballistic Missile Defense Organization, was employing 772 civilian and 154 military personnel in April 1985 (*Army Research Development & Acquisition Magazine*, March/April 1985). Whilst the USASDC reports to the Office of the Chief of Staff of the Army, most of its funds come from the SDIO.

The USASDC has inherited the BMDO's former Washington headquarters and two large field-operating agencies in Huntsville, Alabama. These are the BMD Advanced Technology Center (BMDATC) and the BMD Systems Command (BMDSCOM). BMDSCOM operates the Kwajalein Missile Range in the Marshall Islands, some 4000 kilometres south-west of Hawaii.

Other major Army organizations involved in SDI are the Army Materiel Command's Armament Research and Development Center at Dover, New Jersey, and the Army Materials and

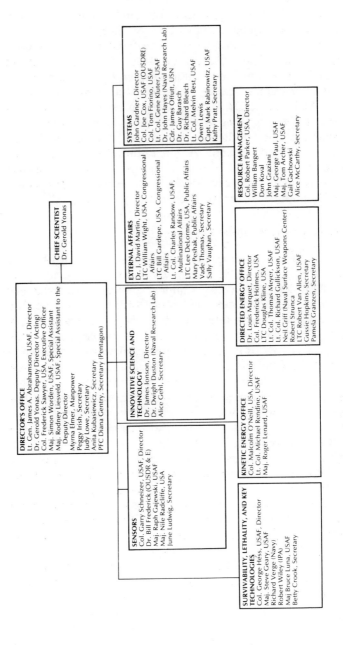

DIRECTOR'S OFFICE
Lt. Gen. James A. Abrahamson, USAF. Director
Dr. Gerold Yonas. Deputy Director (Acting)
Col. Frederick Sawyer, USA, Executive Officer
Maj. Simon Worden, USAF, Special Assistant
Maj. Rodney Liesveld, USAF, Special Assistant to the
     Deputy Director
Myrna Elmer, Manpower
Peggy Irish, Secretary
Judy Lowe, Secretary
Anita Kubasiewicz, Secretary
PFC Diana Gentry, Secretary (Pentagon)

CHIEF SCIENTIST
Dr. Gerold Yonas

SENSORS
Col. Garry Schneizer, USAF, Director
Dr. Bill Frederick (OUSDR & E)
Maj. Raph Gajewski, USAF
Maj. Nile Radcliffe, USA
June Ludwig, Secretary

**INNOVATIVE SCIENCE AND
TECHNOLOGY**
Dr. James Ionson, Director
Dr. Dwight Duston (Naval Research Lab)
Alice Gehl, Secretary

EXTERNAL AFFAIRS
Dr. J. David Martin, Director
LTC William Wight, USA, Congressional
     Affairs
LTC Bill Gardepe, USA, Congressional
     Affairs
Lt. Col. Charles Randow, USAF,
     Multinational Affairs
LTC Lee DeLorme, USA, Public Affairs
Mary Peshak, Public Affairs
Vade Thomas, Secretary
Sally Vaughan, Secretary

SYSTEMS
John Gardner, Director
Col. Joe Cox, USAF (OUSDRE)
Col. Tom Fiorino, USAF
Lt. Col. Gene Kluter, USAF
Dr. John Hayes (Naval Research Lab)
Cdr. James Offutt, USN
Dr. Guy Barasch
Dr. Richard Bleach
Lt. Col. Melvin Best, USAF
Owen Lewis
Capt. Mark Rabinowitz, USAF
Kathy Pratt, Secretary

**SURVIVABILITY, LETHALITY, AND KEY
TECHNOLOGIES**
Col. George Hess, USAF, Director
Maj. Steve Geary, USAF
Richard Verge (Navy)
Robert Wiley (IPA)
Maj Bruce Luna, USAF
Betty Crook, Secretary

KINETIC ENERGY OFFICE
Col. Malcolm O'Neill, USA, Director
Lt. Col. Michael Rendine, USAF
Maj. Roger Lenard, USAF

DIRECTED ENERGY OFFICE
Dr. Louis Marquet, Director
Col. Frederick Holmes, USA
LTC Douglas Kline, USA
Lt. Col. Thomas Meyer, USAF
Lt. Col. Richard Gullickson, USAF
Neil Griff (Naval Surface Weapons Center)
Robert Strunca
LTC Robert Van Allen, USAF
Gussie Hopkins, Secretary
Pamela Granzen, Secretary

RESOURCE MANAGEMENT
Col. Robert Parker, USA, Director
William Bangert
Don Koval
John Graziani
Maj. George Paul, USAF
Maj. Tom Archer, USAF
Gail Gachowski
Alice McCarthy, Secretary

Figure AP2.1   Strategic Defense Initiative, April 1985.
Source: SDI Organization

Mechanics Research Center at Watertown, Massachusetts.

Perhaps the most important *US Air Force* organization working on SDI is the Air Force Systems Command (AFSC), whose headquarters are at Andrews Air Force Base (AFB) in Maryland. A subsidiary of the AFSC is its Space Division at Los Angeles. As a lead agency and integrator for the SDI, Space Division is working in many areas: natural backgrounds, advanced cryocoolers (for super-cooling of sensors), infra-red focal plane, radiation hardening, focal mirrors, space boosters, threat analysis, large optics, atmospheric composition, and space logistics (*Air Force Magazine*, July 1985).

Numerous other Air Force organizations are working on SDI projects. They include: the AFSC's Electronic Systems Division (ESD) at Hanscom AFB, Massachusetts; the ESD's Rome Air Development Center (RADC) at Griffiss AFB, New York; the Aeronautical Systems Division (ASD) and its Wright Aeronautical Laboratories, both at Wright-Patterson AFB, Ohio; the Air Force Armament Division at Eglin AFB, Florida; the Air Force Space Technology Center and its Weapons Laboratory, both at Kirtland AFB, New Mexico; the Air Force Rocket Propulsion Laboratory at Edwards AFB, California; and the Air Force Geophysical Laboratory at Hanscom AFB, Massachusetts.

The Ballistic Missile Office at Norton AFB, California, runs a penetration aids programme and is believed to be working closely with the SDIO. Similar cooperation exists with the long-range planning staff at Air Force Space Command, Colorado Springs, Colorado, who would have the task of integrating SDI weapons into the central coordinating computers of Space Command.

US Navy organizations working on SDI include: the Naval Ocean Systems Center, San Diego, California; the USN Space and Warfare Systems Command, Arlington, Virginia; and the Navy Center for Advanced Research in Artificial Intelligence, which is located at the Naval Research Laboratory, Washington D.C. and run from the Office of Naval Research in Arlington.

Also at Arlington is the *Defense Advanced Research Projects Agency* (DARPA), many of whose staff are employed on SDI research. Members of the *Defense Nuclear Agency* are responsible for the SDI's target lethality and hardening programme. And the *Defense Communications Agency* researches into SDI battle management.

A crucial part in SDI is taken by the *Department of Energy*, with its triad of major laboratories: Lawrence Livermore National Laboratory, Livermore, California; Sandia National Laboratory,

Albuquerque, New Mexico; and Los Alamos National Laboratory, Los Alamos, New Mexico (table AP2.8). These specialize in high-energy optical weapons, especially short-wavelength lasers, and also in 'third generation' nuclear weapons such as the 'nuclear-pumped' X-ray laser. Other DoE laboratories are also doing SDI-related work, such as the Brookhaven Laboratory, which does research into space nuclear reactors.

G: BMD Becomes SDI – Before–and–After Comparisons

In view of the hypothesis suggested in this book (p. 80), that the SDI represents as much a political re-shaping of the campaign for BMD programmes as any really novel strategic departure, it may be useful to give some comparisons beween the size and rate of growth of BMD budgets before and after March 1983.

A Congressional Budget Office report in May 1984 cited testimony from General Abrahamson to the effect that, prior to March 1983, the Department of Defense had planned to make BMD budget requests to Congress for $15 billion for the period 1985–9 (CBO, 1984:5). Other testimony, referred to in a market analysis from the same period, put the figure for the same requests at $10 billion. And one Electronic Industries Association report gave a figure of $18 billion. It follows that no precise figure can be derived for the relationship between the initial $26 billion five-year SDI budget request, and the BMD budget that 'might have been' requested for the same period in the absence of the SDI, still less for that between the actual budgets currently authorized for the SDI and those which might have been authorized for the corresponding programmes in the same period, if the SDI had not been launched.

There is also a more basic problem with trying to answer the before-and-after question about the SDI. Before the creation of the SDIO at the beginning of 1984, those BMD projects which were later gathered together under its major programme elements were distributed unevenly across 27 military service and defence agency programmes. The CBO report already referred to gives a list of programme elements taken over from the FY 1984 budget into the SDIP, which was prepared by the Pentagon. Analysing these, the report shows that such 'SDI-equivalent' programme elements had actually grown by 29 per cent in 1982 and 12 per cent in 1983, and had been planned to grow by a further 20 per cent in 1984. If generalized, such a growth rate should have been sufficient to keep the proportion of BMD expenditure within the Pentagon's overall Research, Development, Testing and Evaluation (RDT&E) budget fairly constant, at around 9 or 10 per cent.

H: The SDI as a Growth Market

As noted above, the total US Department of Defense budget for FY 1986 will be $297.4 billion, of which $2.75 billion has been appropriated for the SDI. In strictly monetary terms, therefore, the SDI accounts for less than 1 per cent of the current DoD budget. But its importance for the major US defence corporations involved (table AP2.8) is far greater than such a figure might suggest.

The SDI was initially projected to run to $26 billion over its first five years. An early market forecast prepared for the Electronic Industries Association (1984) gave a figure of $69 billion for the first ten years, 1985–94. The Pentagon continues to use its original rough estimate of $26 billion over five years as a basis for making up, through future budget requests, for amounts already 'lost' in actual budget authorizations for the SDIP's first two full years. In fact, the Congressional authorization rate, which determines what actually gets spent, is running well below the original request (table AP2.1). At the start of 1986 one investment consultant estimated SDI expenditure to 1989 at $17 billion, but also pointed out that 'SDI is one of the handful of space-related business opportunities that holds a payoff in the short term' (*Aviation Week & Space Technology*, 27 January 1986).

All current estimates or allocations apply only to initial research and development. The 'real' money, or 'gigabucks' as corporation strategists call it, would come with any decision to seek deployment. That would initiate a costly full-scale RDT&E phase across all projects that had been positively assessed in the current, 1985–9 phase. The price of actual system deployment, if the RDT&E work were itself judged 'successful', would of course be even greater. A de-classified DoD study on laser weapons in 1982 put the cost of building an anti-missile defence system with limited capabilities, for 'damage denial', at $500 billion (Senator Larry Pressler to Senate Foreign Relations Committee, 25 April 1984). Other sources have put the cost of an SDI-type deployment at $1 trillion (*Aviation Week & Space Technology*, 2 April 1984). In November 1983 Dr Richard DeLauer, then Under-Secretary of Defense for Research and Engineering, told the House Armed

Services Committee that:

> When the time comes that you deploy any of these technologies,
> you'll be staggered at the cost that they will involve.

It is this long-term prospect that interests the defence corporations. TRW Inc., a major SDI contractor, put it this way in one
recent brochure: 'We're standing on the first rung of a defense
development that will dominate the industry for the next 20 years.'

Thus when the Pentagon announced its readiness to make
'Systems Architecture' awards for studies to define an overall
concept for the SDI, more than 200 companies responded to the
request for proposals. Senior administration officials were
reported to be 'immensely impressed' by the industry's enthusiasm
for SDI, which extended to the investment of company funds well
in excess of those provided by DoD contracts. But there is nothing
disinterested about such zeal. Today's 'architecture' decisions will
determine which special facilities offered by which corporations
will or will not have any chance of competing, tomorrow and the
day after, for the RDT&E and deployment phases of any future
US strategic defence system, in a period when the proportion of
US defence dollars going to BMD would have to be very much
higher than at present.

More recently, there has been some anxiety within the industry
that the enormous budget deficit will end the balmy days of
steadily rising defence expenditure. Senator Sam Nunn, the senior
Democrat on the Senate Armed Services Committee, has predicted that there will probably be no real growth in defence
spending over the next five years:

> Indeed, there could well be decreases in the defense budget where
> annual increases would not even cover inflation. (*Aviation Week &
> Space Technology*, 28 October 1985)

In such an event, the SDI could only continue to grow at the
expense of other military programmes. Advocates of increased
SDI spending might encounter opposition from vested interests
elsewhere in the defence industry. But if the SDI looked like
winning such a contest, it would become all the more crucial to the
'assured survival' strategy of each major corporation that it should
be securely 'on board', whatever the fate of its rivals. (The
Pentagon's wish to allay such anxieties has been evidenced in the

terms of its five recently awarded contracts for major architecture studies, under which each 'winning' corporation is required to avail itself of contributions from the 'losers' also – see table AP2.9.) For these reasons, conservative resistance to 'Star Wars' innovation is perhaps more likely to come from within the armed services themselves, with their strong institutional commitment to a plethora of traditional roles, than it is from industry.

Table AP2.8   Major SDI Contractors 1983–86

Company		Total contracts ($ in millions)
DoE Lawrence Livermore	L	725 000
General Motors	C	579 000
Lockheed	C	521 000
TRW	C	354 000
McDonnell Douglas	C	350 000
Boeing	C	346 000
DoE Los Alamos	L	196 000
Rockwell International	C	188 000
Teledyne Brown	C	180 000
EG&G	C	140 000
GenCorp Inc.	C	135 000
Textron	C	93 000
DoE Sandia	L	91 000
LTV Aerospace	C	90 000
Flow General	C	89 000
Raytheon Co.	C	72 000
Science Applications	C	69 000
Honeywell	C	69 000
Nichols Research	C	63 000
MIT Lincoln Lab.	U	63 000

Key:
C = Commercial
L = Government laboratory
U = University

Note:
Amounts shown include funds already obligated and others to be obligated in future, also some priced contract options not yet exercised. Some contract values are estimates only. Estimates of the proportion of funding likely to go to certain establishments for certain technologies, e.g. to Lawrence Livermore for building a free-electron laser, are sometimes contentious.

Source: Federation of American Scientists in *Aviation Week & Space Technology,* 21 April 1986.

Table AP2.9 The Architecture Contracts

Initial one-year studies	Selected for ongoing major design studies, fall 1985
General Research	
Hughes Aircraft	
Lockheed Missiles & Space	
Martin Marietta Denver Aerospace	Martin Marietta Denver Aerospace
McDonnell Douglas Astronautics	
Rockwell International	Rockwell International
Science Applications	Science Applications
Sparta	Sparta
Teledyne Brown	
TRW Electronics & Defense	TRW Electronics & Defense

Sources: US Department of Defense *Memorandum*, 17 December 1984; Federation of American Scientists.

Appendix 3 Does the Soviet Union Observe the ABM Treaty?

In February 1985 the White House published the unclassified version of President Reagan's second *Report to the Congress on Soviet Noncompliance with Arms Control Agreements*. This was followed in October by an illustrated booklet on *Soviet Strategic Defense Programs* from the Pentagon (US DoD, 1985c). On 23 December 1985 the President published a third report on the subject (US WH, 1985c). And six weeks later the Arms Control and Disarmament Agency published its own findings (US ACDA, 1986a). Except for the declassification, in the President's third report, of two ABM Treaty issues that had been placed in the classified part of his second report, which could then be aired in the ACDA report also, the assertions about Soviet noncompliance with the ABM Treaty are substantially the same, often using identical or closely parallel wording, in all four documents.

They establish that the US government's legal complaints about Soviet space weapons activities are in fact confined to matters arising from the ABM Treaty. (There is therefore no need in this context to examine the administration's frequent application to Soviet actions of the elusive concept of 'violations of its political commitment with respect to . . .' this or that treaty which is not legally in force, often because the United States has declined to ratify it.)

Most previous US administrations have given the Soviet Union high marks for treaty compliance. In general the assessment of Western experts has been that Soviet governments tend to observe such agreements very close to the letter, taking everything they are allowed but seldom trying to get away with even marginal breaches. Where doubtful incidents have occurred, it has usually been possible for United States negotiators to reach a satisfactory resolution of the issue at the SCC. Under the Carter administration a broadly based official commission found no substantive

307

proof of Soviet infringements. And the US representative on the SCC declared that 'the commission has never yet had to deal with a case of real or apparent clear and substantial non-compliance with an existing agreement' (in Mische, 1985: 66).

The unclassified version of President Reagan's first Report to Congress on the issue of Soviet treaty compliance, in January 1984, made no charges of outright Soviet breaches of the ABM Treaty, though there have since been reports that the Americans brought up the matter of the Abalakova radar at the SCC during 1983 (*Washington Times*, 7 August 1985). In April 1984, however, the third edition of the Pentagon's annual booklet *Soviet Military Power* declared that construction of a large phased-array radar 'in Siberia . . . almost certainly violates the 1972 ABM Treaty.' This accusation has been repeated with increasing vigour ever since.

The Abalakova radar is in fact the *only* definite charge of a Soviet breach of the ABM Treaty publicly advanced by the Reagan administration. A second charge, of concurrent testing of SAM and ABM components contrary to the Treaty, is given the status of a 'probable violation'. Soviet disregard for the Treaty is also claimed, however, on four other, less certain counts. Least firmly stated is the charge 'that the USSR may be preparing an ABM defense of its national territory' (US WH, 1985b: 11; US DoD, 1985c: 12). This aspect of the Treaty is inherently difficult to apply, since no definition of what might comprise a 'base' for such a national defence was ever agreed. Furthermore, at what stage does 'providing' begin? Suppose, for example, that the United States may start deploying some form of national anti-missile defence in the year 2003. It might at that point be reasonably judged that work 'to provide a base' for that defence had begun with the first years of the SDI. If so, however, the United States also appears at present to be infringing the Treaty in this respect. Indeed, it could plausibly be argued that the United States is now far more clearly committed to at least *trying* to acquire such a national defence than is the Soviet Union.

In practice, the charge of providing a base for national defence is little more than a restatement of four other more or less tentative US charges of Soviet noncompliance with the Treaty. These concern the concurrent testing of air-defence and anti-missile system components, the development of air-defence missiles (SAMs) with significant ABM capabilities, the development, testing and deployment of mobile ABM radars, and the development and testing of rapidly reloadable ABM launchers.

Concurrent Testing, and SAMs with ABM capability?

These are not really two separate issues. The Treaty and its associated documents do not establish any prohibition whatsoever on testing SAMs, as SAMs, at the same time or in the same area as testing ABMs, as ABMs. However, there was concern on the American side that such concurrent testing might be used to develop covert ABM capabilities in air-defence systems, a process which certainly is prohibited by article VI.a. (Soviet representatives were less liable to feel such a concern, because of the neglect of air-defence systems by the United States during the 1970s.) Accordingly, an informal, unpublished agreement was reached at the SCC in 1978, to the effect that air-defence and ABM systems located at the same test range would no longer be tested concurrently. Any Soviet deviation from that agreement, however, can only form part of a case for supposing the Soviet Union might be in breach of the Treaty's ban on upgrading SAMs to ABM capability, since there exists *within the Treaty* no separate ban on concurrent testing of the two types of system as such.

In the President's February 1985 report, the 'concurrent testing' accusation was further sub-divided into two parts. The finding in respect of possible Soviet concurrent *operations* with air-defence and anti-missile components was that there was insufficient evidence to assess the matter one way or the other (US WH, 1985b: 11). However, it was claimed that air-defence components were probably being *tested* 'in an ABM mode', in violation of article VI.a of the Treaty. The distinction between testing and peacetime operations is hardly self-evident. In the Pentagon booklet the former is explained as 'conducting tests involving the use of SAM air defense radars in ABM-related testing activities' (US DoD, 1985c: 12). But the notion of an 'ABM-related' activity was not explained. The President made matters yet more obscure by stating, in the letter introducing his February report, that the Soviet Union *had* after all 'probably violated the ABM Treaty restriction on concurrent testing of SAM and ABM components' (US WH, 1985b: 2).

These confusions were dispelled in the President's December 1985 report, in which it was made clear that only concurrent testing was at issue, and in which the President's earlier personal evaluation of a 'probable violation' was upheld. On the main question, however, of whether the Soviet Union has been

upgrading its SAMs to a significant level of ABM capability, the
recent US reports have not amended the (classified) finding of
February 1985 that 'the evidence . . . is insufficient to assess
compliance'.

This combination of findings is legally anomalous. The concurrent testing issue is only relevant to Soviet compliance with the
ABM Treaty in so far as it provides evidence for a forbidden effort
towards upgrading SAMs to ABM capability. To find it 'highly
probable' that there has been concurrent testing of the kind which,
though not forbidden, it has been informally agreed should not
occur, but at the same time to find the evidence about SAM
upgrading generally 'insufficient', seems hardly consistent. Certainly, it is not legally possible to regard the former alleged
activities as constituting a 'probable violation' *unless* the charge is
applied with equal confidence to the SAM upgrading issue, since
the latter is the only Treaty provision of which concurrent testing
can be in violation.

Mobile Radars?

Before the ABM Treaty was signed, the Soviet negotiators had
accepted a US clarification of it, which thereby attained the status
of a Common Understanding ((2.C)). This states that the Treaty
prohibits deployment (and by implication from ((V.1)) also
development and testing) of 'ABM . . . radars which were not
permanent fixed types.' At the time, the Americans were
concerned about the future capabilities of a Soviet air-defence
radar which could be disassembled, moved, and reassembled at
another site in two to three days. Today, however, their accusation
refers to the Flat Twin radars being built quite legally in the
Moscow ABM deployment area.

> The *Flat Twin* radar is both transportable and modular in the sense
> that it can be disassembled, moved in component stages, and
> reassembled in a period of months. This, however, assumes that
> extensive advanced preparation of the site upon which it is being
> relocated has occurred. At issue is whether being 'transportable' in
> this instance means the radar is 'mobile' . . . there is no Agreed
> Statement between the U.S. and the Soviet Union explicitly
> defining 'mobile' and/or 'not of a permanent fixed type'. (Long-
> streth et al., 1985: 58)

Besides the Flat Twin system, US anxiety has also been aroused by a smaller Soviet radar now being developed, the Pawn Shop. It is feared that this apparatus could be made mobile quite quickly, if its van-shaped container were placed on a wheeled base – though this arrangement has not yet been observed. The implication here is that it might already be in breach of the Treaty, as such an ABM radar would be illegal in its engineering development phase. The President's February 1985 report, however, found only that:

> Soviet actions with respect to ABM component mobility are ambiguous, but . . . represent a potential violation . . . (US WH, 1985b: 10)

Furthermore, the US Army Ballistic Missile Defense Organization's commander, General Tate, was prepared to testify as follows, five years ago:

> Tate: The LoADS radar we are talking about today is an X band radar and can be packaged in a space 9 feet in diameter and 20 feet or so long and carried on the same type transport vehicle that carries the MX.
>
> Senator Jackson: You can move it right along to whatever place it is going to be?
>
> Tate: Yes. It will play the shell game against that array of shelters [MPS]. (S. Armed Services, 1981: 4142)

Perhaps such work had not then crossed the ill-defined boundary that supposedly separates research from development within the ABM Treaty regime. Nevertheless, American indignation about Soviet air-defence radars which are merely regarded as *potentially* mobile, or which it is feared it might be possible to relocate over several months, appears more than a little contrived.

Rapid Reload?

The fullest account of US anxieties under this heading is given in the recent ACDA report (1986a: 5–6). It is claimed that above-ground Galosh ABM test launchers, and test launchers for the new high-acceleration terminal (SH–08) interceptor, have been observed being reloaded and in the first case refired within a day.

The Galosh launchers in question are presumably of the ABM-1B type, since the new, SH-04 version of Galosh, for the ABM-X-3 system, will be silo-launched.

Article 5.2 of the ABM Treaty forbids the development, testing or deployment of 'automatic or semi-automatic or other similar systems for rapid reload of ABM launchers.' No clarification of the terms in this clause has ever been agreed between the parties. When the Treaty was being negotiated, the chief US negotiator Ambassador Gerard Smith explained to the other side that the United States would regard a reloading test as 'rapid' if it was achieved in a 'strategically significant' time. He also indicated that no changes in the existing Galosh system, believed erroneously by the Nixon administration of the day to be reloadable within 15 minutes, would be required in order to meet this stipulation. The lowest estimate of an apparent Soviet reload interval in recently observed testing is said to have been two hours, well outside the Smith criterion for 'rapid' (Longstreth et al., 1985: 57).

On the US side there were fairly advanced studies in the late 1970s and early 1980s for mobile LoADS Terminal Defense Units (TDUs), designed for concealed deployment alongside MX missiles in their MPS basing mode. Not just the mobility of such launchers would have breached the Treaty. Article 5.2 bans rapid reload systems immediately after, and for the same reasons as, banning ABM launchers capable of launching 'more than one ABM interceptor at a time.' But LoADS TDUs with multiple launch capabilities were regularly discussed and proposed (Moore, 1981). It is unlikely, however, that any were built or tested.

Abalakova?

The strongest and indeed the only substantive American charge of a Soviet breach of the ABM Treaty concerns the large phased-array radar under construction at Abalakova (p. 147). Because of its location and orientation, it will certainly breach the Treaty once it becomes operational at the end of this decade, *if* it is for early-warning ((VI.b; 1.F)). The Soviet Union has declared that it is for space tracking, but the Reagan administration points to its similarity with other Soviet early-warning radars and to the fact that it could fill the last remaining gap in an all-round early-warning coverage, on which the Soviet Union has expended major efforts for several years past (US DoD, 1985c: 10–12). It is also claimed that the design is not 'optimized' for space tracking, and that the radar would add nothing significant to existing Soviet space-tracking capabilities.

If the Abalakova radar is indeed for early-warning, one reason for its improper siting may have been the hostile climate and inadequate economic infra-structure of the north-eastern 'periphery' of Soviet Siberia. Two earlier Soviet radar projects in such latitudes, intended perhaps to provide coverage of the Arctic and Northern Pacific oceans, were apparently abandoned because of the instability of the tundra permafrost. One US official has suggested that 'what they tried to do was to build the radar inland near the Trans-Siberian railway where they could maintain it and hope they could pass it off as a space-tracker' (*Arms Control Reporter*, April 1985). In 1981 and 1982 Soviet representatives at the SCC explained, in response to questions about the relatively inland location of the Pechora LPAR, that radar siting always had to take account of 'technical and practical considerations involved in their placement' (ibid.; also Longstreth et al., 1985: 54). Viewed in retrospect, these may even have been hints about Abalakova. If Soviet representatives have not yet made that reply to the specific US accusation about Abalakova, that could well be because of the hostile tone in which the latter has been delivered, and the domestic public relations campaign to which it has been unstintingly applied.

However well-founded the American objection to Abalakova may turn out to be, it suffers politically from the fact that the US Air Force has long been engaged in constructing one new early-warning radar actually outside US national territory, at Thule, Greenland, and proposes to start building another shortly, at Fylingdales Moor, Britain. The legal arguments arising out of the Treaty's restraints on early-warning radars ((VI.b; 1.F)) are too complex to examine here. But a good case can be made for viewing these US radars as infringements of the Treaty, if nothing stronger (Bulkeley, 1985). This does not of course put the alleged early-warning radar at Abalakova back on the right side of Treaty limits. But if such parallel stretching and perhaps infringement of the Treaty continues on both sides, the entire Treaty regime is bound to be threatened.

On the other hand, the United States and the Soviet Union *could* decide to handle these alleged radar infringements in a spirit of reconciliation aimed at strengthening the Treaty, if they so chose. Amendments to permit increased early-warning capability are conceivable, despite the problems posed by the fact that all modern early-warning radars are almost certain to be potentially powerful 'early-tracking' radars also, and thus of considerable use to ABM systems. Such an approach might require no more than joint re-affirmation of the ABM Treaty's strict restraints on the development of the sort of strategically significant ABM capabilities that might be linked to such radars. With future large-scale BMD deployments ruled out, large radars could have no other military role than early-warning, to enhance each side's old-fashioned 'retaliatory deterrence'. (A *total* asat and ABM ban (p. 156) would of course be even more effective in this respect.) The likelihood of such an accommodation being reached, however, is bound up with the general implications for disarmament of current space weapon programmes.

Glossary

The glossary contains the principal acronyms and abbreviations used in the text, and technical terms if they are not explained at or close to their first occurrence (see index). If explained in the text, acronyms are unpacked here without further comment. The following groups of acronyms have largely been *omitted* from the glossary: those occurring only in appendix 2; those used only for references (chapter 1, note 2, p. 325); and the missile designations explained in chapter 3, note 9 (p. 327).

A-A: Anti-Aircraft artillery.

ablative shield: material applied to objects in order to cool them by evaporating or burning off when heated.

ABM: Anti-Ballistic Missile.

ABRES: US Advanced Ballistic Re-Entry Systems programme (now ASMS).

accuracy: of ballistic missiles, the radius of a circle within which the warhead has a 50 per cent probability of landing (Circular Error Probable), but *not* the probability of that circle being centred on the intended target.

ACDA: US government Arms Control and Disarmament Agency.

ACIS: Arms Control Impact Statements, annual assessment by ACDA of arms control aspects of US weapons development and procurement programmes.

acquisition: detection of a potential target by the sensors of a weapons system.

active countermeasure: device to reduce performance of a BMD system by damaging, destroying or spoofing (by emission of misleading signals) one or more of its components.

active sensor: system emitting radiation (or sound) in order to detect objects by means of the pattern of energy reflected, e.g. radar.

315

adaptive optics: optical systems capable of modification to compensate for distortions in light signals, e.g. by altering the shape of mirrors.

AFB: (US) Air Force Base.

AIDS: Artificial Immune-Deficiency Syndrome.

algorithm: systematic mathematical procedure for problem solving.

A-LV: Soviet Launch Vehicle type 'A', used for Soyuz spacecraft.

AMS: US Advanced Meteorological Satellite.

anti-simulation: making warheads look like decoys to deceive enemy sensors.

arc second: angular measure, ¹⁄₆₀th of a degree.

architecture: overall structure of a multi-layered BMD system.

asat: weapon for destroying satellites (by abbreviation from 'anti-satellite').

ASMS: US Advanced Strategic Missiles Systems programme.

ATBM: Anti-Tactical Ballistic Missile; in arms control parlance, a 'tactical' missile is a ground- or air-launched missile system with a range less than 5500 km. (All SLBMs are bracketed with 'strategic' systems. Sea-launched cruise missiles, however, are a grey area.)

atomic: of nuclear weapons and explosions, those using only fission of atomic nuclei (by contrast with 'thermonuclear'); physical constraints limit such explosions to kiloton sizes.

balance: estimated ratio of military strength between rival states.

BAMBI: BAllistic Missile Boost Intercept.

battle management: instructions and rules embedded in or generated by computer software and associated hardware to control a BMD or other weapon system.

beam divergence: spreading out of electro-magnetic radiation in a directed-energy beam, caused by diffraction, proportional to the ratio between the wavelength and the diameter of the aperture.

beta blackout: interference to radio and radar signals caused by dispersal of electrons from a nucleus explosion within Earth's magnetic field.

betatron: relatively low-power particle accelerator forming a beam of beta particles – electrons spontaneously emitted by radioactive elements.

billion: one thousand million (10^9).

birdcage: satellite constellation using several similar orbits in planes at equal angular intervals.

birth-to-death tracking: tracking of objects from moment of launch

from missile bus until they are destroyed, disabled, or reach
the ground.

BMD: Ballistic Missile Defence; use or capability of using ABMs
to destroy attacking ballistic missiles.

BMEWS: US Ballistic Missile Early-Warning System.

booster: main missile engine/s, or small engine or gas-jet used to
adjust a vehicle trajectory in space or upper atmosphere.

bus: post-boost vehicle containing warheads and penetration aids
on ICBMs and SLBMs.

C^3: Command, Control and Communications.

CBO: US Congressional Budget Office.

Chevaline: nuclear warhead and penetration aids system fitted to
British Polaris submarine-launched missiles in early 1980s.

CIA: US Central Intelligence Agency.

cm: centimetre.

coherence: matching, in space and time, of the wave structure of
different parallel rays of a single frequency of electromagnetic
radiation.

constellation: arrangement of several satellites of the same type
along one or more orbits, whether in a navigation, BMD, or
other system.

Cosmos: generic Soviet satellite designation in use since 1962, but
distinguished by Western analysts into a variety of actual
types, well over half of which have been for military purposes.

counter-city: use or threatened use of nuclear weapons mainly
against people.

counter-force: use or threatened use of weapons to destroy enemy
weapons or forces before they can be used (e.g. missiles in
silos).

countermeasures: actions aimed at reducing the effectiveness of a
weapon system; see 'active c.' and 'passive c.'.

damage limitation: use of offensive or defensive weapons primarily
to reduce damage from enemy weapons.

DARPA: US Defense Advanced Research Projects Agency.

decapitate: of a nuclear 'first strike', to destroy the enemy's
political and military C^3.

decoy: object designed to confuse or 'draw fire' from a BMD
system by its resemblance to a nuclear warhead.

DEW: Directed Energy Weapon, such as laser or particle beam
accelerator, able to deliver destructive energy at or near the
speed of light.

DIA: US Defense Intelligence Agency.

diffraction: mutual interference between rays of light or other electro-magnetic radiation passing through an aperture or leaving the edge of a reflecting body – causes beam divergence in DEW.

digital imaging: scanning and coding of optical or radar images as sequences of binary numbers for radio transmission and computer processing.

directed energy: see DEW.

D-LV: Soviet Launch Vehicle type 'D', used by Salyut spacecraft.

DNA: US Defense Nuclear Agency.

DoE: Department of Energy.

Doppler effect: apparent shift in frequency of electro-magnetic radiation (or sound) due to relative motion between the source and the observer.

DOV: Denial Of Victory, 1980s pro-BMD strategic doctrine advocated by Colin Gray and Keith Payne.

drag: deceleration and deflection of objects passing through atmosphere at high speeds.

DSAT: weapon for or mission of Defending SATellites by destroying or interfering with asats.

DSCS: US Defense Satellite Communications System.

DSP: Defense Support Program (US early-warning satellites).

DTS: Defensive Technology Study, also known as Fletcher Study, conducted for US administration in 1983.

early-warning: early detection of enemy missile launch, usually by means of surveillance satellites or long-range radar.

EDI: 'European Defence Initiative'; so-called, no such programme actually exists.

efficiency: of a machine, ratio of output energy to input energy.

electro-magnetic radiation: energy propagated from electric charges in motion, producing simultaneous wavelike variation of electric and magnetic fields in space; highest frequencies (shortest wavelengths) are gamma rays, originated by processes within atomic nuclei; lower frequencies include X-rays, ultraviolet light, visible light, infra-red light, microwaves, and radio waves.

EMRLD: Excimer, Moderate power, Repetitively pulsed Laser Device.

EMP: Electro-Magnetic Pulse.

encryption: automatic ciphering of digitalized signals, usually employing randomization techniques.

ERIS: Exo-atmospheric Reentry-vehicle Interception System.

ESA: European Space Agency.

F-15: USAF interceptor/strike aircraft, converted to carry air-launched MHV asat.

FAS: Federation of American Scientists.

FLTSATCOM: US FLeet SATellite COMmunications system.

F-LV: Soviet Launch Vehicle type 'F', used in former SS-9 ICBM and current orbital-pursuit asat.

FOBS: Fractional Orbit Bombardment System.

footprint: ground area threatened or defended by a weapon system.

FSSS: Future Security Strategy Study, conducted for US administration in 1983 by the Hoffman and Miller Panels.

FY: US government Fiscal Year, from 1 October of previous year to 30 September of year specified.

g: unit of acceleration, equivalent to earth gravity at sea level, approx. 9.8 metres per second per second.

Gemini: second US manned spaceflight project, 1964–6.

GEO: GEostationary Orbit.

GOES: Geostationary Operational Environmental (meteorological) Satellite.

GPS: Global Positioning System.

grey area: poorly defined and therefore ill-kept part of legal provisions in a treaty or other legal enactment.

ground-station: military or civilian installation for satellite C^3.

hardness: estimated resistance of a possible target to some form of potentially destructive energy; e.g. heating, measured in kJ/cm^2, the amount of heat energy that can be absorbed before damage is sustained, or blast, measured in psi.

HEDI: High Endo-atmospheric Defense Interceptor.

HOE: Homing Overlay Experiment.

homing: combination of on-board sensors, computers and course adjustment mechanisms to enable a vehicle to 'steer itself' towards a sensed object or location.

HQ: headquarters.

hypervelocity: very high speed, usually measured in km/sec; within the SDI, usually defined as at or over 10 km/sec.

ICBM: Inter-Continental Ballistic Missile.

impulse kill: use of directed energy to destroy a target by ablative shock, i.e. by violently and rapidly boiling off the surface of the target and setting up a mechanical shock wave to cause structural failure.

INF: Intermediate-range Nuclear Forces.

IOC: Initial Operating Capability.

ion: 'charged atom' formed by removing or adding one or more negatively charged electrons in an atom, which is normally electrically neutral since charges of component particles balance out.

IONDS: US Integrated Operational Nuclear Detection System.

ionization: process of creating ions.

IRBM: Intermediate-Range Ballistic Missile.

JIC: Joint Intelligence Committee of the British Cabinet.

joule: unit of work or energy, the work done per second by a current of 1 ampere flowing through a resistance of 1 ohm.

KEW: Kinetic Energy (collision) Weapon.

KH: US Key-Hole reconnaissance satellite.

kill assessment: detection and processing of information within a weapon system to determine the status – destroyed/damaged/ undamaged – of object/s it is being used to attack.

kill radius: radius of spherical space within which the effects of a weapon are estimated to have a larger than 0.5 probability of destroying a target.

kilojoule: 1000 joules (kJ).

kiloton: unit of explosive power, equivalent to detonation of 1000 tons of TNT; strictly, 10^{12} calories.

kJ/cm^2: 1000 joules energy acquired within one square centimetre of a surface.

km: kilometre.

km/sec: kilometres per second; 1 km/sec = 3600 kilometres or 2237 miles per hour.

layer: self-contained BMD subsystem, using one method of attack at one stage or successive stages of the attacking missile trajectories.

layered defence: BMD with several layers.

leakage: estimated or actual proportion of attacking missiles, warheads or possible warheads (including undiscriminated decoys) surviving a BMD layer or system.

LEO: Low Earth Orbit.

Limited Test Ban Treaty: multilateral treaty initially signed by Britain, the United States and the Soviet Union in 1963; prohibits all except underground nuclear weapon tests.

LoADS: Low Altitude Defense System.

LPAR: Large Phased-Array Radar.

MAD: Mutual Assured Destruction.

megaton: unit of explosive power, equivalent to detonation of 1 million tons of TNT; strictly, 10^{15} calories.

MHD: Magneto-HydroDynamic, hydro-electric power system developed in Soviet Union for rapid generation of a large pulsed power supply to an experimental high-energy laser.

MHV: Miniature Homing Vehicle.

micron: one millionth of a metre.

microradian: one millionth of a radian angle; lateral (or planar) aiming accuracy can be expressed in terms of this angle, which corresponds effectively to one millionth part of the range – thus at 3000 km, accuracy to one tenth of a microradian equals accuracy to within 30 cm in one dimension.

MilStar: Military Strategic-Tactical And Relay, US satellite communications system, not yet deployed.

MIRACL: Mid-Infra-Red Advanced Chemical Lasers.

MIRV: Multiple Independently-targeted Re-entry Vehicle.

MIT: Massachusetts Institute of Technology.

MJ: megajoule, one million joules.

modular: system of construction by assembling relatively large and complex prefabricated parts.

MOU: Memorandum of Understanding.

MPS: Multiple Protective Shelters.

MW: megawatt, one million watts.

MX: US ICBM (Missile eXperimental) to be deployed in 1986.

NASA: US National Aeronautics and Space Administration, established in 1958.

NASP: US National AeroSpace Plane.

NATO: North Atlantic Treaty Organization.

NORAD: North American Aerospace (formerly Air) Defense Command.

NOSS: US Navy Ocean Surveillance Satellite.

NPG: Nuclear Planning Group.

NTM: National Technical Means of arms control verification, e.g. surveillance satellites.

NUDET: NUclear explosion DETection.

OTA: US Congressional Office of Technology Assessment.

PAR: Phased Array Radar.

passive countermeasure: device to reduce performance of a BMD system by deceiving its sensors without actual spoofing (see under 'active countermeasure'), or by negating the effects of one or more of its components.

passive sensor: device for detecting radiation emitted or reflected by an object.

payload: part of any missile or other vehicle intended to reach the target zone, arrive in orbit, etc.

PBW: Particle Beam Weapon.

PE: Program Element within the SDIP.

penetration aid: object released by a missile bus together with warheads to increase likelihood of warheads penetrating defensive systems.

Pentagon: US Department of Defense main office building in Washington D.C., used by extension to refer to the Department itself.

psi: pounds per square inch, unit of blast overpressure produced from an explosion, hence of hardness in any structure able to survive that amount of blast intact. 1 psi $= 6.89476 \times 10^3$ pascals (or newtons per square metre).

R&D: Research and Development.

radian: angle at centre of a circle subtended by an arc equal to the length of the radius $= 57.296°$.

radio-frequency quadrupole: a radio-frequency field at very high oscillation, e.g. 80 million times a second, forming an electric field with four poles instead of two; used to focus a beam of negative hydrogen ions at the same time as accelerating it.

RAF: British Royal Air Force.

range: distance between points of initiation and of destructive effect in any weapon system; in the case of missiles, measured along the surface of the Earth, i.e. less than the length of trajectory.

red-out: blinding or dazzling of infra-red sensors; can be produced either by air friction around their vehicle, or by high levels of infra-red radiation from a nuclear explosion in the upper atmosphere.

RF: Radio-Frequency.

RORSAT: Soviet Radar ocean-reconnaisance satellite.

RV: Re-entry Vehicle, portion of any system entering space designed to survive friction effects during atmospheric re-entry; in missiles, small container for nuclear warhead, or decoy resembling such; frequently used as euphemism for 'warhead'.

SAINT: US SAtellite INTerception and inspection system.

SALT: Strategic Arms Limitation Talks; hence 'SALT 2', for the 1979 US–Soviet Treaty on Limitation of Strategic Offensive Arms.

salvage-fusing: design of warheads to explode if attacked.

SAM: Surface-to-Air Missile.

SATKA: Surveillance, Acquisition, Tracking and Kill Assessment.

SAMOS: Satellite And Missile Observation System.

SANA: Scientists Against Nuclear Arms

SCC: Standing Consultative Commission (for SALT).

SDI: Strategic Defense Initiative.

SDIO: Strategic Defense Initiative Organization.
SDIP: Strategic Defense Initiative Program.
SDS: US Satellite Data System.
sensors: electronic instruments for detecting radiation; see 'passive
s.' and 'active s.'
shuttle: re-usable vehicle able to enter LEO from and return to the
ground; hence 'the Shuttle', the US Space Transportation
System (STS), now comprising a fleet of three orbital vehicles
(Orbiters), following the loss of the second Orbiter (OV-099)
Challenger on 28 January 1986.
signature: distinctive pattern of radiation emitted or reflected from
an object, enabling identification of its type.
Silicon Valley: region running north-west from San Jose, in Santa
Clara county, California, with a high concentration of firms
designing and manufacturing advanced microchips, computers and software, often for military applications, together
with the Lockheed Space Tracking Center and NASA's Ames
Research Center.
silo: vertical hardened concrete structure for underground housing
of a missile.
simulation: making decoys look like warheads.
SIOP: Single Integrated Operational Plan.
SIPRI: Stockholm International Peace Research Institute.
SLBM: Submarine-Launched Ballistic Missile.
slew-time: time needed for a weapon to re-aim to a new target after
firing at a previous one.
soft: highly vulnerable to nuclear weapon effects.
software: coded instructions and information put into computers to
enable performance of required calculations.
SPADATS: SPAce Detection And Tracking System.
SPADOC: SPAce Defense Operations Center.
SPD: West German Socialist Party.
Sputnik: first Soviet satellite series, 1957–61.
squirt transmission: very brief and rapid radio transmission of
digitalized data, intended to evade hostile monitoring.
State: US Department of State (i.e. Foreign Affairs).
STEW: Space-To-Earth Weapon.
super-hardening: hardening to unusually high levels, several
thousand psi.
survivability: capacity of a system to survive and function despite
direct attack.
TAV: Trans-Atmospheric Vehicle.

TBD: To Be Decided.
TBM: Tactical Ballistic Missile.
TDRS: US Tracking and Data Relay Satellite.
TDU: Terminal Defense Unit.
telemetry: signals returning data from instruments placed on-board a missile or satellite to record details of performance, especially in test or prototype models.
thermonuclear: nuclear weapon/explosion releasing energy by the fusion of light atomic nuclei into heavier ones; term derives from very high temperatures needed to initiate the process, usually obtained from a fission ('atomic') explosion.
throw-weight: total payload capacity of a missile.
trajectory: path followed by ballistic projectile, e.g. missile.
transistor: multi-electrode semiconductor device in which current flowing between two specified electrodes is modulated by the current or voltage applied to one or more specified electrodes. In microelectronic circuits for digital processing, incorporated in such materials as crystalline silicon or gallium arsenide, the transistors function as switches to provide the logical structure of the circuitry.
trillion: one million million (10^{12}).
UCS: Union of Concerned Scientists.
UN: United Nations.
unilateral, -ism: of arms build-up or disarmament, proceeding by independent national decision; of deterrence, seeking the capacity to threaten other states whilst diminishing or removing their capacity for similar threats in return.
USAF: United States Air Force.
warhead: explosive or other kind of weapon carried by a missile; usually uses nuclear or conventional explosives, or direct impact, for destructive effect; can also be chemical or biological weapon – see also 'RV'.
WEU: Western European Union.

Notes

1 Just Another Place

1 The problem of interpreting such Soviet statements is discussed in chapter 9.
2 References to official US sources are organized as follows. Congressional Hearings and Reports are cited by committee and the year in which they occurred, usually with a page reference, and listed in the References under 'House of Representatives' (H.), or 'Senate' (S.). The Congressional Budget Office (CBO) and the (Congressional) Office of Technology Assessment (OTA) are listed separately. Departments and organs of the US Government, including the Departments of Defense (US DoD) and State (US State), the DoD's Strategic Defense Initiative Organization (US SDIO), the White House (US WH), National Security Council (US NSC), and Arms Control and Disarmament Agency (US ACDA), are cited by the abbreviations just given and listed under 'United States Government'. Speeches and articles by leading members of the administration and other officials or advisers are listed under their names. Thus Weinberger, 1985 for a speech to the National Press Club, but US DoD, 1984a for his FY 1985 Report to Congress.

2 The Beginning of Space Weapons

1 On 13 January 1986 *The Guardian* carried a report by Robert Scheer which quotes a recent study, written by physicists Albert Latter and Ernest Martinelli for a Los Angeles think-tank, R & D Associates, as follows:

> in a matter of hours, a laser defense system powerful enough to cope with the ballistic missile threat can also destroy the enemy's major cities by fire. The attack would proceed city by

city, the attack time for each city being only a matter of minutes. Not nuclear destruction, but armageddon all the same.

Scheer also cites other US experts in confirmation of this possibility.

2 In contrast with two short paragraphs in the magazine, the article in the book had a four-page discussion of the bomb-platform idea. In this version, the bombs were to be carried on a separate platform trailing behind the main satellite on the same orbit. Their re-entry into the atmosphere could then be directly guided from on board the space station (figure 2.1). The editor's Introduction warned about confident Soviet claims to be able to build such a space station also, but spoke warmly of the ability of any Western counterpart to put an end to 'Iron Curtains wherever they might be' (Ryan, 1952: xiv).

3 The first US satellite (31 January 1958) was in fact to be the 14 kg Explorer 1. The figure given here for Sputnik 2 is the mass of the complete vehicle placed in orbit. Western journalists and 'experts' have usually preferred to record the only slightly less embarrassing figure of 508 kg, which refers to the apparatus, batteries and capsule, with its live passenger, which were indeed placed in orbit, but still attached to the entire third stage, which also contained some of the instruments, as a single unit (Martin, C-N., 1967: 70).

4 In 1960, a few months after the interception of an American U-2 reconnaissance plane at an unprecedented altitude for anti-aircraft systems, there were in fact to be official Soviet claims of a capability to serve US satellites in the same fashion (Peebles, 1983: 95).

5 This organization keeps a detailed record of the several thousand artificial objects now in orbit, until under the gradual pull of Earth's gravity they re-enter the atmosphere and either burn up completely or in some cases return to the surface as debris. The information is necessary both for safety reasons, for deciding when and where to launch any new space vehicle, and for the strategic warning mission, which has to distinguish any unidentified and potentially threatening space objects from those already up there. The Soviet equivalent, for which there is no name in general use, is interesting for the use it makes of a fleet of more than ten specialized tracking and communications ships, developed since 1960 to compensate for a lack of suitable land bases (Turnill, 1984: 286).

6 Between February 1959 and June 1960 the first twelve Discoverer shots all failed in this or other technical ways.

7 For 'megaton', see glossary. The use of what was only the fourth successful US satellite in a second space-weapons experiment, just ten days before its orbit began to decay, is perhaps symbolic of the importance given from the outset to the development of space weapons.

8 In 1969 the payload in one such test flight remained in orbit, probably

due to a malfunction of some sort, and later re-entered as a shower of fragments, several of which landed in the American mid-West.

9 Names such as 'Galosh' or 'Guild' for Soviet missiles are of NATO origin. The Western numbering system for Soviet missiles, SS-7, SA-X-12, and so on, is produced by the US Department of Defense according to simple rules: SS for Surface-to-Surface, SA for Surface-to-Air, X for Experimental, N for Naval, and so on. Much Soviet missile nomenclature eventually becomes known, but Western intelligence officers and other concerned parties usually need a system for referring to them well before that happens. Code-names like 'Try Add' and 'Hen House' for Soviet radars are bestowed by US intelligence agencies. The alphabetical nomenclature for Soviet space launch vehicles (A-LV, D-LV etc.) was originated by the late C. S. Sheldon II of the Congressional Research Service. By contrast, Western terms for Soviet satellites usually follow the original nomenclature, because they are registered with international bodies and therefore quickly available.

3 The First Anti-Missile Weapons

1 Strictly speaking, the world's first anti-missile defence system was the very effective one deployed by the British Air Defence command against the V-1 'flying bombs', some months before the V-2s came into action. But the V-1 was a cruise missile (box 1.1), with a maximum speed of about 640 kph at a constant altitude of between 600 and 900 metres. The V-2, in contrast, was a ballistic missile, hurled by its initial rocket boost into an elliptical curve the top of which was 80 to 100 kilometres high, on the fringes of outer space. From that point it fell at a speed of up to 1.5 km/sec, slowed by the atmosphere to 1 km/sec at impact. Its arrival took place only about five minutes after launching, by contrast with flight-times of 30 minutes or more for the V-1. Two other firsts for the anti-V-1 defenses should be mentioned. They were the first to employ anti-missile missiles, if the 'two hundred rocket projectors' deployed alongside anti-aircraft batteries may be so described, and the first to use a 'layered' anti-missile defense, with successive independent forces – fighters, A-A, and barrage balloons – seeking to intercept the intruding missiles. (Collier, 1976; McGovern, 1965.)

2 The inaccuracy of early long-range missiles, relative to warhead sizes of the order of those used against Japan, led to some notorious failures to see what was coming, notably that of Dr Vannevar Bush, Truman's senior military scientist, who remarked that ICBMs were:

> not so effective from the standpoint of cost or performance as the airplane with a crew aboard . . .

[A]s long as atomic bombs are scarce and highly expensive in terms of destruction accomplished per dollar disbursed, one does not trust them to a highly complex carrier [the ICBM] of inherently low precision, for lack of precision decidedly increases such costs. (1950: 90, 96–7)

In contrast to Bush, Britain's P. M. S. Blackett (1948) went along with the official technical pessimism of US sources, but wisely and, as it turned out, accurately limited his negative forecast to the next ten years.

3 The precise locations of these two LPARs, and of a third one nearing completion 'at Pushkino', are obscure. Pentagon sketch-maps, on which figure 3.1 is based, regularly place all three Moscow LPARs about 10–15 kilometres away from the places they are said to be 'at' (p. 000). We have not been able to resolve this inconsistency. Figure 3.1. therefore places the 'Pushkino' LPAR, for instance, approximately at Ashukino, 15 kilometres north-east of Pushkino.

4 The widespread assertion that US MIRV technology was developed mainly to counter Soviet anti-missile defenses is not borne out by the historical record, despite Robert McNamara's occasional use of it for purposes of political rationalization. His Director of Research and Engineering, Dr John S. Foster, told a different story:

The MIRV concept was originally generated to increase our targeting capability rather than to penetrate ABM defenses. In 1961–62 planning for targeting the Minuteman force it was found that the total number of aim points exceeded the number of Minuteman missiles. By splitting up the payload of a single missile [deleted] each [deleted] could be programmed [deleted] allowing us to cover these targets with [deleted] fewer missiles. [Deleted.] MIRV was originally born to implement the payload split-up [deleted]. It was found that the previously generated MIRV concept could equally well be used against ABM [deleted]. (S. Appropriations, 1968: 2310)

Dr. Foster's version of events is confirmed in Greenwood's authoritative history of the MIRV programme (1975). See also (p. 000).
5 Davis (1979: 29) puts it a little longer, at four months.

4 Military Satellites and Asats

1 This series has been ended, and the one remaining operational KH-11 was to have been succeeded in 1986 by the first two KH-12s. As this book was being finished, the serious delays to that programme, and others, resulting from the disaster with the *Challenger* Shuttle in

January 1986, were still being evaluated. Five of the fifteen Shuttle missions planned for 1986 were to have been flown by this spacecraft. The second US Tracking and Data Relay Satellite (TDRS), part of a network intended to provide a more or less uninterrupted and secure relay system between satellites, cutting out the vulnerable ground-stations (p. 135), was destroyed on board the *Challenger*. (TDRS-1 was deployed on *Challenger's* maiden flight in April 1983, but owing to a slight malfunction of the system for boosting it to GEO took six months to be placed in orbit at its planned location. It has a predicted life of ten years.) Other Shuttle-launched US military satellites for 1986 would have included two DSCS Phase III communication satellites, two Navstar navigation and nuclear explosion detection satellites, and the Teal Ruby experimental space-borne sensor (p. 298). In 1987, ten more Navstar launches were planned, together with important electronic intelligence and early-warning satellites.

Earlier DSCS and Navstar models had been launched on Titan and Atlas boosters respectively, so it may be possible to adapt the current versions to do the same if, as seems likely, the Shuttle programme is seriously delayed. However, the KH-12 is reported to be totally dependent on the Shuttle, the absence of which could seriously limit US reconnaissance capabilities for some time. The first two satellites in a new British military communications system, Skynet 4, may also be affected unless facilities are obtained from the (previously rejected) French Ariane launch system. The Ariane, however, is almost completely booked until late 1987.

2 Two or three years ago various experts believed the last pair of Vela satellites might still be operational (Stares 1984: 46; Jasani & Lee, 1984: 44). With the passage of time, and in the absence of hard information, it is increasingly improbable that their systems are working reliably, if at all. They have therefore been omitted from table 4.1.

3 Estimates vary. The Union of Concerned Scientists (1984: 195) say over 60 per cent; Stares (1984: 53), over 70 per cent; Jasani (1982: 58) has 80 per cent.

4 However, the Office of Technology Assessment has estimated that the Soviet asat could destroy targets at up to 5000 kilometres (OTA, 1985b).

5 Also known as the Space Detection and Tracking System (SPADATS).

6 It turned out that the satellite was still being used by US military scientists for solar measurements said to be of assistance to the USAF.

5 From 'Deterrence' to 'Star Wars'

1 This view was repeated two years later by USAF Chief of Staff, General Lew Allen (H. Armed Services, 1981: 23).

2 The widely used distinction between two senses of the term 'policy' originated with Mr Nitze thirty years ago:

> In one sense, the *action* sense, it refers to the general guide lines which we believe should and will in fact govern our actions in various contingencies. In the other sense, the *declaratory* sense, it refers to policy statements which have as their aim political and psychological effects. (Nitze, 1956: 187)

3 By the time his talk was published, Mr Weinberger had been appointed to the post of Secretary for Health, Education and Welfare.

4 Newcomers to such discussions should grasp as soon as possible that the term 'balance' in nuclear strategic parlance has nothing to do with parity or equality. It is simply a polite way of referring to the *ratio* between US and Soviet capabilities, a ratio which US force planners have, at least since the late 1960s, been determined to keep *unequal*, in their side's favour. Hence the expression 'favourable balance', now falling somewhat out of use as a little too revealing.

As to how policy-makers should be judged, who simultaneously protested at alleged Soviet first-strike capabilities or contingency plans, did their very best to deny Soviet forces the alternative of a reliable second-strike posture, and tried to secure an effective counter-force first-strike arsenal for the United States, let the reader decide. The inimitable Mr Weinberger simply puts the policy of his Nixon years behind him, together with his own express satisfaction, as one of its leading members, at the certainty that US 'offensive forces' would have 'a commanding lead'. He now affects to believe that the policy he accepted in those days 'was based on the assumption that both sides, Soviet and United States, would stay equal in offensive power' (S. Armed Services, 1984: 55). Except, that is, for days on which he believes it was based on retaining 'significant American nuclear superiority'!

5 For Reagan's own susceptibility to this misperception, from at least 1976, see Barrett, 1983: 305.

6 Eighteen months later Carbaugh resigned his post in the Congressional service, amidst speculation that he had been moved to do so by the media exposure given to his extensive political influence, based neither on election nor on any governmental position. His role in the Group, which continues to play an important role in Washington life, appears to have been assumed by Senator Wallop's aide, Angelo Codevilla.

6 The President's Programme?

1 In what can only be described as an apologia for atmospheric testing and nuclear fall-out, written after the notorious 'Lucky Dragon' incident of 1954, Teller had already claimed that:

Clean, flexible and easily delivered weapons of all sizes would make it possible to use these bombs as we want to use them: as tools of defense. (Teller & Latter, 1958: 172)

2 Some if not all of these Senators were strongly influenced by right-wing Republican staffers in the Madison Group, alias 'The Survivalists' (p. 64).
3 The document is undated, but can be shown from internal evidence to have been written around April 1984.
4 One topic that seems to have been troubling the Chiefs particularly at the time, and with them the great aerospace corporations, was the increasing reluctance of the American public to pay for ever 'bigger and better' offensive missile programmes (Thompson, 1985: 128–37).
5 As many commentators immediately pointed out, for the peoples of East and West to be held hostage, in future, by only a half or a quarter of each other's massively superfluous arsenals, would be rather like being offered an immediate return to the warhead numbers of, say, 1971. But, had the world been safer from nuclear war at that time?
6 Unwelcome, because of their long-standing disagreements with several past US administrations over the ABM issue, due to clearly perceived divergences of national interest (Yost, 1982). See chapter 12.
7 The State Department's June 1985 report, echoing the SDIO's 1984 Interim Charter (note 8 below), declared:

the SDI research program will place its emphasis on options which provide the basis for eliminating the general threat posed by ballistic missiles. Thus, the goal of our research is not, and cannot be, simply to protect our retaliatory forces from attack. (US State, 1985: 4)

Interestingly, the denial is not extended to the possibility that the SDI may produce new offensive capabilities.
8 The course of events between President Reagan's speech and the formation of the Strategic Defense Initiative Organization (appendix 2) a year later was as follows. Immediately after his speech the President signed a directive setting up two parallel studies, the Defensive Technology Study (DTS) into the technological possibilities and most promising avenues for ballistic missile defence, and the Future Security Strategy Study (FSSS), to examine the role of BMD within future US strategy. The former is commonly referred to by the name of its chairperson as the 'Fletcher Study'. The FSSS was carried out by two teams, one made up of government officials, chaired by F. C. Miller, and another of outside experts, chaired by F. S. Hoffman. The 'Hoffman Report', which is unclassified, is the report of the second of these two panels for the FSSS. None of these studies was charged with examining *whether* the President's Initiative should go forward, only how best it should do so.

The three panels reported in October 1983. On 6 January 1984 the President signed another order, directing the Defense Department to proceed with establishing a Strategic Defense Initiative Organization. The organisation was set up in March 1984, with Lieutenant General James Abrahamson, who had headed the Space Shuttle programme at NASA, as its Director. On 24 April 1984 it received its Interim Charter, which specifies that: 'The ultimate goal of the SDI is to eliminate the threat posed by nuclear ballistic missiles . . .' (H. Appropriations, 1984: 696).

7 Weapons for Ballistic Missile Defence

1 Dr Hagengruber, director of system studies at the Sandia National Laboratory, has cautioned against setting too much store by:

> what we call strap-down chicken tests, where you strap the chicken down, blow it apart with a shotgun, and say shotguns kill chickens. But that's quite different from trying to kill a chicken in a dense forest while it's running away from you. (*International Herald Tribune*, 19 December 1985)

The *Tribune* reported that MIRACL had been virtually next door to its target in BMD terms (half a mile, according to another source), that its achievement was thought to have taken 'several seconds', far too long for an effective anti-missile weapon, and that 9000 gallons of water had been needed to cool its dozens of mirrors – an almost insuperable logistics requirement for anything but a ground-based system.

2 A technique for this purpose, known as Stimulated Raman Scattering, has recently been demonstrated in a high-pressure gas both for combining beams from several sources and for 'cleaning up' the combined beam that results.

3 Dr. Allen, a leading British authority on laser physics, writes: 'We do not know what an intense coherent beam of X-rays can do. I suspect that it is likely that they will pass through quite a lot of air; but it remains to be seen' (personal communication).

4 On 24 November 1985 the *Boston Globe* reported that there had only been four test explosions in the X-ray laser programme, which implied that only the first test, 'Dauphin', on 14 November 1980, had been a technical success. Other sources put the number of tests, to the end of 1985, at six.

5 According to one press report (*Aviation Week & Space Technology*, 15 July 1985), Lawrence Livermore Laboratory has demonstrated the theoretical possibility of a method of firing a charged particle beam through the atmosphere. A high-energy annular-shaped laser beam is used to cause gas breakdown and create a cylindrical 'Faraday cage'

channel, through which a charged particle beam can be propagated. Scaling of this physical effect to the size needed for a weapon system appears problematic. If it could be achieved, the technique might be applicable at low atmospheric pressures, at altitudes between 85 and 600 kilometres.

8 The Effectiveness of BMD Systems

1 It would take about two minutes for warning of a Soviet ICBM launch to travel from the DSP-East satellite, via its Australian ground-station, to the United States. The plan to replace such installations with faster and more secure satellite relay stations received a major setback with the destruction of the second such satellite on board the *Challenger* in January 1986 (p. 329).
2 '[M]ost experts now feel four to ten times that amount may be necessary' (*IEEE Spectrum*, September 1985: 42).
3 The continuing relevance of this problem was stressed in the eventual findings of the Eastport Study Group. They reported that:

> The contractors [for architecture studies, see table AP2.9] treated battle management as something that is expected to represent less than 5% of the total cost of the system, and therefore could not significantly affect the system architecture. They have developed their proposed architectures around the sensors and weapons and have paid only lip service to the structure of the software that must control and coordinate the entire system. (*Aviation Week & Space Technology*, 13 January 1986)

4 It is no consolation to be told that 'simulators . . . are capable of 'fooling' the system under test into believing that simulated data comes from real sources' (Martin, J., 1985:45). All too often the weapon and its sensors, on the one hand, and the simulator on the other, will have been designed on shared assumptions, even if the latter has not actually been built with the former in mind. The capacity of a simulator to pass data to a weapons system which the latter cannot discriminate from the real world is just as likely to be caused by limitations in the weapon as by the technical excellence of the simulator.
5 Since the chemical laser programmes were demoted to 'silver' status within the SDI (p. 94), as unlikely ever to yield any feasible weapons, the 'chemical laser architecture' debates of 1983–4 are by now mainly of historical interest. They are summarized here in order to illustrate some fundamental aspects of any space-based BMD. See also note 6.
6 Lord Zuckerman (1986) has traced the history of the arguments about the numbers of low-orbit laser battle-stations that might be needed for

boost-phase interception. Initially abstract and inadequate calculations by UCS critics of the SDI led to an excessive estimate that 2,400
satellites would be necessary. Within five weeks the UCS team had
been the first to point out their own mistakes and to offer more
realistic calculations, producing an estimate of 800 satellites, which
was soon afterwards revised again to 300. This public retraction and
self-correction did not prevent one enthusiast for the SDI, Dr Robert
Jastrow, from publishing several contributions to the 'debate', a year
later, as if the UCS scientists had never issued them.

7 The Shuttle has cost about $3 million per ton lifted into low orbits
towards the east. Putting payloads into polar orbits – the purpose of
the new military Shuttle base at Vandenberg AFB – will be at least
twice as expensive because the momentum of Earth's rotation cannot
be exploited (UCS, 1984: 101). (Any attempt to modify the Shuttle
design for greater safety, after January 1986, will have to bear in mind
the greater accelerations or longer burn-times required to achieve such
orbits.)

8 It used to be thought that the new space-launch system would need a
payload capacity of up to 100 tons. But the axeing of many of the
large-scale space-based elements in the proposed system, and their
replacement, if at all, by multiple smaller satellites, has led to
emphasis on the project for a lighter National Aerospace Plane
(NASP).

9 One scheme is to use 'adaptive optics' to compensate for atmospheric
distortion by predistorting the beam so that it arrives at the relay
mirror as a near-perfect wave-front. A beacon laser near the relay
mirror reveals the pattern of atmospheric distortion as it changes. This
is feasible because atmospheric turbulence effects remain stable for
milliseconds whereas the laser beam takes less than 0.1 millisecond to
deliver its information.

10 Early calculations by the Union of Concerned Scientists suggested that
the electrical generating stations alone, to power such lasers, would
cost between $40 and $110 billion (1984: 109).

11 The original BAMBI study was ended in 1963 once the idea was found
to be impractical with the technologies of the time (Manno, 1984:
101–2; 165).

12 For one critique of the High Frontier concept see Carter, 1984a: 34–5.

13 Carter (1984a) estimates the mass of such a projectile at 420 kg.

14 Not all that is announced about tests of hypervelocity gun technologies
is quite what meets the eye. In one experiment in November 1985 a
solid slug was fired at a one-third size mock-up of a Soviet booster.
This was said by General Abrahamson to have 'demonstrated the
potential' of such a system. It later transpired that the destruction in
question had actually been wrought by an air-gun (*International
Herald Tribune*, 19 December 1985).

9 Current Soviet Space Weapon Programmes

1 Principal sources for this chapter have been: Longstreth et al., 1985; OTA, 1985a; Pike, 1985; Sagdeyev & Kokoshin, 1985; Stevens, 1984; Stirling, 1985; Turnill, 1984; US DoD, 1985a, 1985c.

2 Yonas's use of the term 'strategic' might seem to imply that all the NATO aircraft against which such Soviet air defences have been deployed are now regarded by the Pentagon as 'strategic offensive forces', because of their potential for attacks on the national territory of the Soviet Union. Sensible as such a categorization might appear to be, it is not the one usually accepted by United States representatives in arms control negotiations.

3 A US-Canadian agreement on a new joint network of warning radars to be completed by 1992, the North Warning System, was signed in March 1985. This development is closely related to the rationale of rendering obsolete not just nuclear missiles but all deliverable nuclear *weapons*, such as 'air-breathing' bombers and cruise missiles, which underpins at least the public relations aspect of the SDI.

4 The possibility that 'accurate' counter-force missiles might perform significantly below their claimed capabilities 'on the day', due to poor reliability, or to unpredictable or unfavourable natural or war-related conditions, has been a topic of considerable controversy in recent years. See Anderson, 1982, Cockburn & Cockburn, 1980, Bunn & Tsipis, 1983, Marsh, R. T. 1982.

5 By early 1985 the British Cabinet's Joint Intelligence Committee had apparently accepted an analysis which found that Abalakova was not an 'ABM radar'. The press reports (*Daily Telegraph*, 16 March 1985; *Observer*, 24 April 1985) did not, however, reveal the JIC's finding on the question of its being an *early-warning radar* in breach of the Treaty, which has after all been the principal American accusation (appendix 3). The reticence of the then British Defence Secretary, Mr Heseltine, when questioned on this subject by the House of Commons Defence Committee (5 December 1985), suggests there may be an interesting divergence between the official British and American positions on the matter.

6 One estimate puts the range at 4800 kilometres for objects between 2° and 50° above the horizon, which would be slightly more compatible with the early-warning theory.

7 This summary is based on three sources: a CIA report on *Soviet Directed Energy Weapons,* released in March 1985 but written the previous year; testimony to a joint hearing of the Senate Appropriations and Armed Services Committees on 26 June 1985, on *Soviet Strategic Force Developments*, given by R. M. Gates, Chairman of the National Intelligence Council, and L. K. Gershwin, National Intelligence Officer for Strategic Programs; and the censored testimony of Dr J. K. Sellers, chief of the Strategic Defense branch in the DIA's

Directorate of Scientifics, given to the House Appropriations Committee on 7 and 23 May, 1985. The analysis of the first two sources by Stirling (1985) was particularly helpful.

It is worth noting here that whilst in general these sources originate or strongly support US apprehensions about possible Soviet achievements in directed-energy and other BMD technologies, they are sometimes also far more candid about the extent of US *ignorance* of the actual state of Soviet programmes than are the more sweepingly alarmist hand-outs from the Pentagon, the White House and the ACDA. Thus for example Dr Sellers pointed out that 'in terms of the very advanced optical systems . . . we judge they are probably behind us in that area, but we don't have a lot of specific evidence.' And in respect of neutral particle beams, he explained that strategic evaluation was difficult. 'We see small windows into their program, we happen to have one in that area, but we do not get the full scope of what they are working on.' Asked whether he thought it was a major effort, he could only reply: 'That is difficult to say. I have no context. When we say we have evidence of how many people, how large the effort is' [sic] (H. Appropriations, 1985: 675, 679).

8 US progress with it, however, achieving velocities of up to 40 km/sec, was highlighted in a press briefing given by General Abrahamson in Brussels in December 1985 (*Aviation Week & Space Technology*, 9 December 1985). See also his censored but clearly intelligible testimony of 7 May 1985 (H. Appropriations, 1985: 620).

9 The larger estimate is still given in some sources (*Aviation Week & Space Technology*, 27 May 1985).

10 A subsequent Soviet UN proposal for international cooperation on peaceful uses of space equated 'non-militarization of space' with 'abstention . . . from the production of space strike weapons (including research), their testing and their deployment . . .' (*General Assembly* A/10/192: 4, 16 August 1985).

11 Such a total ABM ban was briefly considered during the negotiations for the ABM Treaty, until the American side thought better of what one participant has recalled as little more than a negotiating tactic (Garthoff, 1984: 305).

10 Who Needs Space Weapons?

1 As shown at several points in this book, the US administration has been profoundly ambivalent and inconsistent on the question of what level of anti-missile defences they expect the SDI to produce. On occasions when the technical problems are admitted, the line now seems to be that missiles are to be rendered 'obsolete' – some day – by the indirect, political powers of BMD systems rather than by their direct, physical ones. But they are not always admitted. It is, for

example, often stated or implied that it is the effectiveness of the ABMs that will produce the looked-for outcome at the negotiating table – a completely inverse relationship.

2 The level of defensive capability postulated for this part of the discussion may be thought of as falling between 'Level 3', in which cities cannot be protected against a determined attacker, and 'Level 4', 'a high level of urban survival' but probably only with 'widespread civil defense', as defined in the recent OTA report (1985a: 108). The former cannot be seriously regarded as population defence. The latter is too far-fetched to merit analysis.

The phrase '*during* a nuclear war' recognizes that the whole discussion leaves entirely out of account the potentially massive disruption of most of Earth's ecological systems, and devastation of their human and non-human participants alike, that could develop *after* a nuclear war from such relatively long-term systemic effects as 'nuclear winter' (Greene et al., 1985). This concession to what passes for 'strategic' thinking about space weapons is made deliberately, in order to tackle its arguments on their own terms.

3 Desirability must of course be assessed in respect of particular ends. The implications of space weapons for human moral and spiritual goals and values are touched on briefly elsewhere. This chapter focuses on the generally accepted goal of strategy and foreign policy, that of securing the physical and social well-being of human communities against such dangers as their various disputes and confrontations may from time to time produce.

4 See for example, from both sides of the 'ABM debate' of the late 1960s, Barnaby, 1969, and Wohlstetter, 1969: 124–8. The point is repeated, with rather little in the way of critical examination, by Drell et al. (1984: 68).

5 The SDI Organization has rejected the characterization of its projected near-term anti-missile systems as 'point defence', on the grounds that they would begin to operate so far out in the 'threat tunnel', both in the mid-course phase and in the early part of the terminal phase, that they would cover 'footprints' too large to be deemed 'points' any longer. The issue is largely semantic, since if the degree of protection within the footprint is still too low for 'soft' civilian targets to survive, and probably too low for all hardened targets to survive either, the end-result will in practice be that of point defence. Wohlstetter's expression (1969) is therefore revived at this point, to focus discussion on the military outcomes of such partially effective defences, rather than on possible but irrelevant alterations in the technique by which they are produced.

6 Its cultural and psychological bases have been examined in an interesting study by Major Nunn (1982).

7 Obviously the force defences would no longer be needed to protect those missiles that had already departed their silos in a pre-emptive

strike. But nuclear war planning in the 1980s, possibly on both sides, is placing increased emphasis on the potential survivability of a large proportion of nuclear offensive forces, together with command-and-control systems, through a prolonged series of 'exchanges' (Branch, 1984).

8 This is also explained by Freedman (1985: 46). But he does not make it clear whether he accepts that this negative effect of force-defence ABMs is enough to make them a poor strategic choice.

9 Any supposedly benign effect of real or apparent increases in the threat to populations seems certain to be extremely small, in view of the enormous penalties which are already well understood as things are. Indeed, the authors are acutely aware, throughout this discussion, of having strayed perhaps too far into the realm of what Raymond Aron called 'strategic fiction', by allowing certain very questionable distinctions between alternative war scenarios to stand unchallenged, for the sake of argument. Readers who find an obscene unreality about much of the reasoning we have chosen to engage with here should be reassured that they are probably right to do so. However, such distinctions are seen as important in the calculations of the nuclear decision-takers, which gives their 'unreality' a purchase on the real world and on its chances of avoiding nuclear war. The frequent use in this section of subjective expressions, such as 'expected', is intended to reflect this aspect of the matter.

10 The OTA's figure 5.1, on the same page, says the exact opposite to the text, and should probably be disregarded.

11 As long as the aspiration for one-way, US-Soviet deterrence is more or less latent and concealed within the arguments of the US BMD lobby, something else is wanted to explain why the threat of massive devastation is felt to be sufficient (though hardly needed of course) to restrain the United States from nuclear war, but not to deter the Soviet Union. For the modern American Right, the historian Richard Pipes has done much to supply this deficiency by explaining why, in 'culturalist' terms, Soviet leaders are so much less bothered by the prospect of a nuclear holocaust than are other mortals. Brennan adopted this line in 1980, rejecting his earlier assimilation of the motivations of Soviet decision-makers to those of their American counterparts, and it can be found in frequent use throughout the BMD lobby:

> Since virtually all US commentators agree that the Soviet Union is not attracted to MAD reasoning, the long-familiar 'instability' case against urban-area BMD and non-marginal civil defense provision, is simply wrong. (Gray, 1979: 83)

However, Gray has referred elsewhere (Gray & Payne, 1980: 25) to there having once been such a thing 'in Soviet perspective' as the 'assured destruction effect [i.e. of US nuclear forces on the Russians] of the late 1960s' after all! He does not seem to have explained convincingly why it is no longer there.

12 The 1983 Defense Guidance dropped only the expressions 'protracted' and 'prevail', but not their substantive meaning, as the distinguished critic Theodore Draper was quick to point out to Mr Weinberger (Draper & Weinberger, 1983).

13 The point was made, in their own terms, in the first paper on space weapons by the 'Committee of Soviet Scientists for Peace, against Nuclear Threat', which referred to 'additional difficulties in assessing the correlation of forces' (Sagdeyev & Kokoshin, 1984: 27).

14 The administration has been unswerving in its strategic philosophy from its very first week in office:

> [I]f this nation becomes a hostage to the strategic nuclear forces of the U.S.S.R., then it becomes difficult if not impossible for the United States to employ or risk using conventional force against Soviet interests, or to conduct its diplomacy successfully. Without full strategic nuclear deterrence, the United States and its allies would always have the spector [sic] of having to do battle at Armageddon. (Weinberger, in S. Armed Services, 1981: 15)

15 The 'argument' that a population-defence solution to the problem of American vulnerability to Soviet nuclear forces simply had to be available, because the President and some other like-minded people wanted it so badly, was by no means an original irrelevancy. It had been around for at least twenty years:

> I am not talking about the Nike-Zeus. I am just talking about the mandatory requirement for this *sort* of weapon, irrespective of how the solution of the technical problem is attempted. Nevertheless, a question is in order: if Nike-Zeus does not fill the bill, where, pray, is the system which does? If there is no effective anti-missile programme, why isn't there? (Possony, 1963: 551)

16 It is perhaps not entirely cynical to suppose that various phases of the nuclear arms race are about whatever intention the superpowers have been most concerned to *deny* at the time. When it was about 'massive retaliation' it was important to deny any wish to destroy the other society except if forced to do so, as a last resort. More recently, it has become necessary to start denying any thought of acquiring military superiority.

17 A decision to proceed with full-scale development of the TAV was reported in November 1985 (*Daily Telegraph*, 26 November 1985).

The status of the programme, however, remains unclear. President Reagan expressed strong support for it in early February 1986, but a senior DARPA official said it had not yet gone into full-scale development and testing (*Aviation Week & Space Technology*, 3 February 1986). Its current budget is very low – $50 million in FY 1986, rising to twice that in FY 1987. The British government announced on 5 February 1986 that it would fund initial studies of a similar vehicle, that might at first be unmanned only, by British Aerospace.

18 The chronology is not absolutely clear from Stares's text, but this appears to be what is meant.

19 Where there is no 'dead ground' there can be no 'high ground' either. But if weapons for combat in space were ever orbited in appreciable numbers, and if super-synchronous orbits became popular, this judgement might in the distant future have to be revised. For the Moon might in such circumstances become a piece of tactically desirable real estate, as a moderately defensible military base for weapons intended to 'cover' orbiting military space platforms at very high altitudes. This would also depend, of course, on the successful development of weapons capable of acting over not just tens but even hundreds of thousands of kilometres in short times, such as lasers or particle accelerators even more powerful than those discussed in this book, which are themselves still very far from having been developed.

20 Ford gives the average for US systems as 'about five contacts per satellite per day' (1985: 65).

11 Arms Control or Space Control?

1 Few experts believe that any East-West agreements to date have done much to restrain the arms race, let alone reverse it (Goldblat, 1982: 355).

2 Mr Weinberger's grasp of the Treaty has continued to be inadequate. Besides repeating this trivial error (since the 1974 Vladivostok Protocol, only *one* ABM site each is permitted) at least once, he single-handedly anticipated the extreme 'McFarlane Version' (p. 223) of the Treaty at the time of the President's March 1983 speech, by stating that:

> There is no violation involved in the study, the research, the development . . . the Treaty goes only to block deployment. (*New York Times*, 25 March 1983)

But see also note 12.

3 The Treaty is in fact of unlimited duration, and does not have to be renewed or re-ratified through the review process. However, though either party could table substantive amendments or serve notice of

abrogation at any time, it is probable that a party which felt unable to continue with the Treaty in its present form might for political advantage prefer to use the review process as the occasion to express its dissatisfaction, and then to renounce the Treaty if necessary, provided of course that the review date was not too far off when it was deciding how to proceed.

4 The earlier welcome for the project may have assumed that 'traditional', that is, nuclear-armed point-defence ABMs would still be politically acceptable. That might also explain why the software taken out of the system in the mid-1970s cannot now simply be restored. But the dangers and difficulties of patchwork tinkering with very long and complex software should also not be underestimated (Lin, 1985).

5 A detailed account of the extremely confused, and confusing, US policy-making process with respect to ATBM, till early 1985, can be found in Brauch, 1985b.

6 The question of what sort of 'interference' with one another's military satellites the United States and the Soviet Union may from time to time get up to would be sensitive enough even without the matter of the legal protection afforded to 'national technical means of verification' (NTM) by the SALT 1 agreements of which the ABM Treaty was a part. Apart from straightforward monitoring of the coded signals from such satellites to their ground-control stations, it seems at least probable that a certain amount of active probing of the opponent's systems may from time to time take place. This might be done in two ways, either by sending signals to which it is assumed the *sensors* of the object under investigation are sensitive and then monitoring its emitted signals, or by sending signals to probe the on-board satellite *control systems* to see what can be gleaned thereafter both from its signal output and from its other responses, such as actual manœuvres. The second of these would obviously come closer to a breach of the NTM provision than the first. However, it should also be realized that the two superpowers are unlikely to accept *all* each other's reconnaisance satellites as legitimate NTM, particularly in the case of those designed to gather sensitive electronic intelligence on all military, political and economic activities of the opponent, not just those covered by strategic arms control agreements. And there are other important military satellite systems, such as those for communications and navigation, not protected by the NTM rule at all. However, few if any of such discreetly unpublicized mutual electronic probings would seem to amount to 'jamming' of the other side's system, let alone to damaging it.

7 In the former option, which was adopted by the Soviet Union, there may be up to six ABM radar complexes within that area, none of which may be more than three kilometres across. In the latter, chosen by the United States, there may be only two large phased-array radars (box 3.1), of a power-aperture comparable to the large ABM radars

which the United States had developed for its system by the end of the 1960s, and 18 smaller ABM radars, all within the deployment area ((III)). Large phased-array radars for ballistic missile early-warning may be deployed in any number, but are restricted to the periphery of the party's national territory and should face outwards ((VI.b; 1.F)). (References in double brackets are to the articles of the Treaty, if wholly or partly in Roman numerals, otherwise to its three supporting documents: 1 – Agreed Statements; 2 – Common Understandings; 3 – Unilateral Statements, within which clauses are identified by capital letters.)

8 Its reasons for so doing must, in the language of the Treaty, be based on 'extraordinary events'. The Nixon administration indicated at the time what this might mean, in a Unilateral Statement ((3.A)) affirming that the purpose of the follow-on SALT talks 'should be to constrain and reduce on a long-term basis threats to the survivability of our respective strategic retaliatory forces', and that, 'if an agreement providing for more complete strategic arms limitations were not achieved within five years, U.S. supreme interests could be jeopardized. Should that occur, it would constitute a basis for withdrawal from the ABM Treaty.' There is no lack of voices in Washington today to say that it has indeed occurred, and that the United States should withdraw accordingly. Many of those now lamenting the absence of offensive arms limitations are of course the same people who acted successfully in 1979 and 1980 to prevent US ratification of the SALT 2 Treaty, which whatever its shortcomings was certainly a 'more complete' arms control measure than SALT 1.

9 Goldblat (1985) expresses the more positive opinion that BMD '[r]esearch alone . . . is not unlawful *per se*.'

10 The term 'currently' was inserted immediately after the definition of ABM systems in article II during a joint revision by both teams of an already agreed text, in order to make quite sure the definition could not be read as applying to existing ABM technologies alone. But see the discussion of Judge Sofaer's reading of the Treaty (p. 225ff.).

11 The confusion persists. On 31 October 1985 a BBC correspondent in Moscow, Kevin Ruane, reported the Soviet view that programmes within the SDI would be in breach of the Treaty, by referring to article V's prohibition of the 'creation, testing and deployment' of space-based systems. Mr Ruane does not seem to have asked his informant what the official Soviet explanation was for statement D's apparent willingness to contemplate that novel systems might be 'created in the future', *before* they became subject to negotiation.

12 This had long been the Pentagon's preferred solution for treaty-compliance problems with its BMD programmes:

An Army official said testing of LoADS in 3–5 years will not be prohibited, since it is only a 'demonstration of the generic

capability' of LoADS and not specifically a demonstration of the
mobile system. (*Aviation Week & Space Technology*, 9 February
1981)
13 Officially, that is. Unofficially this tendentious reading of the Treaty
had been hanging about on the right-wing fringe in Washington for
some time, and is mentioned, only to be dismissed, in at least one
study of the Treaty from before Kunsberg and Sofaer were put to work
(Schneiter, 1984: 228).
14 The McCloy-Zorin Agreement, named after the principal negotiators
on the two sides, was reached between the United States and the
Soviet Union in September 1961 and unanimously welcomed by the
UN General Assembly later that year. It establishes a set of agreed
principles and an overall framework for the process of international
disarmament which have remained valid and generally accepted ever
since. The text appears as appendix C in Huzzard & Meredith, 1985.

12 A Tour of the Battlefields

1 Administration officials are frequently reduced to phrases such as: 'the
President has said that, yes, and . . .', to precede some more or less
fudging reply to a question on this point.
2 To students of the history of the nuclear arms race, there is a striking
parallel between talk of the United States some day 'sharing' the SDI
with the Soviet Union and the 1946 'Baruch Plan', an American
proposal tabled at the United Nations whereby, as the sole possessor
of atomic bombs at the time, the United States would have given them
up only after an intrusive and totally foolproof system for international
control over all fissile materials and nuclear facilities in the Soviet
Union and elsewhere had been established.
3 The NORAD Commander is always a US officer, and combines the
post with those of US C-in-C Space and Commander, USAF Space
Command.
4 Outright opponents of nuclear weapons are often naturally, if
unwisely, indifferent to disputes between supporters of nuclear
deterrence as to just why they are sure it is a good idea.
5 Much of the narrative in this section is based on Hans Günter Brauch's
invaluable and comprehensively documented account (1985a).
6 Thatcher restated her position, referring to the 'restraints' that had
been accepted for the SDI, in a speech to the Lord Mayor's Banquet
on 12 November 1984, shortly before Gorbachev's visit to London.
7 There is confusion over when the Reagan administration first made
clear its willingness for allied participation in SDI research. Brauch
(1985a: 8) mentions an invitation for West German participation as
coming from Weinberger on 7 November 1984, but does not supply a
reference. The same author then records a Bonn spokesperson as

complaining, on 13 February 1985, of the absence of any concrete US offer, after remarks from Chancellor Kohl at the Wehrkunde Conference which implied an initial lack of US consideration for European economic interests (1985a: 9–10). Whatever the facts, the Germans certainly repaid the Americans in kind, by taking a full year to agree to a 'strictly civilian' participation.

8 This point was echoed by Jacques Huntzinger, the French strategic analyst and senior adviser to the then ruling Socialist Party, at an SPD symposium in October 1985. Western BMD would probably both license and provoke a large-scale Soviet counterpart, he thought. Against such defences the existing NATO nuclear doctrine of 'flexible response' would become a non-starter, because small nuclear strikes could no longer reliably penetrate. Hence only large attacks would make sense, and NATO's long-established [if unrealistic – authors] intention of controlling escalation within the first stages of nuclear conflict would have to be abandoned.

9 Even the North Atlantic Assembly's resolution in support of the SDI, at its October 1985 meeting in San Francisco, was qualified by a successful amendment from a West German Christian Democrat, to the effect that any decisions to move beyond the research phase would have to depend on consultations within the alliance and on negotiations with the Soviet Union.

10 Not to be confused with 'Eureca', the European Space Agency's Shuttle-launched retrievable carrier, designed to place a cargo of experiments into higher and therefore longer-lasting orbits for about six months, after which it descends for recovery by the Shuttle. Eureca is intended to enter service by 1988.

11 In February 1986, shortly before the election at which it was defeated, there were signs that the Socialist government in Paris had had a change of heart, in favour of French companies participating in the SDI after all. Its conservative successor may therefore seek to reverse the previous policy of 'Eureka and only Eureka' before long.

12 Within weeks of its signature purported paragraphs of the never-to-be-published MOU began to appear in the US press. On 27 January 1986, *Aviation Week & Space Technology* provided the following:

> To facilitate programs set in implementation of this MOU, the governments understand that subject to their laws, regulations, established policies and procedures and subject to privately owned proprietary rights, each government will, so far as it is able, release to the other and to its agents information and technology necessary to implement such programs . . .
>
> In considering each cooperative project carried out under this MOU, [the governments] will pay specific attention to exercising control both bilaterally and within multinational bodies over the transfer to proscribed countries of technologies and associated manufacturing processes involved in the defense program.

If this process of gradual and unofficial publication continues, the remarks in the text about the MOU's secrecy may soon be of more historical than contemporary relevance.

13 When US Vice President George Bush visited London in July 1985, 78 computer scientists from British university departments presented a petition against the SDI which dwelt largely on what they saw as the insuperable obstacles, in computing terms, to its technical feasibility. Scientists from Edinburgh and Cambridge Universities, and from Imperial College, London, continued their protests with letters to Prime Minister Thatcher in the months leading up to the signing of the MOU, with computing science, physics, and the space sciences all being prominently represented. Dr Richard Ennals, Reader in Computing Science at Imperial College, resigned from his posts as a Departmental Research Manager and as Director of Alvey, the British National Logic Programmes Initiative on fifth-generation computers, over the issue.

14 By the end of 1985 both Dornier and Messerschmitt-Bölkow-Blohm had teamed up with major US corporations (Sperry and Boeing respectively), without waiting for any participation agreement, in rival bids for a $10 million contract for an infra-red telescope experiment, scheduled for a Shuttle flight in 1987. Before the *Challenger* disaster, that contract had seemed likely to be awarded to one or other of the competing US–German partnerships in the spring of 1986, even though both bids had been criticized for 'technical deficiencies' (Pike, 1986a).

15 The Initiative was organized by Parliamentarians for World Order and launched in May 1984 by the Heads of State or Government of Argentina, Greece, India, Mexico, Sweden and Tanzania.

13 Wasting Space

1 Those seeking further light on this remark should read the closing chapters of de Riencourt's neglected masterpiece *The Coming Cæsars* (1958). The point is not that the United States has the answers to the world's problems, many of which she takes a leading part in creating, but that whatever the contributions made by others, from the Soviet Politburo to the French Chamber of Deputies, it is primarily in and through the United States that such answers could, in our time, be given decisive effect.

2 The Pentagon's initial response to the Nitze criteria, if cool, was not exactly negative. On 21 February 1985, the day after Nitze's public exposition of views that would already have been familiar to many of his administration colleagues, Fred Iklé told a Subcommittee of the Senate Armed Services Committee that 'in a sense he was stating the obvious', a remark that leaves Mr Weinberger's subsequent incomprehension (p. 266) looking somewhat strained (*New York Times*, 22 February 1985).

3 It will be recalled that if *f* is too small to produce a low enough value for *NP* in the equation $NP = NL(1-f)^n$ (p. 88), one simply increases *n*.
4 The penchant of the US defence community for terminological fixes is at least as great as its hankering after the technological variety, but far more easily satisfied.
5 The other 10 per cent is reserved for a few very immature technologies, the 'platinum' class, where investment means taking a high risk for a chance of achieving some very effective but also very improbable technique.
6 The President has the power to shelter the SDI from the effects of the Act, and has chosen to do so for FY 1986. Since he is also doing this for military pay, all other Department of Defense budgets are being cut by 4.9 per cent, instead of the 4.3 per cent which the Act imposes elsewhere. However, if the SDI appears 'Gramm-Rudman-free', this may lead Congress to treat its annual budget request more harshly than other programmes.
7 Broad's interviews were conducted in May 1984. The Hertz Foundation has operated out of an office at Livermore for some years, and its principal talent scouts are drawn from senior managers in the Laboratory's directed-energy weapons development teams.
8 The large but poorly acknowledged benefit that we draw from our uneven and all too frequently immoral association with the other intelligent but unfortunately cultureless species of our own planet is but the faintest foreshadowing of what such encounters may bring.
9 There are striking resemblances between much of the language and doctrine of space programmes today, in both East and West, and the declarations of the *State Gazette* in Yevgeny Zamyatin's satire on totalitarianism, written in 1920:

> The great, the historic hour is near when the first Integral shall soar into universal space . . . Now a . . . glorious deed lies before you: that of integrating, by means of the glazed, electrified, fire-breathing Integral, the endless equalization of all Creation. There lies before you the subjugation of unknown creatures to the beneficent yoke of reason – creatures inhabiting other planets, perhaps still in the savage state of freedom. (1970: 23)

References

Entries in bold type below, thus: **Union of Concerned Scientists 1984: The Fallacy of Star Wars,** provide a list of recommended further reading.

Official US sources are organized in the References as follows. Congressional Hearings are listed under 'House of Representatives' (referred to by 'H.') and 'Senate.' (referred to by 'S'), by committee and year. The Congressional Budget Office and the Office of Technology Assessment (OTA) are listed separately. Departments and organs of the US Government, including the Central Intelligence Agency, the Departments of Defense and State, the DoD's Strategic Defense Organization, the White House, National Security Council, and Arms Control and Disarmament Agency, are listed under 'United States'. Speeches, articles etc by leading members of the administration and other officials are listed under their names. Thus Weinberger, 1985 for a speech to the National Press Club, but US DoD, 1984a for his FY 1985 Report to Congress.

Abshire, D. M. 1985: 'New Technology and Intra-Alliance Relationships: New Strengths, New Strains', in *New Technology and Western Security Policy, Pt. III*, Adelphi Paper 199. London: IISS.
Abshire, D. M. & Allen, R. V. (eds) 1963: *National Security*. New York: Praeger.
Allen, L. & Dombey, N. 1985: 'X-ray Lasers to Shoot Holes in the Test-Ban Treaty?', *New Scientist*, 19 September.
Anderson, J. E. 1982: 'Strategic Missiles Debated: Missile Vulnerability – What You Can't Know!', *Strategic Review*, Spring.
Ball, D. 1974: *Déja Vu: The Return to Counterforce in the Nixon Administration*. California Seminar on Arms Control and Foreign Policy.
1983: *Targeting for Strategic Deterrence*, Adelphi Paper 185. London: IISS.
Barnaby, C. F. 1969: 'Arguments for and against the Deployment of Anti-Ballistic Missile Systems'. In Barnaby & Boserup (eds), 1969.
Barnaby, C. F. & Boserup, A. (eds) 1969: *Implications of Anti-Ballistic Missile Systems*. London: Souvenir.

348   *Space Weapons*

Barnes, F. 1985: 'Why Reagan must blaze the summit trail'. United Features Syndicate, in *The Times*, 15 October.

Barrett, L. I. 1983: *Gambling With History – Reagan in the White House*. New York: Doubleday.

Bekefi, G., Feld, B. T., Parmentola, J. & Tsipis, K. 1980: 'Particle Beam Weapons – a technical assessment', *Nature*, 20 March.

Bethe, H. A. & Garwin, R. L. 1985: 'New BMD Technologies', Appendix A in *Weapons In Space – Volume II, Daedalus*, Summer.

Bethe, H. A., Garwin, R. L., Gottfried, K. & Kendall, H. W. 1984: 'Space-based Ballistic Missile Defense', *Scientific American*, October.

Blackett, P. M. S. 1948: *Military and Political Consequences of Atomic Energy*. London: Turnstile.

Bosma, J. 1984: 'A Proposed Plan for Project on BMD and Arms Control (Final)'. Unpublished High Frontier discussion document.

Boylan, E. S., Brennan, D. G. & Kahn, H. 1972: *Alternatives to Assured Destruction*. New York: Hudson Institute.

Bracken, P. 1983: *The Command and Control of Nuclear Forces*. New Haven: Yale University Press.

Branch, C. I. 1984: *Fighting A Long Nuclear War*, National Security Affairs Monograph Series 84–5. Washington: National Defense University.

Brauch, H. G. 1985a: *Antitactical Missile Defense: will the European version of SDI undermine the ABM Treaty?*, AFES paper 1. Stuttgart: Institute for Political Science.

1985b: *Militärische Komponenten einer Europäischen Verteidigungsinitiative*, AFES paper 3. Stuttgart: Institute for Political Science.

Brennan, D. G. 1969a: 'The Case for the ABM'. In Brennan et al., 1969.

1969b: 'The Case for Missile Defense', *Foreign Affairs 43*, April.

1969c: 'The Case for Population Defense'. In Holst & Schneider (eds), 1969.

1980: 'BMD Policy Issues for the 1980s'. In Schneider (ed.), 1980.

Brennan, D. G., Johnson, L. W., McGovern, G. S. & Wiesner, J. B. 1969: *Anti-Ballistic Missile: Yes or No?*. New York: Hill and Wang.

Broad, W. J. 1985: *Star Warriors*. New York: Simon & Schuster.

Brodie, B. 1959: *Strategy in the Missile Age*. Princeton: Princeton University.

Buedeler, W. 1957: *Operation Vanguard*. Burke: London.

Bulkeley, R. I. P. 1985: 'RAF Fylingdales and the Anti-Ballistic Missile Treaty', background paper accompanying 'European Aspects of the Strategic Defence Initiative', unpublished evidence submitted to the House of Commons Select Committee on Defence, November.

Bunn, M. 1984: 'Technology of Ballistic Missile Reentry Vehicles', M. I. T. Program in Science and Technology for International Security. Cambridge, Mass.: MIT.

Bunn, M. & Tsipis, K. 1983: 'The Uncertainties of a Preemptive Nuclear Attack', *Scientific American*, November.

Burt, R. 1980: 'Arms and the Man'. In H. Smith et al., 1980.
Bush, V. 1950: Modern Arms and Free Men. London: Heinemann.
Butler, S. M., Sanera, M. & Weinrod, W. B. (eds) 1984: *Mandate for Leadership II – Continuing the Conservative Revolution*. Washington: Heritage Foundation.
Cable, J. 1981: *Gunboat Diplomacy 1919–1979*. London: Macmillan.
Caidin, M. 1960: *Race for the Moon*. London: Kimber.
Campbell, D. 1984: *The Unsinkable Aircraft Carrier*. London: Michael Joseph.
Canan, J. W. 1982: *War In Space*. New York: Harper and Row.
— 1984: 'Bold New Missions in Space', *Air Force Magazine*, June.
Carter, A. B. 1984a: *Directed Energy Missile Defense in Space*, OTA Background Paper. Washington: Office of Technology Assessment.
— 1984b: 'Introduction to the BMD Question'. In Carter & Schwartz (eds), 1984.
— 1986: 'Satellites and Anti-Satellites: the Limits of the Possible', *International Security*, 10, pt 4, Spring.
Carter, A. B. & Schwartz, D. (eds) 1984: *Ballistic Missile Defense*. Washington: Brookings Institution.
Charlton, M. 1985: 'Red Alert: Paul Nitze and the Present Danger', Part 3 of *The Star Wars History*, BBC Radio 3, 20 October.
Chomsky, N. 1981: 'The Cold War and the Superpowers', *The Guardian* 15 June. (Also in *Monthly Review* 33, pt 6, November.)
Christofilos, N. C. 1959: 'The Argus Experiment', paper to Symposium on Scientific Effects of Artificially Introduced Radiations at High Altitudes, April. In Jastrow (ed.), 1960.
Christy, R. 1984: 'Analysis of Soviet launchers'. Pp. 267–271 in Turnill, 1984.
Clemens, W. C. 1973: *The Superpowers and Arms Control*. Lexington: D. C. Heath.
Cockburn, A. & Cockburn, A. 1980: 'The Myth of Missile Accuracy', *New York Review of Books*, 20 November.
Coffin, T. 1964: *The Passion of the Hawks*. New York: Macmillan.
Collier, B. 1976: *The Battle of the V-Weapons 1944–1945*. London: Elmfield.
Collins, J. M. & Cordesman, A. H. 1978: *Imbalance of Power* (Report to Senate Armed Services Committee). San Rafael: Presidio.
Congressional Budget Office, 1984: *Analysis of the Costs of the Strategic Defense Initiative 1985–89*, paper for the Subcommittee on Arms Control, Oceans, International Operations and Environment, of the Senate Foreign Relations Committee. Washington: CBO.
Cunningham, C. T. 1984: *Critique of Systems Analysis in the OTA Study of Directed Energy Missile Defense in Space*. Livermore: Lawrence Livermore National Laboratory, (DDV–84–0007).
Davis, W. A. 1979: 'Current Technical Status of U.S. BMD Programs'. In Nacht (ed.), 1979.

Douglas Aircraft Company Inc. 1946: *Preliminary Design of an Experimental World-Circling Spaceship*, Abstract SM–11827, 2 May.

Draper, T. & Weinberger, C. 1983: 'On Nuclear War: an exchange with the Secretary of Defense', *New York Review of Books*, 18 August.

Drell, S. D., Farley, P. J. & Holloway, D. 1984: *The Reagan Strategic Defense Initiative: A Technical, Political and Arms Control Assessment*. Stanford University: Center for International Security and Arms Control.

Durch, W. J. 1984: 'Steps Into Space'. In Durch (ed.), 1984.

Durch, W. J. (ed.) 1984: *National Interests and the Military Use of Space*. Cambridge, Mass.: Ballinger.

Englebardt, S. L. 1966: *Strategic Defenses*. New York: Thomas Crowell.

Ferguson, C. 1985: Open Letter to a Japanese friend, no title, 9 May. Photocopy, publication details unknown.

Fischer, H. 1985: 'Europas Antwort auf die Militarisierung des Weltraums', Address to SPD Colloquium on Weapons In Space, October. Bonn: SPD.

Fletcher, J. C., 1984: 'The Technologies for Ballistic Missile Defense', *Issues in Science and Technology*, Fall.

Ford, D. 1985: *The Button: the Nuclear Trigger, Does It Work?*. London: Allen & Unwin.

Fossedal, G. 1982: 'Exploring the High Frontier', *Conservative Digest*, June.

Freedman, L. 1985: 'The "Star Wars" Debate: the Western Alliance and Strategic Defence: Part II'. In *New Technology and Western Security Policy III*, Adelphi Paper 191. London: IISS.

Frei, D. 1983: *Risks of Unintentional Nuclear War*. London: Croom Helm.

French, S. 1985: *'Star Wars'*. London: SANA.

Frye, A. 1974: 'US Decision Making for SALT'. In Willrich & Rhinelander (eds), 1974.

Garthoff, R. L. 1984: 'BMD and East–West Relations'. In Carter & Schwartz (eds), 1984.

Garwin, R. 1985: 'How many orbiting lasers for boost-phase intercept?', *Nature*, 23 May.

Gavin, J. M. 1958: *War and Peace in the Space Age*. New York: Harper Bros.

Gehlen, M. P. 1967: *The Politics of Coexistence*. Bloomington: Indiana University.

Glasstone, S. & Dolan, P. J. 1979: *The Effects of Nuclear Weapons*. Washington: Departments of Defense & Energy.

Goldblat, J. 1982: *Agreements for Arms Control*. London: Taylor & Francis.

   1985: 'Space Weapons and Arms Control', SIPRI Fact Sheet, October. Stockholm: SIPRI.

Golovine, M. N. 1962: *Conflict In Space*. London: Temple Press.

Gorbachev, M. 1985: Letter to the Union of Concerned Scientists, Boston. Extracts in OTA, 1985a: 315.

Goulden, J. A. & Singer, M. 1972: 'AT&T and the ABM'. In Pursell (ed.), 1972.

Graham, D. O. 1982: *High Frontier: A New National Strategy*. Washington: Heritage Foundation.

Gray, C. S. 1979: 'Nuclear Strategy: the case for a Theory of Victory', *International Security* 4.

1981: 'A New Debate on Ballistic Missile Defence', *Survival*, March/April.

1983: *American Military Space Policy*. Cambridge, Mass.: Abt Associates.

Gray, C. S. & Payne, K. B. 1980: 'Victory Is Possible', *Foreign Policy* 39, Summer.

1981: *SALT: Deep Force Level Reductions*, prepared for the SALT/Arms Control Support Group, Office of the Assistant to the Secretary of Defense (Atomic Energy). New York: Hudson Institute.

Greene, O., Percival, I. & Ridge, I. 1985: *Nuclear Winter*. Cambridge: Polity Press.

Greenwood, T. 1975: *Making the MIRV*. Cambridge, Mass.: Ballinger.

Gumble, B. 1985: 'Air Force Upgrading Defenses at NORAD', *Defense Electronics*, August.

Gunston, B. 1979: *Encyclopedia of Rockets and Missiles*. London: Salamander.

Harndt, R. 1985: 'Die Militärische Nutzung des Weltraums und ihre völkerrechtlichen Grenzen', *Military Law and Law of War Review*, 69–121.

Hecht, J. 1984: *Beam Weapons*. New York: Plenum Publishing Corp.

1985: 'Star Wars – an astronomical bribe for scientists', *New Scientist*, 20 June.

Hobbes, T. (ed. Oakeshott) 1947: *Leviathan*. Oxford: Basil Blackwell.

Hoffman, F. 1983: *Ballistic Missile Defenses and U.S. National Security*, Report of a team of outside experts for the Future Security Strategy Study. Washington: Institute for Defense Analyses.

1985: 'The SDI in U.S. Nuclear Strategy', Statement before the Subcommittee on Strategic and Theater Nuclear Forces of the Senate Armed Services Committee, *International Security* 10, Summer.

Holloway, D. 1983: *The Soviet Union and the Arms Race*. New Haven: Yale University Press.

Holst, J. J. 1969: 'Missile Defense: Implications for Europe'. In Holst & Schneider (eds), 1969.

Holst, J. J. & Schneider, W. (eds) 1969: *Why ABM?*. New York: Pergamon.

House of Representatives (US Congress):
Appropriations

1984: *Authorization of DoD Budget for FY 1985.*
1985: *Authorization of DoD Budget for FY 1986.*
Armed Services
 1981: *Continental Air Defense.*
 1982: *Authorization of DoD Budget for FY 1983.*
Science and Astronautics
 1959: *Nuclear Explosions in Space.*
Howe, G. 1985: 'Defence and Security in the Nuclear Age', speech to the
 Royal United Services Institute, 15 March, London Press Service
 VSO26/85. Also in *Arms Control & Disarmament Newsletter* 23,
 January/March.
Huzzard, R. & Meredith, C. (eds) 1985: *World Disarmament: an idea
 whose time has come.* Nottingham: Spokesman.
Jasani, B. 1982: 'Military space technology and its implications'. In Jasani
 (ed.), 1982.
 1984: 'The Arms Control Dilemma – An Overview'. In Jasani (ed.),
 1984.
Jasani, B. (ed.) 1982: *Outer Space – A New Dimension of the Arms Race.*
 London: Taylor & Francis.
 **(ed.) 1984: *Space Weapons – The Arms Control Dilemma.* London: Taylor
 & Francis.**
**Jasani, B. & Lee, C. 1984: *Countdown to Space War.* London: Taylor &
 Francis.**
Jastrow, R. 1984: 'The War Against "Star Wars"', *Commentary*,
 December.
Jastrow, R. (ed.) 1960: *The Exploration of Space.* New York: Macmillan.
Kaldor, M. 1982: *The Baroque Arsenal.* London: Andre Deutsch.
Kaplan, F. 1983: *The Wizards of Armageddon.* New York: Simon and
 Schuster.
**Karas, T. 1983: *The New High Ground: Strategies and Weapons of Space
 Age War.* New York: Simon & Shcuster.**
Kerr, D. 1984: 'Implications of Anti-Satellite Weapons for ABM Issues'.
 In Jasani (ed.), 1984.
Keyworth, G. 1985a: Remarks to the American Security Council
 (excerpts), 4 June. London: USIS.
 1985b: 'The Impact of the Strategic Defense Initiative on the Western
 Alliance', Address to the European Atlantic Group, 17 June.
Kistiakowsky, G. 1976: *A Scientist at the White House.* Cambridge, Mass.:
 Harvard University Press.
Lapp, R. E. 1968: *The Weapons Culture.* New York: Norton.
Lenzer, C. 1985: 'WEU and the Strategic Defence Initiative', report from
 WEU Committee on Scientific, Technological and Aerospace Ques-
 tions. WEU Document 1036, 6 November.
**Lin, H. 1985: *Software for Ballistic Missile Defense.* Boston: MIT, Center
 for International Studies.**
Longstreth, T. K., Pike, J. & Rhinelander, J. B. 1985: *The Impact of U.S.*

and Soviet Ballistic Missile Defense Programs on the ABM Treaty.
Washington: National Campaign to Save the ABM Treaty.
McDougall, W. 1985: . . .*The Heavens and the Earth.* New York: Basic
  Books.
McGovern, J. 1965: *Crossbow and Overcast.* London: Hutchinson.
Manno, J. 1984: *Arming the Heavens: The Hidden Military Agenda for
  Space, 1945–1995.* **New York: Dodd, Mead & Co.**
Marsh, P. 1985: *The Space Business.* **London: Penguin.**
Marsh, R. T. 1982: 'Strategic Missiles Debated: Missile Accuracy – We
  Do Know!', *Strategic Review*, Spring.
Martin, C-N. 1967: *Satellites Into Orbit.* London: Harrap.
Martin, J. 1985: 'Answering the Software Critics', *Defense Science &
  Electronics*, October.
Martin, L. (ed.) 1979: *Strategic Thought in the Nuclear Age.* London:
  Heinemann.
Meyer, S. M. 1983: 'Soviet Military Programmes and the "New High
  Ground"', *Survival*, Sept./Oct.
  1985: 'Soviet Strategic Programmes and the US SDI', *Survival*,
  Nov./Dec.
Mische, P. 1985: *Star Wars and the State of our Souls.* Minneapolis:
  Winston Press.
Moore, J. T. 1981: 'Low Altitude Defense: an analysis of its effect on MX
  survivability', unpublished M.Sc. thesis, USAF Institute of Technol-
  ogy, March.
Nacht, M. (ed.) 1979: *U.S. Arms Control Objectives and the Implications
  for Ballistic Missile Defense.* Cambridge, Mass.: Center for Science
  and International Affairs, Harvard University.
Newell, H. E. 1955: 'The Satellite Project', *Scientific American*, 193 pt 6.
Nitze, P. 1956: 'Atoms, Strategy and Policy', *Foreign Affairs* 34, January.
  1985: 'On the Road to a More Stable Peace', Speech to the Philadelphia
  World Affairs Council, 20 February. London: USIS.
Nunn, J. H. 1982: *The Soviet First Strike Threat: the U.S. perspective.* New
  York: Praeger.
Oberth, H. 1929: *Wege zur Raumschifffahrt (The Way to Space Travel).*
  Munich: R. Oldenbourg.
Office of Technology Assessment 1985a: *Ballistic Missile Defense Technolo-
  gies,* **OTA-ISC-254, September. Washington: O.T.A.**
  1985b: *Anti-Satellite Weapons, Countermeasures, and Arms Control,*
  OTA-ISC-281, September. Washington: O.T.A.
Openshaw, S., Steadman, P. & Greene, O. 1983; *Doomsday – Britain
  after Nuclear Attack.* Oxford: Basil Blackwell.
Paine, C. 1982: 'Arms Buildup', *Bulletin of the Atomic Scientists,* 38, pt 8,
  October.
Parry, A. 1960: *Russia's Rockets and Missiles.* London: Macmillan.
Parson, N. A. 1962: *Missiles and the Revolution in Warfare.* Cambridge,
  Mass.: Harvard University Press.

Payne, K. B. 1981: 'Deterrence, Arms Control, and U.S. Strategic Doctrine', *Orbis*, 25, pt 3, Fall.

Peebles, C. 1983: *Battle for Space*. Poole: Blandford Press.

Perry, G. E., 1985: 'Extent of Militarisation of Outer Space'. In Rotblat & Hellman (eds), 1985.

Pike, J. 1985: 'Assessing the Soviet ABM Programme'. In Thompson et al., 1985.

   1986a: 'SDI Contracts: Will U.S. Allies Be Sold Short?', *International Herald Tribune*, 16 January.

   1986b: *The Emperor's Newest Clothing – Changes to the SDI as a Result of Phase 1 Architecture Studies,* February. Washington: F.A.S.

Polanyi, J. C. 1985: 'Time for More NO in NORAD?', *Globe & Mail*, 2 December.

Possony, S. T. 1963: 'Toward a Strategy of Supremacy'. In Abshire & Allen (eds), 1963.

Possony, S. T. & Rosenzweig, L. 1955: 'The Geography of the Air', *Annals of the American Academy of Political and Social Science*, May.

Pretty, R. T. 1985: 'Will SDI Work?', *Jane's Defence Weekly*, 7 December.

Pursell, C. W. (ed.) 1972: *The Military-Industrial Complex*. New York: Harper and Row.

Ranger, R. 1981: *The Implications of the Possible U.S. Introduction of Ballistic Missile Defence into the North American Air Defence System.* Ottawa: Operational Research & Analysis Establishment, Dept of National Defence.

Reagan, R. 1983: 'Address on National Security', 23 March. London: USIS.

Reich, R. 1984: 'High Technology, Defense, and International Trade'. In Tirman (ed.), 1984.

Rhinelander, J. B. 1974: 'The SALT 1 Agreements'. In Willrich & Rhinelander (eds), 1974.

Riencourt, A. de 1958: *The Coming Cæsars*. London: Jonathan Cape.

Rogers, P. 1986: 'The Effect of the SDI Programme on Soviet Strategic Nuclear Forces with reference to Arms Control', unpublished evidence to House of Commons Select Committee on Defence, January.

Rosenberg, D. A. 1983: 'The Origins of Overkill', *International Security*, Spring.

Rotblat, J. & Hellman, S. (eds) 1985: *Nuclear Strategy and World Security*. London: Macmillan.

Rowen, H. 1979: 'The Evolution of Strategic Nuclear Doctrine'. In Martin, L. (ed.), 1979.

Ryan, C. (ed.) 1952: *Across the Space Frontier*. London: Sidgwick & Jackson.

Sagdeyev, R. Z. & Kokoshin, A. A. 1984: *A Space-Based Anti-Missile System with Directed Energy Weapons: Strategic, Legal and Political*

Implications, xeroxed typescript of English translation. Moscow: Committee of Soviet Scientists for Peace, Against the Nuclear Threat.

1985: *Space-Strike Arms and International Security* (abridged). Moscow: Committee of Soviet Scientists for Peace, Against the Nuclear Threat.

Sanders, J. W. 1983: *Peddlers of Crisis*. London: Pluto.

Scheer, R. 1982: *With Enough Shovels*. New York: Random.

Schneider, W. (ed.) 1980: *U.S. Strategic-nuclear Policy and Ballistic Missile Defense: the 1980s and beyond*. Washington: Institute for Foreign Policy Analysis.

Schneiter, G. R. 1984: 'The ABM Treaty Today'. In Carter & Schwartz (eds), 1984.

1985: 'Implications of the Strategic Defense Initiative for the ABM Treaty', *Survival*, Sept./Oct.

Scoville, H. 1981: *MX – Prescription for Disaster*. Cambridge, Mass. : MIT Press.

Scowcroft, B. 1983: *Report of the President's Commission on Strategic Forces*, April. Washington: White House.

Senate (US Congress):

Aeronautical and Space Sciences

1969: *Authorization of NASA FY 1970 Budget*.

Appropriations

1968: *Authorization of DoD FY 1969 Budget*.

1981: *MX Missile Basing Mode*.

1985: *Authorization of DoD FY 1986 Budget*.

Armed Services

1972: *Military Implications of the Treaty on the Limitation of Anti-Ballistic Missiles and the Interim Agreement on Limitation of Strategic Offensive Arms*.

1979: *Military Implications of the SALT 2 Treaty*.

1981: *Authorization of DoD FY 1982 Budget*.

1982a: *Authorization of DoD FY 1983 Budget*.

1982b: *Nato: Can The Alliance Be Saved?*, Report by Senator Sam Nunn.

1983: *Authorization of DoD FY 1984 Budget*.

1984: *Authorization of DoD FY 1985 Budget*.

1985: *Soviet Strategic Force Developments* (Joint Hearing with Appropriations).

Foreign Relations

1963: *The Limited Test Ban Treaty*.

1979: *The SALT 2 Treaty*.

1981: *Foreign Policy and Arms Control Implications of President Reagan's Strategic Weapons Proposals*.

1982: Subcommittee on Arms Control etc., *Arms Control and the Militarization of Space*.

1984: *Strategic Defense and Anti-Satellite Weapons*.

SIPRI (Stockholm International Peace Research Institute) 1980: *Year-*

book: World Armaments and Disarmament. London: Taylor & Francis.

Skolnik, M. I. 1981: *Introduction to Radar Systems.* Tokyo: McGraw–Hill.

Smith, H., Clymer, A., Silk, L., Lindsey, R. & Burt, R. 1980: *Reagan the Man, the President.* New York: Macmillan.

Smith, R. J. 1982: 'Carter's Plan for MX Lives On', *Science*, 216, pt 4545, 30 April.

1983: 'The Search for a Nuclear Sanctuary', *Science,* 221, pts 4605/6, 1/8 July.

Sofaer, A. 1985: Statement to the House of Representatives Foreign Affairs Committee, 22 October. London: USIS.

Sokolovskiy, V. D. (ed. Scott) 1975: *Soviet Military Strategy.* London: Macdonald & Jane's.

Stares, P.B. 1984: 'Space and U.S. National Security'. In Durch (ed.), 1984.

1985: *Space Weapons and US Strategy – origins and development.* London: Croom Helm.

Starsman, R. E. 1981: *Ballistic Missile Defense and Deceptive Basing*, National Security Affairs Monograph Series 81–1. Washington: National Defense University.

Stevens, S. 1984: 'The Soviet BMD Program'. In Carter & Schwartz (eds), 1984.

Stirling, A. (1985): 'Soviet Strategic Defence Technologies', unpublished draft, November.

Sun Tzu (ed. Phillips), 1953: *The Art of War.* Harrisburg: Military Service Publishing Co.

Swift, J. 1985: 'Strategic Superiority Through SDI', *Defense & Foreign Affairs*, December.

Talensky, N. 1964: 'Anti-Missile Systems and Disarmament', *International Affairs (Moscow)*, October.

Teller, E. 1982: 'Dangerous Myths About Nuclear Arms', *Reader's Digest*, November (US edn). Also in Von Hippel, 1983.

Teller, E. & Latter, A. 1958: *Our Nuclear Future.* London: Secker & Warburg.

Thompson, E. P. 1985: 'Folly's Comet'. In Thompson et al., 1985.

Thompson, E. P., Bulkeley, R., Pike, J. & Thompson, B. 1985: *Star Wars.* London: Penguin.

Tirman, J. (ed.) 1984: *The Militarization of High Technology.* Cambridge, Mass.: Ballinger.

Tsipis, K. 1984: 'Why Reagan's Star Wars Plan Won't Work', *Playboy*, June.

Turnill, R. 1984: *Jane's Spaceflight Directory.* London: Jane's.

Union of Concerned Scientists (ed. Tirman, J.) 1984: *The Fallacy of Star Wars.* New York: Vintage.

United States Government:
Arms Control and Disarmament Agency
  Arms Control Impact Statements (ACIS), annual series:
  1978: ACIS for FY 1979
  1981: ACIS for FY 1982
  1982: ACIS for FY 1983
  1984: ACIS for FY 1985
  1985: ACIS for FY 1986
  1986a: *Soviet Noncompliance*, 1 February.
  1986b: ACIS for FY 1987
Central Intelligence Agency
  1985: *Soviet Directed Energy Weapons,* March.
Department of Defense
  1979: *Annual Report to the Congress: FY 1980*
  1984a: *Annual Report to the Congress: FY 1985*
  1984b: *Five Year Defense Program (FYDP) Program Structure,*
  Office of the Assistant Secretary of Defense (Comptroller).
  1985a: *Soviet Military Power* (4th edn), April.
  1985b: *Report to the Congress on the Strategic Defense Initiative,* 18
  April.
  1985c: *Soviet Strategic Defense Programs,* October.
  1986: *Soviet Military Power* (5th edn), March.
Department of State
  1985: *The Strategic Defense Initiative.* Bureau of Public Affairs,
  Special Report 129, June.
National Security Council
  1958: *Preliminary U.S. Policy on Outer Space,* NSC 5814/1, July.
Strategic Defense Initiative Organization
  1985: *Introduction to the SDIO Innovative Science and Technology
  Office,* March.
White House
  1982: *Fact Sheet on National Space Policy,* 4 July. In Gray, 1983.
  1984: *The President's Report to the Congress on U.S. Policy on ASAT
  Arms Control,* 31 March.
  1985a: *The President's Strategic Defense Initiative,* January.
  1985b: *The President's Unclassified Report to the Congress on Soviet
  Noncompliance with Arms Control Agreements,* 1 February.
  1985c: *The President's Unclassified Report to the Congress on Soviet
  Noncompliance with Arms Control Agreements,* 23 December.
Van Cleave, W. R. & Thompson, W. S. 1979: *Strategic Options for the
  Early Eighties: what can be done?.* White Plains: Automated Graphic
  Systems.
Viotti, P., Swan, P. A., & Friedenstein, C. D. 1981: *The Great Frontier –
  Military Space Doctrine,* Report of Military Space Doctrine Sympo-
  sium, 1–3 April. Washington: USAF Academy.
von Braun, W. 1952a: 'Crossing the Last Frontier', *Collier's,* 22 March.
  1952b: 'Prelude to Space Travel'. In Ryan (ed.), 1952.

von Hippel, F. 1983: 'The myths of Edward Teller', *Bulletin of the Atomic Scientists*, March.

Walbridge, E. 1984: 'Angle constraint for nuclear-pumped X-ray laser weapons', *Nature*, 19 July.

Wallop, M. 1979: 'Opportunities and Imperatives of Ballistic Missile Defense', *Strategic Review*, Fall.

Weinberger, C. W. 1972: 'The Defense Budget: Perspectives and Priorities'. In Weinberger et al., 1972.

1985: 'What Is Our Defense Strategy', Remarks to the National Press Club, 9 October. DoD Public Affairs Office: News Release.

Weinberger, C. W., Weidenbaum, M. L. & La Rocque, G. R. 1972: *The Defense Budget*. Washington: American Enterprise Institute.

Weiner, S. 1984: 'Systems and Technology'. In Carter & Schwartz (eds), 1984.

Wilkening, D. A. 1984: 'Space-Based Weapons'. In Durch (ed.), 1984.

Willrich, M. & Rhinelander, J. B. (eds) 1974: *SALT: the Moscow Agreements and Beyond*. New York: Free Press.

Wohlstetter, A. 1969: 'The Case for Strategic Force Defense'. In Holst & Schneider (eds), 1969.

Wolfe, T. 1979: *The Right Stuff*. London: Jonathan Cape.

Wolfrum, R. 1984: 'The Problems of Limitation and Prohibition of Military Use of Outer Space', *Zeitschrift für ausländisches öffentliches Recht und Volkerrecht*, 784–806.

Wörner, M. 1986: 'A Missile Defense for NATO Europe', *Strategic Review*, Winter.

Yonas, G. 1985: 'Strategic Defense Initiative: The politics and science of weapons in space', *Physics Today*, June.

Yost, D. S. 1982: 'Ballistic Missile Defense and the Atlantic Alliance', *International Security*, Fall.

1985: 'Soviet ballistic missile defence and NATO', *NATO Review*, October.

Zamyatin, Y. 1970: *We*. London: Cape.

Zuckerman, S. 1986: 'The Strange Story of SDI', *New York Review of Books*, 30 January.

Acknowledgements

This being about the only page of the book he has not seen, the authors gladly seize the opportunity to express their gratitude for the contribution made by Christopher Meredith, a founding member and former Secretary of Scientists Against Nuclear Arms (SANA) and an active participant in SANA's Space Weapons Group. The title of 'editor' is simply inadequate to the case. It was he who originally conceived the book and who urged us, who had not then met, to write it. He found us our helpful and long-suffering publisher, and played a major role in the constant gathering in of source materials without which the enterprise could not have gone forward. His editorial help was painstaking and unstinted, and an immense contribution to the book in its final form. It takes not just ability, but also a true generosity to give and to go on giving in such a perfect spirit of cooperation, even though not all of one's suggestions are accepted. Not only our book, but we ourselves, have benefited greatly by the collaboration.

In another sense, this book arose out of the general ongoing work of the SANA Space Weapons Group. Too many SANA colleagues to be listed here have sent us material or otherwise contributed. But we should thank Les Allen and Denis Hall in particular, for their help with revising chapters 7 and 8 and with part of chapter 13, Roger Hutton for his with appendix 2, and Fran Bagenal and Alan Walker for other generous assistance.

The authors are especially grateful to Brian Aldiss for a positive response to our request for his participation, and to Ed Reiss for letting himself be persuaded to write appendix 2. John Pike, of the Federation of American Scientists, was always ready to help clear up problems of information or interpretation. Andrew Stirling of Greenpeace International generously allowed us to draw on the first, unpublished draft of a forthcoming paper on Soviet BMD. And we can only hope that Mr Craig Ferguson's excellent 'open letter' on the significance of the SDI within the US economy, already widely circulated in 'samizdat' form, has by now appeared in print. Though we could not reach him with a request, we trust the use made of it in chapter 13 will not offend. Other friends who have helped in numerous ways include Dan Baruch, Jane Bulkeley, Donald Mackenzie, Sheena Phillips, Paul Rogers, Malcolm Spaven and Edward Thompson. And the alphabetically senior author here records his debt for materials, insights and information to many colleagues in the international Pugwash organization, whose Symposium on Strategic Defences he attended in London at the end of 1985. Pugwash conventions forbid the direct attribution of anything to anyone, but Horst Fischer, Ray Garthoff and John Polanyi are names that spring particularly to mind.

We are grateful to the Union of Concerned Scientists, Boston, Mass., for permission to reproduce or adapt figures 4.4 and 8.1, for figure 9.1 to the Brookings Institution, Washington D.C., and to our publisher, as well as to the SANA colleagues in whose book it first appeared, for figure 5.1. Five other figures have been gratefully received from Steven French's 'Star Wars', another SANA production.

Many a librarian has come to our aid in time of need. We are especially indebted to staff at the following: Bodleian Library, British Library of Economics and Political Science, British Newspaper Library, International Institute of Strategic Studies, Rhodes House, the Science Museum, and the USIS Reference Center, London.

Apart from those whose names can be found with ours at the other end of the book, none of the above is even partly or indirectly responsible for such deficiencies as, surely, remain.

RB & GS

Index

The index covers the main text, appendices 2 and 3, and the notes. Individuals are indexed only if mentioned or themselves directly quoted in the text, not for every reference in brackets. Multi-authored quotations are indexed by first author. Publications are likewise indexed only if named in the text proper, not for every reference. Satellites mentioned only in table 4.1, and organizations, places and projects mentioned only in appendix 2, have not been indexed. **Bold type** indicates major topic headings, also page numbers for principal explanations or discussions.

368   *Index*

Trans-Atmospheric Vehicle (TAV) 6, 193, 194, 196, 339
Trans-Siberian railway 313
Tremaine, Stanley 193
Trident *see* missiles, ballistic
Trident submarines: 113, 164
Truman, Harry 26, 327
TRW Inc. 305–6
Try Add *see* radar
Tsiolkovsky, Konstantin 25
Tsipis, Kosta 271
Turkey 77, 241
Turnill, Reginald 42
Tyuratam 15, 48

U-2 27, 326
UK Atomic Energy Authority 253
Union of Concerned Scientists (UCS) 6, 48, 157, 198, 329, 334
United Nations 12, 22, 155, 225–8, 249, 260, 279, 336, 343; Charter 227; Committee on Disarmament 248, 277; General Assembly 19–20, 343; Special Sessions on Disarmament (UNSSD) 261
United States, Congress 32, 61, 153, 184, 190, 194, 207, 209, 212–14, 225, 233, 268–9, 271, 290, 299, 303, 346; House 8, 32–3, 303–4; Senate 64, 207–8, 211, 212, 236–7, 266, 303, 304, 335, 346
United States, forces 7, 305; Space Command 195, 239, 269
Air Force (USAF): 6, 11, 13, 17, 18, 21, 22–4, 32, 42, 44, 48–9, 57, 69–71, 193–5, 207, 299–301, 329; Armament Division 121; Ballistic Missile Division 192; Strategic Air Command 239
Army: 11, 17, 22, 32–3, 69–72, 198, 210, 250, 298, 299, 301, 311, 342; BMD Organization 70–2, 205–6, 210, 299

Navy: 16, 17, 49, 93, 131, 195, 267, 299, 301; Naval Research Lab. 50, 301
see also Central Intelligence Agency; Strategic Defense Initiative
United States, government 8, 13, 14, 20, 21
ACDA: 204, 211, 218–19, 231, 274–5, 307, 311, 336
budget: 11, 189, 212, 268–70, 346
Department of Defense: 16–17, 19, 21–2, 52, 61, 64–5, 66, 72, 98, 119, 128, 147, 149–50, 153, 155, 164, 166, 170, 181, 184–5, 188–9, 192, 209–11, 214, 219–20, 223–5, 239, 250, 252, 264–6, 272, 274, 308–9, 328, 335, 336, 342–3, 345–6; budget 59–60, 74–5, 212, 270, 290, 298, 302–5, 340, 346
Department of State: 177, 185, 187–8, 264–5, 331
National Aeronautical and Space Administration (NASA): 21, 36, 93, 195, 257, 332
National Security Council: 14, 266
see also Reagan administration
United States, public opinion 11, 14–16, 18, 191, 203–4, 232–3
Univac 24
Ural Mountains 147
Utah 69, 71

V weapons 10; V-1 170, 327; V-2 25–6, 327
Van Allen, James 7
Van Cleave, William 67
van Reuth, Edward 132
Vandenberg AFB 23, 124, 334
Vanguard satellites 153; Vang.1 15
Vela satellites 44, 329
Vienna 227, 256
Vienna Convention *see* arms control